Respiratory Disease in Agricultural Workers: Mortality and Morbidity Statistics

February 2007

DEPARTMENT OF HEALTH AND HUMAN SERVICES
Centers for Disease Control and Prevention
National Institute for Occupational Safety and Health

DISCLAIMER

Mention of any company or product does not constitute endorsement by the National Institute for Occupational Safety and Health (NIOSH). In addition, citations to Web sites do not constitute NIOSH endorsement of the sponsoring organizations or their programs or products. Futhermore, NIOH is not responsible for the content of these Web sites.

All Web addresses referenced in this report were accessible as of the date this manuscript was prepared for publication.

ORDERING INFORMATION

To receive documents or other information about occupational safety and health topics, contact NIOSH at

NIOSH
Publications Dissemination
4676 Columbia Parkway
Cincinnati, OH 45226-1998

Telephone: 1-800-35-NIOSH (1-800-356-4674)
Fax (513) 533-8573
e-mail: pubstaft@cdc.gopbv
or visit the NIOSH web at **www.cdc.gov/niosh**

This document is in the public domain and may be freely copied or reprinted.

DHHS (NIOSH) Publication Number 2007-106

February 2007

Table of Contents

Preface .... *iv*

Acknowledgements .... *v*

Abbreviations .... *vi*

List of Tables and Figures .... *vii*

Highlights and Limitations .... *xxiii*

        Highlights .... *xxv*

        Limitations .... *xxix*

Section 1. Demographics 1

Section 2. Mortality 9

Section 3. Morbidity 177

        NHIS 179

        NHANES 195

        SOII 256

Section 4. Recommendations for Future Studies 259

Appendix A. Sources of Data A-1

Appendix B. Methods B-1

Appendix C. ICD-9 Codes and Descriptions for Respiratory Diseases Included in the Mortality Analysis C-1

Appendix D. States and Years with Industry and Occupation Codes from Death Certificates Used in the Mortality Analysis, 1988–1998 .... D-1

Appendix E. Agricultural Groups Used in the Mortality Analysis and Their Derivation from the U.S. Bureau of Census Industry and Occupation Codes E-1

Appendix F. Agricultural Groups Used in the Morbidity Analysis and Their Derivation from the National Health and Interview Survey (NHIS) Industry and Occupation Codes . F-1

Appendix G. Agricultural Groups Used in the Morbidity Analysis and Their Derivation from the Third National Health and Nutrition Examination Survey (NHANES III) Industry and Occupation Codes.. .... G-1

Preface

Respiratory Disease in Agriculture: Mortality and Morbidity Statistics presents summary tables and figures of occupational respiratory disease surveillance data focusing on various occupationally relevant respiratory diseases for the Agriculture, Forestry, and Fishing industries. The report has seven major sections that provide the following data: (1) highlights and data usage limitations; (2) demographic statistics for agricultural workers; (3) mortality statistics for agricultural workers, including by sex and race/ethnicity; (4) morbidity statistics for agricultural workers, including by sex, race/ethnicity, smoking status, and source of data; (5) recommendations to fill research gaps for respiratory disease in agriculture; and (6) appendices with descriptions of data sources, methods, and other supplementary information.

Data contained in this report originate from various publications, reports, data files, and tabulations provided by the National Center for Health Statistics (NCHS) and the Bureau of Labor Statistics (BLS). Details on the major data sources and on the methods used to compute specific statistics can be found in Appendices A and B, respectively.

Interpreted with appropriate caution, the information contained in this report can help to establish priorities for research and respiratory disease prevention in agriculture. To increase the utility of future surveillance of occupational respiratory disease in agriculture, comments on the report, descriptions of how the information could be used, and suggestions of other data for inclusion in future reports are invited.

Send comments, suggestions, and other correspondence to:

Respiratory Disease in Agricultural Workers
Public Health Surveillance Team
Surveillance Branch
Division of Respiratory Disease Studies
NIOSH
1095 Willowdale Road
Morgantown, WV 26505-2888

Phone: 304-285-5754
FAX: 304-285-6111

Acknowledgements

This report was prepared under contract (#200-2001-08009) by Michael Koontz and Laura Niang of GEOMET Technologies, LLC, a subsidiary of Versar, Inc. Marc Schenker at the University of California-Davis provided helpful consultation.

Special appreciation is expressed to Gregory R. Wagner, former Director, Division of Respiratory Disease Studies (DRDS), and Frank J. Hearl, former Deputy Director, DRDS for initiating the effort that resulted in this report.

The following NIOSH staff contributed to this report: Michael D. Attfield, Ki Moon Bang, Robert M. Castellan, Mark F. Greskevitch, Bret L. Jackson, Gregory J. Kullman, Jacek Mazurek, Stephen A. Olenchock, Teri Palermo, and Cathy J. Rotunda.

Drafts of this report were provided for review and comment to epidemiologists, physicians, industrial hygienists, agricultural health experts, and representatives of industry and labor associations. Their comments have been considered in preparing the final version of this report.

Abbreviations

BLS	Bureau of Labor Statistics
CDC	Centers for Disease Control and Prevention
DHHS	Department of Health and Human Services
FEV_1	forced expiratory volume in one second
FVC	forced vital capacity
ICD	International Classification of Diseases
L	liters
LCL	lower confidence limit
LLN	lower limit of normal
L/sec	liters per second
NCHS	National Center for Health Statistics
NHANES	National Health and Nutrition Examination Survey
NHIS	National Health Interview Survey
NIOSH	National Institute for Occupational Safety and Health
PEF	peak expiratory flow
PMR	proportionate mortality ratio
PR	prevalence ratio
SD	standard deviation
SOII	Survey of Occupational Injuries and Illnesses
UCL	upper confidence limit

List of Tables and Figures

Demographics

Table 1-1. Demographic characteristics of employed U.S. agricultural workers by agricultural group and occupation, 2002 3

Figure 1-1. Distribution of employed U.S. agricultural workers by sex, race, and ethnicity in comparison to all U.S. workers, 1997 and 2002 4

Figure 1-2. Distribution of employed U.S. agricultural groups by sex, race, and ethnicity, 1997 5

Figure 1-3. Distribution of employed U.S. agricultural groups by sex, race, and ethnicity, 2002 6

Table 1-2. Distribution of employed U.S. agricultural workers by state, 2002. 7

Mortality

Mortality by Disease Category within Agricultural Group

Table 2-1. Crop farm workers: Proportionate mortality ratio (PMR) adjusted for age, sex, and race/ethnicity by disease category, U.S. residents age 15 and over, selected states, 1988–1998 11

Table 2-2. Livestock farm workers: Proportionate mortality ratio (PMR) adjusted for age, sex, and race/ethnicity by disease category, U.S. residents age 15 and over, selected states, 1988–1998. 12

Table 2-3. Farm managers: Proportionate mortality ratio (PMR) adjusted for age, sex, and race/ethnicity by disease category, U.S. residents age 15 and over, selected states, 1988–1998. 13

Table 2-4. Landscape and horticultural workers: Proportionate mortality ratio (PMR) adjusted for age, sex, and race/ethnicity by disease category, U.S. residents age 15 and over, selected states, 1988–1998 14

Table 2-5. Forestry workers: Proportionate mortality ratio (PMR) adjusted for age, sex, and race/ethnicity by disease category, U.S. residents age 15 and over, selected states, 1988–1998. 15

Table 2-6. Fishery workers: Proportionate mortality ratio (PMR) adjusted for age, sex, and race/ethnicity by disease category, U.S. residents age 15 and over, selected states, 1988–1998. 16

Mortality by Agricultural Group within Disease Category

Figure 2-1. Tuberculosis: Proportionate mortality ratio (PMR) adjusted for age, sex, and race/ethnicity by agricultural group, U.S. residents age 15 and over, selected states, 1988–1998 17

Figure 2-2. Mycoses: Proportionate mortality ratio (PMR) adjusted for age, sex, and race/ethnicity by agricultural group, U.S. residents age 15 and over, selected states, 1988–1998 18

Figure 2-3. Sarcoidosis: Proportionate mortality ratio (PMR) adjusted for age, sex, and race/ethnicity by agricultural group, U.S. residents age 15 and over, selected states, 1988–1998 19

Figure 2-4. Malignant neoplasms of trachea/bronchus/lung/pleura: Proportionate mortality ratio (PMR) adjusted for age, sex, and race/ethnicity by agricultural group, U.S. residents age 15 and over, selected states, 1988–1998 20

Figure 2-5. Acute respiratory infections: Proportionate mortality ratio (PMR) adjusted for age, sex, and race/ethnicity by agricultural group, U.S. residents age 15 and over, selected states, 1988–1998 21

Figure 2-6. Other diseases of upper respiratory tract: Proportionate mortality ratio (PMR) adjusted for age, sex, and race/ethnicity by agricultural group, U.S. residents age 15 and over, selected states, 1988–1998 .. 22

List of Tables and Figures

Figure 2-7. Pneumonia and influenza: Proportionate mortality ratio (PMR) adjusted for age, sex, and race/ethnicity by agricultural group, U.S. residents age 15 and over, selected states, 1988–1998 23

Figure 2-8. Chronic obstructive pulmonary disease and allied conditions: Proportionate mortality ratio (PMR) adjusted for age, sex, and race/ethnicity by agricultural group, U.S. residents age 15 and over, selected states, 1988–1998. 24

Figure 2-9. Pneumoconioses and other lung diseases–external agents: Proportionate mortality ratio (PMR) adjusted for age, sex, and race/ethnicity by agricultural group, U.S. residents age 15 and over, selected states, 1988–1998 25

Figure 2-10. Other diseases of respiratory system: Proportionate mortality ratio (PMR) adjusted for age, sex, and race/ethnicity by agricultural group, U.S. residents age 15 and over, selected states, 1988–1998 26

Mortality by Disease Category within Agricultural and Sex Group

Table 2-7. Crop farm workers, males: Proportionate mortality ratio (PMR) adjusted for age and race/ethnicity by disease category, U.S. residents age 15 and over, selected states, 1988–1998. 27

Table 2-8. Livestock farm workers, males: Proportionate mortality ratio (PMR) adjusted for age and race/ethnicity by disease category, U.S. residents age 15 and over, selected states, 1988–1998. 28

Table 2-9. Farm managers, males: Proportionate mortality ratio (PMR) adjusted for age and race/ethnicity by disease category, U.S. residents age 15 and over, selected states, 1988–1998. 29

Table 2-10. Landscape and horticultural workers, males: Proportionate mortality ratio (PMR) adjusted for age and race/ethnicity by disease category, U.S. residents age 15 and over, selected states, 1988–1998 30

Table 2-11. Forestry workers, males: Proportionate mortality ratio (PMR) adjusted for age and race/ethnicity by disease category, U.S. residents age 15 and over, selected states, 1988–1998 31

Table 2-12. Fishery workers, males: Proportionate mortality ratio (PMR) adjusted for age and race/ethnicity by disease category, U.S. residents age 15 and over, selected states, 1988–1998 32

Table 2-13. Crop farm workers, females: Proportionate mortality ratio (PMR) adjusted for age and race/ethnicity by disease category, U.S. residents age 15 and over, selected states, 1988–1998. 33

Table 2-14. Livestock farm workers, females: Proportionate mortality ratio (PMR) adjusted for age and race/ethnicity by disease category, U.S. residents age 15 and over, selected states, 1988–1998. 34

Table 2-15. Farm managers, females: Proportionate mortality ratio (PMR) adjusted for age and race/ethnicity by disease category, U.S. residents age 15 and over, selected states, 1988–1998 35

Table 2-16. Landscape and horticultural workers, females: Proportionate mortality ratio (PMR) adjusted for age and race/ethnicity by disease category, U.S. residents age 15 and over, selected states, 1988–1998 36

Table 2-17. Forestry workers, females: Proportionate mortality ratio (PMR) adjusted for age and race/ethnicity by disease category, U.S. residents age 15 and over, selected states, 1988–1998. 37

Table 2-18. Fishery workers, females: Proportionate mortality ratio (PMR) adjusted for age and race/ethnicity by disease category, U.S. residents age 15 and over, selected states, 1988–1998. 38

Mortality by Agricultural Group and Sex within Disease Category

Figure 2-11. Tuberculosis: Proportionate mortality ratio (PMR) adjusted for age and race/ethnicity by agricultural group and sex, U.S. residents age 15 and over, selected states, 1988–1998. 39

List of Tables and Figures

Figure 2-12. Mycoses: Proportionate mortality ratio (PMR) adjusted for age and race/ethnicity by agricultural group and sex, U.S. residents age 15 and over, selected states, 1988–1998. 40

Figure 2-13. Sarcoidosis: Proportionate mortality ratio (PMR) adjusted for age and race/ethnicity by agricultural group and sex, U.S. residents age 15 and over, selected states, 1988–1998. 41

Figure 2-14. Malignant neoplasms of trachea/bronchus/lung/pleura: Proportionate mortality ratio (PMR) adjusted for age and race/ethnicity by agricultural group and sex, U.S. residents age 15 and over, selected states, 1988–1998. 42

Figure 2-15. Acute respiratory infections: Proportionate mortality ratio (PMR) adjusted for age and race/ethnicity by agricultural group and sex, U.S. residents age 15 and over, selected states, 1988–1998. 43

Figure 2-16. Other diseases of upper respiratory tract: Proportionate mortality ratio (PMR) adjusted for age and race/ethnicity by agricultural group and sex, U.S. residents age 15 and over, selected states, 1988–1998 . 44

Figure 2-17. Pneumonia and influenza: Proportionate mortality ratio (PMR) adjusted for age and race/ethnicity by agricultural group and sex, U.S. residents age 15 and over, selected states, 1988–1998. 45

Figure 2-18. Chronic obstructive pulmonary disease and allied conditions: Proportionate mortality ratio (PMR) adjusted for age and race/ethnicity by agricultural group and sex, U.S. residents age 15 and over, selected states, 1988–1998. 46

Figure 2-19. Pneumoconioses and other lung diseases–external agents: Proportionate mortality ratio (PMR) adjusted for age and race/ethnicity by agricultural group and sex, U.S. residents age 15 and over, selected states, 1988–1998. 47

Figure 2-20. Other diseases of respiratory system: Proportionate mortality ratio (PMR) adjusted for age and race/ethnicity by agricultural group and sex, U.S. residents age 15 and over, selected states, 1988–1998... 48

Mortality by Disease Category within Agricultural and Race/Ethnicity Group

Table 2-19. Crop farm workers, white, non-Hispanic: Proportionate mortality ratio (PMR) adjusted for age and sex by disease category, U.S. residents age 15 and over, selected states, 1988–1998 49

Table 2-20. Livestock farm workers, white, non-Hispanic: Proportionate mortality ratio (PMR) adjusted for age and sex by disease category, U.S. residents age 15 and over, selected states, 1988–1998 50

Table 2-21. Farm managers, white, non-Hispanic: Proportionate mortality ratio (PMR) adjusted for age and sex by disease category, U.S. residents age 15 and over, selected states, 1988–1998. 51

Table 2-22. Landscape and horticultural workers, white, non-Hispanic: Proportionate mortality ratio (PMR) adjusted for age and sex by disease category, U.S. residents age 15 and over, selected states, 1988–1998 .. 52

Table 2-23. Forestry workers, white, non-Hispanic: Proportionate mortality ratio (PMR) adjusted for age and sex by disease category, U.S. residents age 15 and over, selected states, 1988–1998. 53

Table 2-24. Fishery workers, white, non-Hispanic: Proportionate mortality ratio (PMR) adjusted for age and sex by disease category, U.S. residents age 15 and over, selected states, 1988–1998. 54

Table 2-25. Crop farm workers, black, non-Hispanic: Proportionate mortality ratio (PMR) adjusted for age and sex by disease category, U.S. residents age 15 and over, selected states, 1988–1998 55

Table 2-26. Livestock farm workers, black, non-Hispanic: Proportionate mortality ratio (PMR) adjusted for age and sex by disease category, U.S. residents age 15 and over, selected states, 1988–1998 56

List of Tables and Figures

Table 2-27. Farm managers, black, non-Hispanic: Proportionate mortality ratio (PMR) adjusted for age and sex by disease category, U.S. residents age 15 and over, selected states, 1988–1998 57

Table 2-28. Landscape and horticultural workers, black, non-Hispanic: Proportionate mortality ratio (PMR) adjusted for age and sex by disease category, U.S. residents age 15 and over, selected states, 1988–1998 .. 58

Table 2-29. Forestry workers, black, non-Hispanic: Proportionate mortality ratio (PMR) adjusted for age and sex by disease category, U.S. residents age 15 and over, selected states, 1988–1998 59

Table 2-30. Fishery workers, black, non-Hispanic: Proportionate mortality ratio (PMR) adjusted for age and sex by disease category, U.S. residents age 15 and over, selected states, 1988–1998 60

Table 2-31. Crop farm workers, Hispanic: Proportionate mortality ratio (PMR) adjusted for age and sex by disease category, U.S. residents age 15 and over, selected states, 1988–1998. 61

Table 2-32. Livestock farm workers, Hispanic: Proportionate mortality ratio (PMR) adjusted for age and sex by disease category, U.S. residents age 15 and over, selected states, 1988–1998 62

Table 2-33. Farm managers, Hispanic: Proportionate mortality ratio (PMR) adjusted for age and sex by disease category, U.S. residents age 15 and over, selected states, 1988–1998. 63

Table 2-34. Landscape and horticultural workers, Hispanic: Proportionate mortality ratio (PMR) adjusted for age and sex by disease category, U.S. residents age 15 and over, selected states, 1988–1998 64

Table 2-35. Forestry workers, Hispanic: Proportionate mortality ratio (PMR) adjusted for age and sex by disease category, U.S. residents age 15 and over, selected states, 1988–1998. 65

Table 2-36. Fishery workers, Hispanic: Proportionate mortality ratio (PMR) adjusted for age and sex by disease category, U.S. residents age 15 and over, selected states, 1988–1998. 66

Mortality by Agricultural Group and Race/Ethnicity within Disease Category

Figure 2-21. Tuberculosis: Proportionate mortality ratio (PMR) adjusted for age and sex by agricultural group and race/ethnicity, U.S. residents age 15 and over, selected states, 1988–1998 67

Figure 2-22. Mycoses: Proportionate mortality ratio (PMR) adjusted for age and sex by agricultural group and race/ethnicity, U.S. residents age 15 and over, selected states, 1988–1998 68

Figure 2-23. Sarcoidosis: Proportionate mortality ratio (PMR) adjusted for age and sex by agricultural group and race/ethnicity, U.S. residents age 15 and over, selected states, 1988–1998 69

Figure 2-24. Malignant neoplasms of trachea/bronchus/lung/pleura: Proportionate mortality ratio (PMR) adjusted for age and sex by agricultural group and race/ethnicity, U.S. residents age 15 and over, selected states, 1988–1998. 70

Figure 2-25. Acute respiratory infections: Proportionate mortality ratio (PMR) adjusted for age and sex by agricultural group and race/ethnicity, U.S. residents age 15 and over, selected states, 1988–1998. 71

Figure 2-26. Other diseases of upper respiratory tract: Proportionate mortality ratio (PMR) adjusted for age and sex by agricultural group and race/ethnicity, U.S. residents age 15 and over, selected states, 1988–1998. . 72

Figure 2-27. Pneumonia and influenza: Proportionate mortality ratio (PMR) adjusted for age and sex by agricultural group and race/ethnicity, U.S. residents age 15 and over, selected states, 1988–1998. 73

List of Tables and Figures

Figure 2-28. Chronic obstructive pulmonary disease and allied conditions: Proportionate mortality ratio (PMR) adjusted for age and sex by agricultural group and race/ethnicity, U.S. residents age 15 and over, selected states, 1988–1998 74

Figure 2-29. Pneumoconioses and other lung diseases–external agents: Proportionate mortality ratio (PMR) adjusted for age and sex by agricultural group and race/ethnicity, U.S. residents age 15 and over, selected states, 1988–1998 75

Figure 2-30. Other diseases of respiratory system: Proportionate mortality ratio (PMR) adjusted for age and sex by agricultural group and race/ethnicity, U.S. residents age 15 and over, selected states, 1988–1998 ... 76

Tuberculosis Mortality within and by Agricultural Group

Table 2-37. Crop farm workers: Proportionate mortality ratio (PMR) adjusted for age, sex, and race/ethnicity for tuberculosis, U.S. residents age 15 and over, selected states, 1988–1998 77

Table 2-38. Livestock farm workers: Proportionate mortality ratio (PMR) adjusted for age, sex, and race/ethnicity for tuberculosis, U.S. residents age 15 and over, selected states, 1988–1998 78

Table 2-39. Farm managers: Proportionate mortality ratio (PMR) adjusted for age, sex, and race/ethnicity for tuberculosis, U.S. residents age 15 and over, selected states, 1988–1998 79

Table 2-40. Landscape and horticultural workers: Proportionate mortality ratio (PMR) adjusted for age, sex, and race/ethnicity for tuberculosis, U.S. residents age 15 and over, selected states, 1988–1998 80

Table 2-41. Forestry workers: Proportionate mortality ratio (PMR) adjusted for age, sex, and race/ethnicity for tuberculosis, U.S. residents age 15 and over, selected states, 1988–1998 81

Table 2-42. Fishery workers: Proportionate mortality ratio (PMR) adjusted for age, sex, and race/ethnicity for tuberculosis, U.S. residents age 15 and over, selected states, 1988–1998 82

Figure 2-31. Pulmonary tuberculosis: Proportionate mortality ratio (PMR) adjusted for age, sex, and race/ethnicity by agricultural group, U.S. residents age 15 and over, selected states, 1988–1998 83

Figure 2-32. Other respiratory tuberculosis: Proportionate mortality ratio (PMR) adjusted for age, sex, and race/ethnicity by agricultural group, U.S. residents age 15 and over, selected states, 1988–1998 84

Figure 2-33. Tuberculosis of meninges and central nervous system: Proportionate mortality ratio (PMR) adjusted for age, sex, and race/ethnicity by agricultural group, U.S. residents age 15 and over, selected states, 1988–1998 85

Figure 2-34. Tuberculosis of bones and joints: Proportionate mortality ratio (PMR) adjusted for age, sex, and race/ethnicity by agricultural group, U.S. residents age 15 and over, selected states, 1988–1998 86

Figure 2-35. Tuberculosis of other organs: Proportionate mortality ratio (PMR) adjusted for age, sex, and race/ethnicity by agricultural group, U.S. residents age 15 and over, selected states, 1988–1998 87

Figure 2-36. Miliary tuberculosis: Proportionate mortality ratio (PMR) adjusted for age, sex, and race/ethnicity by agricultural group, U.S. residents age 15 and over, selected states, 1988–1998 88

Mycoses Mortality within and by Agricultural Group

Table 2-43. Crop farm workers: Proportionate mortality ratio (PMR) adjusted for age, sex, and race/ethnicity for mycoses, U.S. residents age 15 and over, selected states, 1988–1998 89

List of Tables and Figures

Table 2-44. Livestock farm workers: Proportionate mortality ratio (PMR) adjusted for age, sex, and race/ethnicity for mycoses, U.S. residents age 15 and over, selected states, 1988–1998 90

Table 2-45. Farm managers: Proportionate mortality ratio (PMR) adjusted for age, sex, and race/ethnicity for mycoses, U.S. residents age 15 and over, selected states, 1988–1998. 91

Table 2-46. Landscape and horticultural workers: Proportionate mortality ratio (PMR) adjusted for age, sex, and race/ethnicity for mycoses, U.S. residents age 15 and over, selected states, 1988–1998 92

Table 2-47. Forestry workers: Proportionate mortality ratio (PMR) adjusted for age, sex, and race/ethnicity for mycoses, U.S. residents age 15 and over, selected states, 1988–1998. 93

Table 2-48. Fishery workers: Proportionate mortality ratio (PMR) adjusted for age, sex, and race/ethnicity for mycoses, U.S. residents age 15 and over, selected states, 1988–1998. 94

Figure 2-37. Candidiasis: Proportionate mortality ratio (PMR) adjusted for age, sex, and race/ethnicity by agricultural group, U.S. residents age 15 and over, selected states, 1988–1998 95

Figure 2-38. Histoplasmosis: Proportionate mortality ratio (PMR) adjusted for age, sex, and race/ethnicity by agricultural group, U.S. residents age 15 and over, selected states, 1988–1998 96

Figure 2-39. Blastomycotic infection: Proportionate mortality ratio (PMR) adjusted for age, sex, and race/ethnicity by agricultural group, U.S. residents age 15 and over, selected states, 1988–1998 97

Figure 2-40. Other mycoses: Proportionate mortality ratio (PMR) adjusted for age, sex, and race/ethnicity by agricultural group, U.S. residents age 15 and over, selected states, 1988–1998 98

Malignant Neoplasm of Trachea/Bronchus/Lung/Pleura Mortality within and by Agricultural Group

Table 2-49. Crop farm workers: Proportionate mortality ratio (PMR) adjusted for age, sex, and race/ethnicity for malignant neoplasms of trachea/bronchus/lung/pleura, U.S. residents age 15 and over, selected states, 1988–1998 99

Table 2-50. Livestock farm workers: Proportionate mortality ratio (PMR) adjusted for age, sex, and race/ethnicity for malignant neoplasms of trachea/bronchus/lung/pleura, U.S. residents age 15 and over, selected states, 1988–1998. 100

Table 2-51. Farm managers: Proportionate mortality ratio (PMR) adjusted for age, sex, and race/ethnicity for malignant neoplasms of trachea/bronchus/lung/pleura, U.S. residents age 15 and over, selected states, 1988–1998 101

Table 2-52. Landscape and horticultural workers: Proportionate mortality ratio (PMR) adjusted for age, sex, and race/ethnicity for malignant neoplasms of trachea/bronchus/lung/pleura, U.S. residents age 15 and over, selected states, 1988–1998. 102

Table 2-53. Forestry workers: Proportionate mortality ratio (PMR) adjusted for age, sex, and race/ethnicity for malignant neoplasms of trachea/bronchus/lung/pleura, U.S. residents age 15 and over, selected states, 1988–1998 103

Table 2-54. Fishery workers: Proportionate mortality ratio (PMR) adjusted for age, sex, and race/ethnicity for malignant neoplasms of trachea/bronchus/lung/pleura, U.S. residents age 15 and over, selected states, 1988–1998 104

List of Tables and Figures

Figure 2-41. Malignant neoplasm of trachea, bronchus, and lung: Proportionate mortality ratio (PMR) adjusted for age, sex, and race/ethnicity by agricultural group, U.S. residents age 15 and over, selected states, 1988–1998 105

Figure 2-42. Malignant neoplasm of pleura: Proportionate mortality ratio (PMR) adjusted for age, sex, and race/ethnicity by agricultural group, U.S. residents age 15 and over, selected states, 1988–1998 106

Acute Respiratory Infection Mortality within and by Agricultural Group

Table 2-55. Crop farm workers: Proportionate mortality ratio (PMR) adjusted for age, sex, and race/ethnicity for acute respiratory infections, U.S. residents age 15 and over, selected states, 1988–1998 107

Table 2-56. Livestock farm workers: Proportionate mortality ratio (PMR) adjusted for age, sex, and race/ethnicity for acute respiratory infections, U.S. residents age 15 and over, selected states, 1988–1998 108

Table 2-57. Farm managers: Proportionate mortality ratio (PMR) adjusted for age, sex, and race/ethnicity for acute respiratory infections, U.S. residents age 15 and over, selected states, 1988–1998 109

Table 2-58. Landscape and horticultural workers: Proportionate mortality ratio (PMR) adjusted for age, sex, and race/ethnicity for acute respiratory infections, U.S. residents age 15 and over, selected states, 1988–1998 110

Table 2-59. Forestry workers: Proportionate mortality ratio (PMR) adjusted for age, sex, and race/ethnicity for acute respiratory infections, U.S. residents age 15 and over, selected states, 1988–1998 111

Table 2-60. Fishery workers: Proportionate mortality ratio (PMR) adjusted for age, sex, and race/ethnicity for acute respiratory infections, U.S. residents age 15 and over, selected states, 1988–1998 112

Figure 2-43. Acute laryngitis and tracheitis: Proportionate mortality ratio (PMR) adjusted for age, sex, and race/ethnicity by agricultural group, U.S. residents age 15 and over, selected states, 1988–1998 113

Figure 2-44. Acute upper respiratory infections of multiple or unspecified sites: Proportionate mortality ratio (PMR) adjusted for age, sex, and race/ethnicity by agricultural group, U.S. residents age 15 and over, selected states, 1988–1998 114

Figure 2-45. Acute bronchitis and bronchiolitis: Proportionate mortality ratio (PMR) adjusted for age, sex, and race/ethnicity by agricultural group, U.S. residents age 15 and over, selected states, 1988–1998 115

Other Diseases of Upper Respiratory Tract Mortality within and by Agricultural Group

Table 2-61. Crop farm workers: Proportionate mortality ratio (PMR) adjusted for age, sex, and race/ethnicity for other diseases of upper respiratory tract, U.S. residents age 15 and over, selected states, 1988–1998 .. 116

Table 2-62. Livestock farm workers: Proportionate mortality ratio (PMR) adjusted for age, sex, and race/ethnicity for other diseases of upper respiratory tract, U.S. residents age 15 and over, selected states, 1988–1998 117

Table 2-63. Farm managers: Proportionate mortality ratio (PMR) adjusted for age, sex, and race/ethnicity for other diseases of upper respiratory tract, U.S. residents age 15 and over, selected states, 1988–1998 118

Table 2-64. Landscape and horticultural workers: Proportionate mortality ratio (PMR) adjusted for age, sex, and race/ethnicity for other diseases of upper respiratory tract, U.S. residents age 15 and over, selected states, 1988–1998 119

List of Tables and Figures

Table 2-65. Forestry workers: Proportionate mortality ratio (PMR) adjusted for age, sex, and race/ethnicity for other diseases of upper respiratory tract, U.S. residents age 15 and over, selected states, 1988–1998 .. 120

Table 2-66. Fishery workers: Proportionate mortality ratio (PMR) adjusted for age, sex, and race/ethnicity for other diseases of upper respiratory tract, U.S. residents age 15 and over, selected states, 1988–1998 .. 121

Figure 2-46. Chronic sinusitis: Proportionate mortality ratio (PMR) adjusted for age, sex, and race/ethnicity by agricultural group, U.S. residents age 15 and over, selected states, 1988–1998. 122

Figure 2-47. Other diseases of upper respiratory tract: Proportionate mortality ratio (PMR) adjusted for age, sex, and race/ethnicity by agricultural group, U.S. residents age 15 and over, selected states, 1988–1998 . 123

Pneumonia and Influenza Mortality within and by Agricultural Group

Table 2-67. Crop farm workers: Proportionate mortality ratio (PMR) adjusted for age, sex, and race/ethnicity for pneumonia and influenza, U.S. residents age 15 and over, selected states, 1988–1998 124

Table 2-68. Livestock farm workers: Proportionate mortality ratio (PMR) adjusted for age, sex, and race/ethnicity for pneumonia and influenza, U.S. residents age 15 and over, selected states, 1988–1998 125

Table 2-69. Farm managers: Proportionate mortality ratio (PMR) adjusted for age, sex, and race/ethnicity for pneumonia and influenza, U.S. residents age 15 and over, selected states, 1988–1998 126

Table 2-70. Landscape and horticultural workers: Proportionate mortality ratio (PMR) adjusted for age, sex, and race/ethnicity for pneumonia and influenza, U.S. residents age 15 and over, selected states, 1988–1998. 127

Table 2-71. Forestry workers: Proportionate mortality ratio (PMR) adjusted for age, sex, and race/ethnicity for pneumonia and influenza, U.S. residents age 15 and over, selected states, 1988–1998 128

Table 2-72. Fishery workers: Proportionate mortality ratio (PMR) adjusted for age, sex, and race/ethnicity for pneumonia and influenza, U.S. residents age 15 and over, selected states, 1988–1998 129

Figure 2-48. Viral pneumonia: Proportionate mortality ratio (PMR) adjusted for age, sex, and race/ethnicity by agricultural group, U.S. residents age 15 and over, selected states, 1988–1998. 130

Figure 2-49. Pneumococcal pneumonia: Proportionate mortality ratio (PMR) adjusted for age, sex, and race/ethnicity by agricultural group, U.S. residents age 15 and over, selected states, 1988–1998 131

Figure 2-50. Other bacterial pneumonia: Proportionate mortality ratio (PMR) adjusted for age, sex, and race/ethnicity by agricultural group, U.S. residents age 15 and over, selected states, 1988–1998 132

Figure 2-51. Pneumonia due to other specified organism: Proportionate mortality ratio (PMR) adjusted for age, sex, and race/ethnicity by agricultural group, U.S. residents age 15 and over, selected states, 1988–1998. . . . 133

Figure 2-52. Bronchopneumonia, organism unspecified: Proportionate mortality ratio (PMR) adjusted for age, sex, and race/ethnicity by agricultural group, U.S. residents age 15 and over, selected states, 1988–1998 134

Figure 2-53. Pneumonia, organism unspecified: Proportionate mortality ratio (PMR) adjusted for age, sex, and race/ethnicity by agricultural group, U.S. residents age 15 and over, selected states, 1988–1998 135

Figure 2-54. Influenza: Proportionate mortality ratio (PMR) adjusted for age, sex, and race/ethnicity by agricultural group, U.S. residents age 15 and over, selected states, 1988–1998 136

List of Tables and Figures

COPD Mortality within and by Agricultural Group

Table 2-73. Crop farm workers: Proportionate mortality ratio (PMR) adjusted for age, sex, and race/ethnicity for chronic obstructive pulmonary disease and allied conditions, U.S. residents age 15 and over, selected states, 1988–1998 137

Table 2-74. Livestock farm workers: Proportionate mortality ratio (PMR) adjusted for age, sex, and race/ethnicity for chronic obstructive pulmonary disease and allied conditions, U.S. residents age 15 and over, selected states, 1988–1998 138

Table 2-75. Farm managers: Proportionate mortality ratio (PMR) adjusted for age, sex, and race/ethnicity for chronic obstructive pulmonary disease and allied conditions, U.S. residents age 15 and over, selected states, 1988–1998 139

Table 2-76. Landscape and horticultural workers: Proportionate mortality ratio (PMR) adjusted for age, sex, and race/ethnicity for chronic obstructive pulmonary disease and allied conditions, U.S. residents age 15 and over, selected states, 1988–1998 140

Table 2-77. Forestry workers: Proportionate mortality ratio (PMR) adjusted for age, sex, and race/ethnicity for chronic obstructive pulmonary disease and allied conditions, U.S. residents age 15 and over, selected states, 1988–1998 141

Table 2-78. Fishery workers: Proportionate mortality ratio (PMR) adjusted for age, sex, and race/ethnicity for chronic obstructive pulmonary disease and allied conditions, U.S. residents age 15 and over, selected states, 1988–1998 142

Figure 2-55. Bronchitis, not specified as acute or chronic: Proportionate mortality ratio (PMR) adjusted for age, sex, and race/ethnicity by agricultural group, U.S. residents age 15 and over, selected states, 1988–1998 143

Figure 2-56. Chronic bronchitis: Proportionate mortality ratio (PMR) adjusted for age, sex, and race/ethnicity by agricultural group, U.S. residents age 15 and over, selected states, 1988–1998 144

Figure 2-57. Emphysema: Proportionate mortality ratio (PMR) adjusted for age, sex, and race/ethnicity by agricultural group, U.S. residents age 15 and over, selected states, 1988–1998 145

Figure 2-58. Asthma: Proportionate mortality ratio (PMR) adjusted for age, sex, and race/ethnicity by agricultural group, U.S. residents age 15 and over, selected states, 1988–1998 146

Figure 2-59. Bronchiectasis: Proportionate mortality ratio (PMR) adjusted for age, sex, and race/ethnicity by agricultural group, U.S. residents age 15 and over, selected states, 1988–1998 147

Figure 2-60. Hypersensitivity pneumonitis: Proportionate mortality ratio (PMR) adjusted for age, sex, and race/ethnicity by agricultural group, U.S. residents age 15 and over, selected states, 1988–1998 148

Figure 2-61. Chronic airway obstruction, not elsewhere classified: Proportionate mortality ratio (PMR) adjusted for age, sex, and race/ethnicity by agricultural group, U.S. residents age 15 and over, selected states, 1988–1998 149

Pneumoconiosis and Other Lung Disease Mortality within and by Agricultural Group

Table 2-79. Crop farm workers: Proportionate mortality ratio (PMR) adjusted for age, sex, and race/ethnicity for pneumoconiosis and other lung diseases–external agents, U.S. residents age 15 and over, selected states, 1988–1998 150

List of Tables and Figures

Table 2-80. Livestock farm workers: Proportionate mortality ratio (PMR) adjusted for age, sex, and race/ethnicity for pneumoconiosis and other lung diseases–external agents, U.S. residents age 15 and over, selected states, 1988–1998 151

Table 2-81. Farm managers: Proportionate mortality ratio (PMR) adjusted for age, sex, and race/ethnicity for pneumoconiosis and other lung diseases–external agents, U.S. residents age 15 and over, selected states, 1988–1998 152

Table 2-82. Landscape and horticultural workers: Proportionate mortality ratio (PMR) adjusted for age, sex, and race/ethnicity for pneumoconiosis and other lung diseases–external agents, U.S. residents age 15 and over, selected states 153

Table 2-83. Forestry workers: Proportionate mortality ratio (PMR) adjusted for age, sex, and race/ethnicity for pneumoconiosis and other lung diseases–external agents, U.S. residents age 15 and over, selected states, 1988–1998 154

Table 2-84. Fishery workers: Proportionate mortality ratio (PMR) adjusted for age, sex, and race/ethnicity for pneumoconiosis and other lung diseases–external agents, U.S. residents age 15 and over, selected states, 1988–1998 155

Figure 2-62. Coal workers' pneumoconiosis: Proportionate mortality ratio (PMR) adjusted for age, sex, and race/ethnicity by agricultural group, U.S. residents age 15 and over, selected states, 1988–1998 156

Figure 2-63. Asbestosis: Proportionate mortality ratio (PMR) adjusted for age, sex, and race/ethnicity by agricultural group, U.S. residents age 15 and over, selected states, 1988–1998 157

Figure 2-64. Pneumoconiosis due to other silica or silicates: Proportionate mortality ratio (PMR) adjusted for age, sex, and race/ethnicity by agricultural group, U.S. residents age 15 and over, selected states, 1988–1998. 158

Figure 2-65. Pneumoconiosis, unspecified: Proportionate mortality ratio (PMR) adjusted for age, sex, and race/ethnicity by agricultural group, U.S. residents age 15 and over, selected states, 1988–1998 159

Figure 2-66. Pneumonitis due to solids and liquids: Proportionate mortality ratio (PMR) adjusted for age, sex, and race/ethnicity by agricultural group, U.S. residents age 15 and over, selected states, 1988–1998 . 160

Figure 2-67. Respiratory conditions due to other and unspecified external agents: Proportionate mortality ratio (PMR) adjusted for age, sex, and race/ethnicity by agricultural group, U.S. residents age 15 and over, selected states, 1988–1998. 161

Other Diseases of Respiratory System Mortality within and by Agricultural Group

Table 2-85. Crop farm workers: Proportionate mortality ratio (PMR) adjusted for age, sex, and race/ethnicity for other diseases of respiratory system, U.S. residents age 15 and over, selected states, 1988–1998 162

Table 2-86. Livestock farm workers: Proportionate mortality ratio (PMR) adjusted for age, sex, and race/ethnicity for other diseases of respiratory system, U.S. residents age 15 and over, selected states, 1988–1998 163

Table 2-87. Farm managers: Proportionate mortality ratio (PMR) adjusted for age, sex, and race/ethnicity for other diseases of respiratory system, U.S. residents age 15 and over, selected states, 1988–1998 164

Table 2-88. Landscape and horticultural workers: Proportionate mortality ratio (PMR) adjusted for age, sex, and race/ethnicity for other diseases of respiratory system, U.S. residents age 15 and over, selected states, 1988–1998 165

List of Tables and Figures

Table 2-89. Forestry workers: Proportionate mortality ratio (PMR) adjusted for age, sex, and race/ethnicity for other diseases of respiratory system, U.S. residents age 15 and over, selected states, 1988–1998 166

Table 2-90. Fishery workers: Proportionate mortality ratio (PMR) adjusted for age, sex, and race/ethnicity for other diseases of respiratory system, U.S. residents age 15 and over, selected states, 1988–1998 167

Figure 2-68. Empyema: Proportionate mortality ratio (PMR) adjusted for age, sex, and race/ethnicity by agricultural group, U.S. residents age 15 and over, selected states, 1988–1998 168

Figure 2-69. Pleurisy: Proportionate mortality ratio (PMR) adjusted for age, sex, and race/ethnicity by agricultural group, U.S. residents age 15 and over, selected states, 1988–1998 169

Figure 2-70. Pneumothorax: Proportionate mortality ratio (PMR) adjusted for age, sex, and race/ethnicity by agricultural group, U.S. residents age 15 and over, selected states, 1988–1998 170

Figure 2-71. Abscess of lung and mediastinum: Proportionate mortality ratio (PMR) adjusted for age, sex, and race/ethnicity by agricultural group, U.S. residents age 15 and over, selected states, 1988–1998 171

Figure 2-72. Pulmonary congestion and hypostasis: Proportionate mortality ratio (PMR) adjusted for age, sex, and race/ethnicity by agricultural group, U.S. residents age 15 and over, selected states, 1988–1998 . 172

Figure 2-73. Postinflammatory pulmonary fibrosis: Proportionate mortality ratio (PMR) adjusted for age, sex, and race/ethnicity by agricultural group, U.S. residents age 15 and over, selected states, 1988–1998 . 173

Figure 2-74. Other alveolar and parietoalveolar pneumonopathy: Proportionate mortality ratio (PMR) adjusted for age, sex, and race/ethnicity by agricultural group, U.S. residents age 15 and over, selected states, 1988–1998 174

Figure 2-75. Other diseases of the lung: Proportionate mortality ratio (PMR) adjusted for age, sex, and race/ethnicity by agricultural group, U.S. residents age 15 and over, selected states, 1988–1998 175

Figure 2-76. Other diseases of respiratory system: Proportionate mortality ratio (PMR) adjusted for age, sex, and race/ethnicity by agricultural group, U.S. residents age 15 and over, selected states, 1988–1998 176

Morbidity

Morbidity by Agricultural Group within Respiratory Condition–NHIS

Table 3-1. Hayfever (past year): Estimated prevalence and prevalence ratio (PR) adjusted for age, sex, race/ethnicity, and smoking status by agricultural group and survey year, U.S. residents age 18 and over, 1997–1999 179

Table 3-2. Sinusitis (past year): Estimated prevalence and prevalence ratio (PR) adjusted for age, sex, race/ethnicity, and smoking status by agricultural group and survey year, U.S. residents age 18 and over, 1997–1999 180

Table 3-3. Chronic bronchitis (past year): Estimated prevalence and prevalence ratio (PR) adjusted for age, sex, race/ethnicity, and smoking status by agricultural group and survey year, U.S. residents age 18 and over, 1997–1999 181

Table 3-4. Emphysema (ever): Estimated prevalence and prevalence ratio (PR) adjusted for age, sex, race/ethnicity, and smoking status by agricultural group and survey year, U.S. residents age 18 and over, 1997–1999 182

List of Tables and Figures

Table 3-5. Asthma (ever): Estimated prevalence and prevalence ratio (PR) adjusted for age, sex, race/ethnicity, and smoking status by agricultural group and survey year, U.S. residents age 18 and over, 1997–1999 183

Table 3-6. Lung cancer (ever): Estimated prevalence and prevalence ratio (PR) adjusted for age, sex, race/ethnicity, and smoking status by agricultural group and survey year, U.S. residents age 18 and over, 1997–1999 184

Figure 3-1. Respiratory conditions: Prevalence ratio (PR) adjusted for age, sex, race/ethnicity, and smoking status by agricultural group, U.S. residents age 18 and over, 1997–1999 185

Morbidity by Respiratory Condition and Sex within Agricultural Group–NHIS

Figure 3-2. Farm workers: Prevalence ratio (PR) adjusted for age, race/ethnicity, and smoking status by respiratory condition and sex, U.S. residents age 18 and over, 1997–1999 186

Figure 3-3. Farm managers: Prevalence ratio (PR) adjusted for age, race/ethnicity, and smoking status by respiratory condition and sex, U.S. residents age 18 and over, 1997–1999 187

Figure 3-4. Forestry/fishery workers: Prevalence ratio (PR) adjusted for age, race/ethnicity, and smoking status by respiratory condition and sex, U.S. residents age 18 and over, 1997–1999 188

Morbidity by Respiratory Condition and Race/Ethnicity within Agricultural Group–NHIS

Figure 3-5. Farm workers: Prevalence ratio (PR) adjusted for age, sex, and smoking status by respiratory condition and race/ethnicity, U.S. residents age 18 and over, 1997–1999 189

Figure 3-6. Farm managers: Prevalence ratio (PR) adjusted for age, sex, and smoking status by respiratory condition and race/ethnicity, U.S. residents age 18 and over, 1997–1999 190

Figure 3-7. Forestry/fishery workers: Prevalence ratio (PR) adjusted for age, sex, and smoking status by respiratory condition and race/ethnicity, U.S. residents age 18 and over, 1997–1999 191

Morbidity by Respiratory Condition and Smoking Status within Agricultural Group–NHIS

Figure 3-8. Farm workers: Prevalence ratio (PR) adjusted for age, sex, and race/ethnicity by respiratory condition and smoking status, U.S. residents age 18 and over, 1997–1999 192

Figure 3-9. Farm managers: Prevalence ratio (PR) adjusted for age, sex, and race/ethnicity by respiratory condition and smoking status, U.S. residents age 18 and over, 1997–1999 193

Figure 3-10. Forestry/fishery workers: Prevalence ratio (PR) adjusted for age, sex, and race/ethnicity by respiratory condition and smoking status, U.S. residents age 18 and over, 1997–1999 194

Morbidity by Agricultural Group within Respiratory Condition–NHANES III

Table 3-7. Wheezing, apart from a cold (current): Estimated prevalence and prevalence ratio (PR) adjusted for age, sex, race/ethnicity, and smoking status by agricultural group, U.S. residents age 17 and over, 1988–1994 195

Table 3-8. Cough (current): Estimated prevalence and prevalence ratio (PR) adjusted for age, sex, race/ethnicity, and smoking status by agricultural group, U.S. residents age 17 and over, 1988–1994 196

Table 3-9. Phlegm (current): Estimated prevalence and prevalence ratio (PR) adjusted for age, sex, race/ethnicity, and smoking status by agricultural group, U.S. residents age 17 and over, 1988–1994 197

List of Tables and Figures

Table 3-10. Shortness of breath (current): Estimated prevalence and prevalence ratio (PR) adjusted for age, sex, race/ethnicity, and smoking status by agricultural group, U.S. residents age 17 and over, 1988–1994 . 198

Table 3-11. Stuffy, itchy, runny nose (past year): Estimated prevalence and prevalence ratio (PR) adjusted for age, sex, race/ethnicity, and smoking status by agricultural group, U.S. residents age 17 and over, 1988–1994. 199

Table 3-12. Cold or flu (past year): Estimated prevalence and prevalence ratio (PR) adjusted for age, sex, race/ethnicity, and smoking status by agricultural group, U.S. residents age 17 and over, 1988–1994 200

Table 3-13. Sinusitis (past year): Estimated prevalence and prevalence ratio (PR) adjusted for age, sex, race/ethnicity, and smoking status by agricultural group, U.S. residents age 17 and over, 1988–1994 201

Table 3-14. Pneumonia (past year): Estimated prevalence and prevalence ratio (PR) adjusted for age, sex, race/ethnicity, and smoking status by agricultural group, U.S. residents age 17 and over, 1988–1994 202

Table 3-15. Wheezing (past year): Estimated prevalence and prevalence ratio (PR) adjusted for age, sex, race/ethnicity, and smoking status by agricultural group, U.S. residents age 17 and over, 1988–1994 203

Table 3-16. Asthma (ever): Estimated prevalence and prevalence ratio (PR) adjusted for age, sex, race/ethnicity, and smoking status by agricultural group, U.S. residents age 17 and over, 1988–1994 204

Table 3-17. Chronic bronchitis (ever): Estimated prevalence and prevalence ratio (PR) adjusted for age, sex, race/ethnicity, and smoking status by agricultural group, U.S. residents age 17 and over, 1988–1994 205

Table 3-18. Emphysema (ever): Estimated prevalence and prevalence ratio (PR) adjusted for age, sex, race/ethnicity, and smoking status by agricultural group, U.S. residents age 17 and over, 1988–1994 206

Table 3-19. Hayfever (ever): Estimated prevalence and prevalence ratio (PR) adjusted for age, sex, race/ethnicity, and smoking status by agricultural group, U.S. residents age 17 and over, 1988–1994 207

Figure 3-11. Respiratory conditions (current): Prevalence ratio (PR) adjusted for age, sex, race/ethnicity, and smoking status by agricultural group, U.S. residents age 17 and over, 1988–1994. 208

Figure 3-12. Respiratory conditions (past year): Prevalence ratio (PR) adjusted for age, sex, race/ethnicity, and smoking status by agricultural group, U.S. residents age 17 and over, 1988–1994 209

Figure 3-13. Respiratory conditions (ever): Prevalence ratio (PR) adjusted for age, sex, race/ethnicity, and smoking status by agricultural group, U.S. residents age 17 and over, 1988–1994. 210

Morbidity by Sex within Respiratory Condition and Agricultural Group–NHANES III

Figure 3-14. Respiratory conditions (current), farm workers: Prevalence ratio (PR) adjusted for age, race/ethnicity, and smoking status by sex, U.S. residents age 17 and over, 1988–1994 211

Figure 3-15. Respiratory conditions (current), farm managers: Prevalence ratio (PR) adjusted for age, race/ethnicity, and smoking status by sex, U.S. residents age 17 and over, 1988–1994 212

Figure 3-16. Respiratory conditions (current), other agricultural workers: Prevalence ratio (PR) adjusted for age, race/ethnicity, and smoking status by sex, U.S. residents age 17 and over, 1988–1994 213

Figure 3-17. Respiratory conditions (past year), farm workers: Prevalence ratio (PR) adjusted for age, race/ethnicity, and smoking status by sex, U.S. residents age 17 and over, 1988–1994 214

Figure 3-18. Respiratory conditions (past year), farm managers: Prevalence ratio (PR) adjusted for age, race/ethnicity, and smoking status by sex, U.S. residents age 17 and over, 1988–1994 215

List of Tables and Figures

Figure 3-19. Respiratory conditions (past year), other agricultural workers: Prevalence ratio (PR) adjusted for age, race/ethnicity, and smoking status by sex, U.S. residents age 17 and over, 1988–1994 216

Figure 3-20. Respiratory conditions (ever), farm workers: Prevalence ratio (PR) adjusted for age, race/ethnicity, and smoking status by sex, U.S. residents age 17 and over, 1988–1994 217

Figure 3-21. Respiratory conditions (ever), farm managers: Prevalence ratio (PR) adjusted for age, race/ethnicity, and smoking status by sex, U.S. residents age 17 and over, 1988–1994 218

Figure 3-22. Respiratory conditions (ever), other agricultural workers: Prevalence ratio (PR) adjusted for age, race/ethnicity, and smoking status by sex, U.S. residents age 17 and over, 1988–1994 219

Morbidity by Race/Ethnicity within Respiratory Condition and Agricultural Group–NHANES III

Figure 3-23. Respiratory conditions (current), farm workers: Prevalence ratio (PR) adjusted for age, sex, and smoking status by race/ethnicity, U.S. residents age 17 and over, 1988–1994 220

Figure 3-24. Respiratory conditions (current), farm managers: Prevalence ratio (PR) adjusted for age, sex, and smoking status by race/ethnicity, U.S. residents age 17 and over, 1988–1994 221

Figure 3-25. Respiratory conditions (current), other agricultural workers: Prevalence ratio (PR) adjusted for age, sex, and smoking status by race/ethnicity, U.S. residents age 17 and over, 1988–1994 222

Figure 3-26. Respiratory conditions (past year), farm workers: Prevalence ratio (PR) adjusted for age, sex, and smoking status by race/ethnicity, U.S. residents age 17 and over, 1988–1994 223

Figure 3-27. Respiratory conditions (past year), farm managers: Prevalence ratio (PR) adjusted for age, sex, and smoking status by race/ethnicity, U.S. residents age 17 and over, 1988–1994 224

Figure 3-28. Respiratory conditions (past year), other agricultural workers: Prevalence ratio (PR) adjusted for age, sex, and smoking status by race/ethnicity, U.S. residents age 17 and over, 1988–1994 225

Figure 3-29. Respiratory conditions (ever), farm workers: Prevalence ratio (PR) adjusted for age, sex, and smoking status by race/ethnicity, U.S. residents age 17 and over, 1988–1994 226

Figure 3-30. Respiratory conditions (ever), farm managers: Prevalence ratio (PR) adjusted for age, sex, and smoking status by race/ethnicity, U.S. residents age 17 and over, 1988–1994 227

Figure 3-31. Respiratory conditions (ever), other agricultural workers: Prevalence ratio (PR) adjusted for age, sex, and smoking status by race/ethnicity, U.S. residents age 17 and over, 1988–1994 228

Morbidity by Smoking Status within Respiratory Condition and Agricultural Group–NHANES III

Figure 3-32. Respiratory conditions (current), farm workers: Prevalence ratio (PR) adjusted for age, sex, and race/ethnicity by smoking status, U.S. residents age 17 and over, 1988–1994 229

Figure 3-33. Respiratory conditions (current), farm managers: Prevalence ratio (PR) adjusted for age, sex, and race/ethnicity by smoking status, U.S. residents age 17 and over, 1988–1994 230

Figure 3-34. Respiratory conditions (current), other agricultural workers: Prevalence ratio (PR) adjusted for age, sex, and race/ethnicity by smoking status, U.S. residents age 17 and over, 1988–1994 231

Figure 3-35. Respiratory conditions (past year), farm workers: Prevalence ratio (PR) adjusted for age, sex, and race/ethnicity by smoking status, U.S. residents age 17 and over, 1988–1994 232

List of Tables and Figures

Figure 3-36. Respiratory conditions (past year), farm managers: Prevalence ratio (PR) adjusted for age, sex, and race/ethnicity by smoking status, U.S. residents age 17 and over, 1988–1994 233

Figure 3-37. Respiratory conditions (past year), other agricultural workers: Prevalence ratio (PR) adjusted for age, sex, and race/ethnicity by smoking status, U.S. residents age 17 and over, 1988–1994 234

Figure 3-38. Respiratory conditions (ever), farm workers: Prevalence ratio (PR) adjusted for age, sex, and race/ethnicity by smoking status, U.S. residents age 17 and over, 1988–1994 235

Figure 3-39. Respiratory conditions (ever), farm managers: Prevalence ratio (PR) adjusted for age, sex, and race/ethnicity by smoking status, U.S. residents age 17 and over, 1988–1994 236

Figure 3-40. Respiratory conditions (ever), other agricultural workers: Prevalence ratio (PR) adjusted for age, sex, and race/ethnicity by smoking status, U.S. residents age 17 and over, 1988–1994 237

Morbidity by Agricultural Group within Spirometry Index: FEV_1, FVC, PEF–NHANES III

Table 3-20. Spirometry: Forced expiratory volume in one second (FEV_1), forced vital capacity (FVC), and peak expiratory flow (PEF) by agricultural group, U.S. residents age 17 and over, 1988–1994.......... 238

Table 3-21. Spirometry: Percent predicted forced expiratory volume in one second (FEV_1), forced vital capacity (FVC), and peak expiratory flow (PEF) by agricultural group, U.S. residents age 17 and over, 1988–1994 239

Table 3-22a. Obstructive abnormality: Estimated prevalence and prevalence ratio (PR) adjusted for age, sex, race/ethnicity, and smoking status by agricultural group, U.S. residents age 17 and over, 1988–1994 240

Table 3-22b. Restrictive abnormality: Estimated prevalence and prevalence ratio (PR) adjusted for age, sex, race/ethnicity, and smoking status by agricultural group, U.S. residents age 17 and over, 1988–1994 240

Morbidity by Agricultural Group and Sex within Spirometry Index: FEV_1, FVC, PEF–NHANES III

Figure 3-41. Percent predicted forced expiratory volume in one second (FEV_1) by agricultural group and sex, U.S. residents age 17 and over, 1988–1994... 241

Figure 3-42. Percent predicted forced vital capacity (FVC) by agricultural group and sex, U.S. residents age 17 and over, 1988–1994... 242

Figure 3-43. Percent predicted peak expiratory flow (PEF) by agricultural group and sex, U.S. residents age 17 and over, 1988–1994... 243

Morbidity by Agricultural Group and Race/Ethnicity within Spirometry Index: FEV_1, FVC, PEF–NHANES III

Figure 3-44. Percent predicted forced expiratory volume in one second (FEV_1) by agricultural group and race/ethnicity, U.S. residents age 17 and over, 1988–1994 244

Figure 3-45. Percent predicted forced vital capacity (FVC) by agricultural group and race/ethnicity, U.S. residents age 17 and over, 1988–1994 245

Figure 3-46. Percent predicted peak expiratory flow (PEF) by agricultural group and race/ethnicity, U.S. residents age 17 and over, 1988–1994 246

List of Tables and Figures

Morbidity by Agricultural Group and Smoking Status within Spirometry Index: FEV$_1$, FVC, PEF–NHANES III

Figure 3-47. Percent predicted forced expiratory volume in one second (FEV$_1$) by agricultural group and smoking status, U.S. residents age 17 and over, 1988–1994 247

Figure 3-48. Percent predicted forced vital capacity (FVC) by agricultural group and smoking status, U.S. residents age 17 and over, 1988–1994 248

Figure 3-49. Percent predicted peak expiratory flow (PEF) by agricultural group and smoking status, U.S. residents age 17 and over, 1988–1994 249

Morbidity by Agricultural Group and Sex within Spirometry Index: Obstructive and Restrictive Abnormality–NHANES III

Figure 3-50. Spirometry: Percent of workers with obstructive abnormality by agricultural group and sex, U.S. residents age 17 and over, 1988–1994 250

Figure 3-51. Spirometry: Percent of workers with restrictive abnormality by agricultural group and sex, U.S. residents age 17 and over, 1988–1994 251

Morbidity by Agricultural Group and Race/Ethnicity within Spirometry Index: Obstructive and Restrictive Abnormality–NHANES III

Figure 3-52. Spirometry: Percent of workers with obstructive abnormality by agricultural group and race/ethnicity, U.S. residents age 17 and over, 1988–1994 252

Figure 3-53. Spirometry: Percent of workers with restrictive abnormality by agricultural group and race/ethnicity, U.S. residents age 17 and over, 1988–1994 253

Morbidity by Agricultural Group and Smoking Status within Spirometry Index: Obstructive and Restrictive Abnormality–NHANES III

Figure 3-54. Spirometry: Percent of workers with obstructive abnormality by agricultural group and smoking status, U.S. residents age 17 and over, 1988–1994 254

Figure 3-55. Spirometry: Percent of workers with restrictive abnormality by agricultural group and smoking status, U.S. residents age 17 and over, 1988–1994 255

Morbidity by Agricultural Group within Dust Diseases of the Lung and Respiratory Conditions Due to Toxic Agents–SOII

Table 3-23. Dust diseases of the lung: Estimated incidence per 10,000 workers by agricultural group, 1995–2001 256

Table 3-24. Respiratory conditions due to toxic agents: Estimated incidence per 10,000 workers by agricultural group, 1995–2001 257

Highlights and Limitations

Highlights

These selected highlights summarize the major findings in the report, including a description of results that were statistically elevated. Mortality statistics were derived from deaths from 24 states for 1988–1998, while morbidity data came from two large population-based surveys of the U.S. undertaken in 1997–1999 and 1988–1994.

- Decedents whose death certificate indicated that they worked as *crop workers* had significantly elevated mortality for a number of respiratory conditions, including hypersensitivity pneumonitis (proportionate mortality more than 10 times higher than expected), asthma, bronchitis, histoplasmosis, tuberculosis, pneumonia, and influenza. (Tables H-1 and H-2)

- Decedents whose death certificate indicated that they worked as *livestock farm workers* had significantly elevated mortality for several respiratory conditions, including hypersensitivity pneumonitis (proportionate mortality more than 50 times higher than expected), asthma, tuberculosis, and influenza. (Tables H-1 and H-2)

- Decedents whose death certificate indicated that they worked as *landscape or horticultural workers* had significantly elevated mortality for chronic obstructive pulmonary diseases (COPD), including chronic airways obstruction, and for abscesses of the lung and mediastinum. (Tables H-1 and H-2)

- Decedents whose death certificate indicated that they worked as *forestry workers* had significantly elevated mortality for tuberculosis, COPD, including chronic airways obstruction, and for pneumonia. (Tables H-1 and H-2)

- Decedents whose death certificate indicated that they worked as *fishery workers* had significantly elevated mortality for COPD, including chronic airways obstruction. (Tables H-1 and H-2)

- At least two of the agricultural groups studied in this report were noted to have significantly elevated mortality for several respiratory diseases, including tuberculosis, hypersensitivity pneumonitis, asthma, COPD, pneumonia, and influenza. Significantly elevated COPD mortality was noted in three agricultural groups (*landscape and horticultural workers*, *forestry workers*, and *fishery workers*). (Table H-2)

- Individuals who reported that their longest job held was *farm worker* had elevated prevalence of phlegm production compared to all non-agricultural workers. Prevalence of wheeze was elevated for female *farm workers* and shortness of breath was elevated for *farm workers* who had 'ever smoked.' (Table H-3)

- *Farm workers* had a prevalence ratio (PR) of 173 for obstructive abnormality. (Table 3-22a)

Highlights

Table H-1. Mortality: Significantly elevated proportionate mortality ratios (PMRs) by agricultural group

Disease (ICD-9 Code)	Number of Deaths	PMR	For more detail see: Table	For more detail see: Figure
Crop Farm Workers				
Hypersensitivity pneumonitis (495)	23	1,228	2-73	2-60
Blastomycotic infection (116)	14	245	2-43	2-39
Histoplasmosis (115)	27	183	2-43	2-38
Bronchitis, not specified as acute or chronic (490)	269	134	2-73	2-55
Abscess of lung and mediastinum (513)	153	120	2-85	2-71
Pulmonary congestion & hypostasis (514)	1,830	113	2-85	2-72
Asthma (493)	813	111	2-73	2-58
Tuberculosis (010–018)	522	148	2-1	2-1
Miliary tuberculosis (018)	35	196	2-37	2-36
Pulmonary tuberculosis (011)	437	152	2-37	7-31
Acute respiratory infections (460–466)	329	124	2-1	2-5
Acute upper respiratory infections of multiple or unspecified sites (465)	87	160	2-55	2-44
Acute bronchitis and bronchiolitis (466)	126	117	2-55	2-45
Pneumonia and influenza (480–487)	26,114	109	2-1	2-7
Influenza (487)	232	142	2-67	2-54
Other bacterial pneumonia (482)	955	120	2-67	2-50
Pneumonia, organism unspecified (486)	23,135	109	2-67	2-53
Livestock Farm Workers				
Hypersensitivity pneumonitis (495)	31	5,563	2-74	2-60
Other respiratory tuberculosis (012)	5	675	2-38	2-32
Tuberculosis of meninges and central nervous system (013)	5	546	2-38	2-33
Asthma (493)	276	150	2-74	2-58
Influenza (487)	73	150	2-68	2-54
Landscape and Horticulture Workers				
Abscess of lung and mediastinum (513)	13	190	2-88	2-71
Chronic obstructive pulmonary disease and allied conditions (COPD) (490–496)	799	109	2-4	2-8
Chronic airway obstruction, nec (496)	624	111	2-76	2-61
Forestry Workers				
Pulmonary tuberculosis (011)	41	143	2-41	2-31
Chronic obstructive pulmonary disease and allied conditions (COPD) (490–496)	2,318	122	2-5	2-8
Chronic airway obstruction, nec (496)	1,890	127	2-77	2-61
Pneumonia and influenza (480–487)	1,771	116	2-5	2-7
Pneumonia, organism unspecified (486)	1,564	117	2-71	2-53
Fishery Workers				
Chronic obstructive pulmonary disease and allied conditions (COPD) (490–496)	568	113	2-6	2-8
Chronic airway obstruction, nec (496)	455	116	2-78	2-61

nec - not elsewhere classified   ICD - International Classification of Diseases
NOTE: PMRs are adjusted for age, sex, and race, U.S. residents age 15 and over, selected states (see Appendix D), 1988–1998. PMRs are significantly different from 100 ($p<0.05$).
SOURCE: National Center for Health Statistics multiple-cause-of-death data

Highlights

Table H-2. Mortality: Disease and disease categories with significantly elevated proportionate mortality ratios (PMRs) in two or more agricultural groups

Disease (ICD-9 Code)	Crop Farm Workers	Livestock Farm Workers	Landscape and Horticulture Workers	Forestry Workers	Fishery Workers
Pulmonary tuberculosis (011)	✓			✓	
Abscess of lung and mediastinum (513)	✓		✓		
Pneumonia/influenza (480–487)	✓			✓	
Pneumonia, organism unspecified (486)	✓			✓	
Influenza (487)	✓	✓			
Chronic obstructive pulmonary disease (490–496)				✓	✓
Asthma (493)	✓	✓			
Hypersensitivity pneumonitis (495)	✓	✓			
Chronic airway obstruction, nec (496)			✓	✓	✓

nec - not elsewhere classified
NOTE: *Crop farm workers* had 10, *livestock farm workers* had 2, and *landscape and horticultural workers* had 1 other respiratory diseases or disease categories with significantly elevated PMRs. See Table H-1. PMRs are adjusted for age, sex, and race, U.S. residents age 15 and over, selected states (see Appendix D), 1988–1998. PMRs are significantly different from 100 ($p<0.05$).
SOURCE: National Center for Health Statistics multiple-cause-of-death data

Table H-3. Morbidity: Significantly elevated prevalence ratios (PRs) by agricultural group

Respiratory Condition	PR	For more details see: Table	For more details see: Figure
Farm Workers			
Phlegm (current)	133	3-9	3-11
Females	226		3-14
Ever smoked	156		3-32
Wheezing (apart from a cold), females	155		3-20
Wheezing (past year), females	146		3-17
Shortness of breath (current), ever smoked	130		3-32

NOTE: PRs are adjusted for age, sex, race, and smoking (except for smoking-specific analyses), U.S. residents age 17 and over, 1988–1994. PRs are significantly different from 100 ($p<0.05$).
SOURCE: National Center for Health Statistics, Third National Health and Nutrition Examination Survey (NHANES III)

Limitations

In addition to the following cautions, readers should see Appendix A for other limitations relating to specific sources of data presented in this report.

General

- In this report, the data are drawn from the major existing databases. However, other data may exist that would improve the completeness and reliability of the findings presented in this report. Readers who are aware of other data that should be considered for inclusion in future editions are encouraged to make their suggestions known (see Preface for contact information).

- Statistics in many tables and figures in this report are based on small numbers. Readers are cautioned that these can be unstable. Hence, inferences should be drawn with care, and should take the numerical basis into account.

- A decedent's or survey respondent's usual or current industry/occupation is not always indicative of the industry and occupation associated with the exposure that may be responsible for that individual's disease even when that disease is work-related. Readers are therefore cautioned not to make definitive causative inferences about industries and occupations based solely on the various mortality and morbidity tables and figures presented in this report.

Mortality Data

- Data from only 24 states were used in the mortality analysis since reliable information on industry and occupation was not available for every state. These 24 states collectively account for 32 percent of the U.S. agricultural worker population (Table 1-2); they do not include the three states having the most agricultural employment (California, Texas, and Florida). Although the information presented is believed to be reasonably representative of health outcomes among all agricultural workers, it may not provide a fully accurate picture.

- Individuals affected by chronic diseases with long latency have much more time to change residences prior to death than individuals affected by acute diseases with short latency. Thus, state of residence at death does not necessarily represent the location of a decedent's occupational exposure, even for a death that results directly from occupational respiratory disease. However, unlike many other occupations, farmers often continue to work well beyond 65 years of age and 18% of the U.S. farm operators are over age 65[1], indicating that farmers are less likely to change residences before death than other occupation.

- Work-related respiratory diseases are often chronic, may also have long latencies, but often may not be reported as the underlying cause of death. This led to a decision to consider both underlying and contributing causes of death in the mortality summary tables and figures in this report.

- Certifying physicians typically do not list all of a decedent's diseases on the death certificate. Therefore, even though contributing causes of death are considered, the mortality data presented in this report probably underestimate the occurrence of some or most respiratory diseases.

- As with any analysis based on death certificate data, there is undoubtedly some misclassification of cause of death. A treating physician may not correctly diagnose a particular disease during a patient's life or, as mentioned above, a certifying physician may fail to list a correctly diagnosed disease on the death certificate, particularly if another disease was directly responsible for the individual's death. In addition, the diagnoses listed on the death certificate sometimes are miscoded.

---

[1]U.S. Department of Commerce [1992]. Census of Agriculture. Washington, DC: U.S. Government Printing Office.

Limitations

- Data that depend, either directly or indirectly, on physician reporting or recording of occupational disease diagnoses can be influenced significantly by the physician's ability or willingness to suspect and evaluate a relationship between work and health. These, in turn, are influenced by evolving medical and scientific information, and by the legal, political and social environment. Some factors may lead to increased diagnosis and recording and reporting whereas others may reduce occupational disease recognition or reporting by physicians.

- The PMRs in this report have not been adjusted for smoking or any other confounding exposure because of lack of these data. Note that PMRs are vulnerable to difficulties in interpretation in that an elevated PMR may reflect an excess in a particular disease mortality or may simply arise from deficits in mortality from other diseases. The PMRs in this report are derived from data reported by only 24 states and omit data from some of the major agricultural states (e.g., California). They therefore may not be representative of the mortality patterns for the whole country. In addition, they may fail to indicate risks for some agricultural operations and situations not, or poorly, represented in the 24 states.

Morbidity Data

- Data from both the NHIS and NHANES surveys are restricted to a sample of household-based respondents in the U.S. A typical round of NHIS or NHANES has about 30,000 respondents. Although weights reflecting the probability of selection for each survey respondent are provided (and were used in the analysis) to enable national estimates, the actual number of respondents is especially small when the data are disaggregated into groups (e.g., agricultural workers). For certain conditions such as emphysema and lung cancer, the numbers are especially small. Hence, the cautions given above for mortality data, against making broad inferences or generalizations from the data provided in this report, apply even more strongly here. In the case of the NHIS data, an attempt was made to compensate for small numbers by summing estimates from the most recent three years (1997–1999) for which survey data were available at the time of the analysis.

- Some of the conditions about which respondents were asked in these surveys relate to the individual's lifetime (e.g., has a doctor ever told you that you have asthma?), whereas others relate to a more recent period (e.g., during the past 12 months, have you had pneumonia?). Hence, the relationship between work and health may be conditional on the time frame of reference for the question, the individual's age, and whether the industry/occupation codes used in the analysis relate to the respondent's current or usual industry/occupation. For the NHANES data, the industry/occupation in which the respondent worked longest was used in the analysis, whereas for the NHIS data only the current industry/occupation was asked of the respondent.

- The questions asked about conditions in the NHANES and NHIS surveys are sensitive to the respondent's ability to recognize such conditions and to correctly answer the questions. Thus, there are potential reporting biases that may be associated, for example, with respondent age or socioeconomic status. The spirometric data from NHANES do not share this limitation, as they are objective measures of respiratory health derived from lung function tests.

- The method used to calculate confidence intervals for prevalence ratios assumes an underlying Poisson distribution and is strictly applicable to outcomes that are rare. Some of the outcomes reported in the survey (e.g., asthma) are not rare, and as a consequence the reported confidence intervals should be regarded as approximate.

Limitations

- Unlike the NHIS and NHANES data, public-use data files were not available for the BLS injury and illness data. Only incidence rates summarized by industry for *dust diseases of the lung* and *respiratory conditions due to toxic agents* are publicly available, and it was not possible to adjust the survey results for factors such as age, sex, race/ethnicity, or smoking status. In the BLS data, work-related diseases are generally under-recognized and under-reported by employers. (Note: BLS confidential microdata for non-fatal injuries and illnesses is available for research, but users may access this data only at the BLS national office in Washington, D.C.)

- The agricultural occupation and industry coding systems for the source data employed in the presentation of the demographic, morbidity, and mortality data are broadly similar but differ in detail, preventing exact comparisons between them. See Appendices E, F, and G for descriptions of the industry and occupation codes relevant to this report.

Section 1

Demographics

Demographics

Table 1-1. Demographic characteristics of employed U.S. agricultural workers by agricultural group and occupation, 2002

Occupation (Census Occupation Code)	Number (in thousands)	Percent Female	Black	Hispanic
Farm operators and managers	1,168	24.5	1.2	4.2
Farmers, except horticultural (473)	898	25.5	0.7	2.5
Horticultural specialty farmers (474)	76	13.5	5.0	17.4
Managers, farms, except horticultural (475)	169	22.4	0.7	6.4
Other agricultural and related occupations	2,181	19.2	6.9	36.4
Farm workers (479)	716	21.0	4.7	45.4
Supervisors, related agricultural occupations (485)	188	7.7	5.5	19.1
Groundskeepers and gardeners, except farm (486)	973	7.8	10.0	35.4
Animal caretakers, except farm (487)	170	68.1	4.3	4.8
Graders and sorters, agricultural products (488)	68	67.7	2.8	71.1
Timber cutting and logging (496)	54	1.6	7.5	6.0
All Farming, Forestry, and Fishing Occupations	**3,480**	**20.6**	**4.9**	**24.4**

NOTE: Data for occupations with fewer than 50,000 employed are not published separately but are included in the total. See Appendices E, F, and G for occupations included in the analyses.
SOURCE: U.S. Bureau of Labor Statistics: *Current Population Survey* (ftp://ftp.bls.gov/pub/special.requests/lf/aa2002/aat11.txt)

Demographics

Figure 1-1. Distribution of employed U.S. agricultural workers by sex, race, and ethnicity in comparison to all U.S. workers, 1997 and 2002

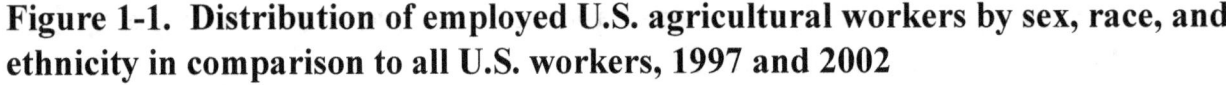

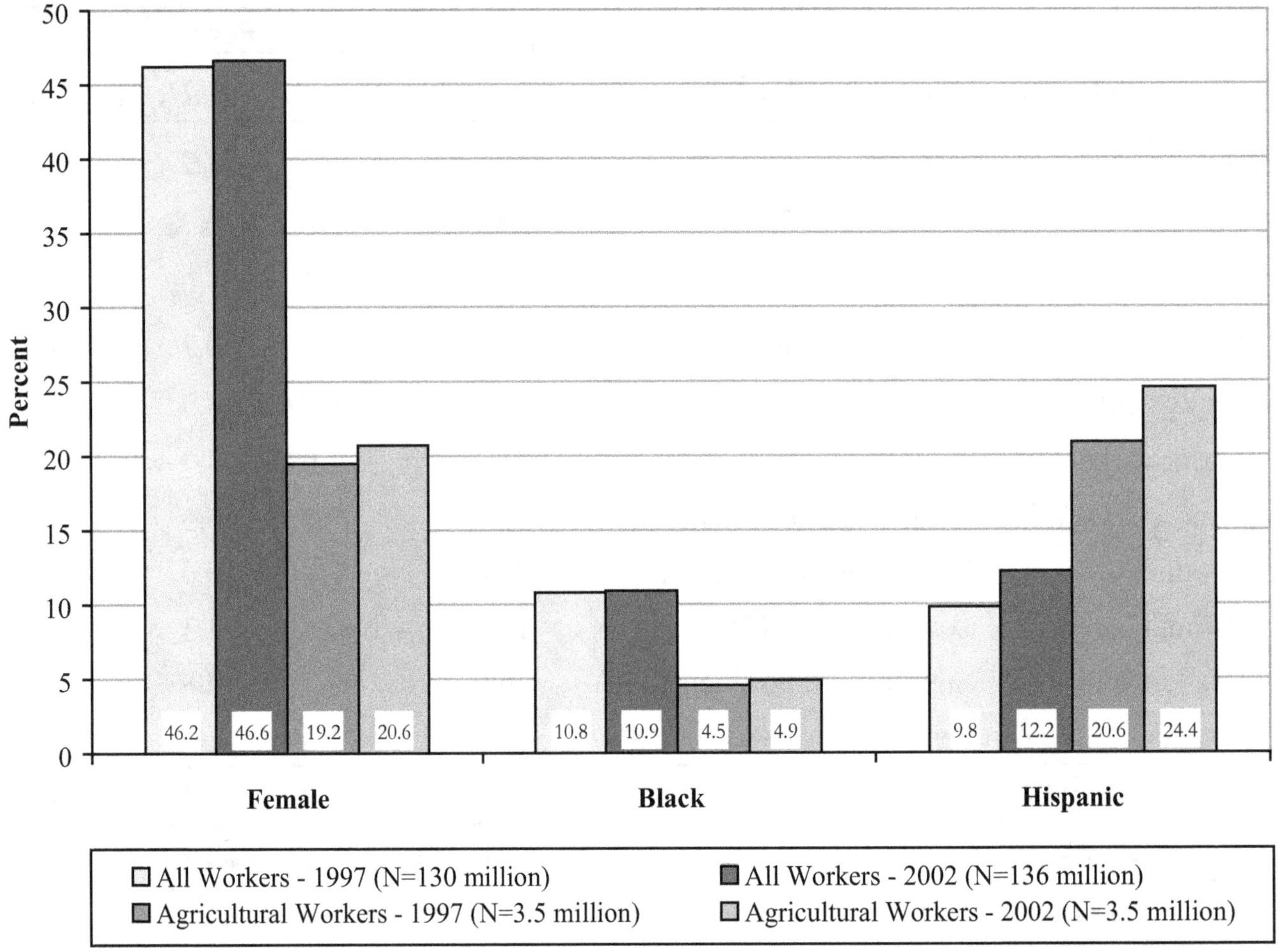

	Female	Black	Hispanic
All Workers - 1997 (N=130 million)	46.2	10.8	9.8
All Workers - 2002 (N=136 million)	46.6	10.9	12.2
Agricultural Workers - 1997 (N=3.5 million)	19.2	4.5	20.6
Agricultural Workers - 2002 (N=3.5 million)	20.6	4.9	24.4

SOURCE: U.S. Bureau of Labor Statistics: *Current Population Survey* (ftp://ftp.bls.gov/pub/special.requests/lf/aa97/aat11.txt) (ftp://ftp.bls.gov/pub/special.requests/lf/aa2002/aat11.txt)

Demographics

Figure 1-2. Distribution of employed U.S. agricultural groups by sex, race, and ethnicity, 1997

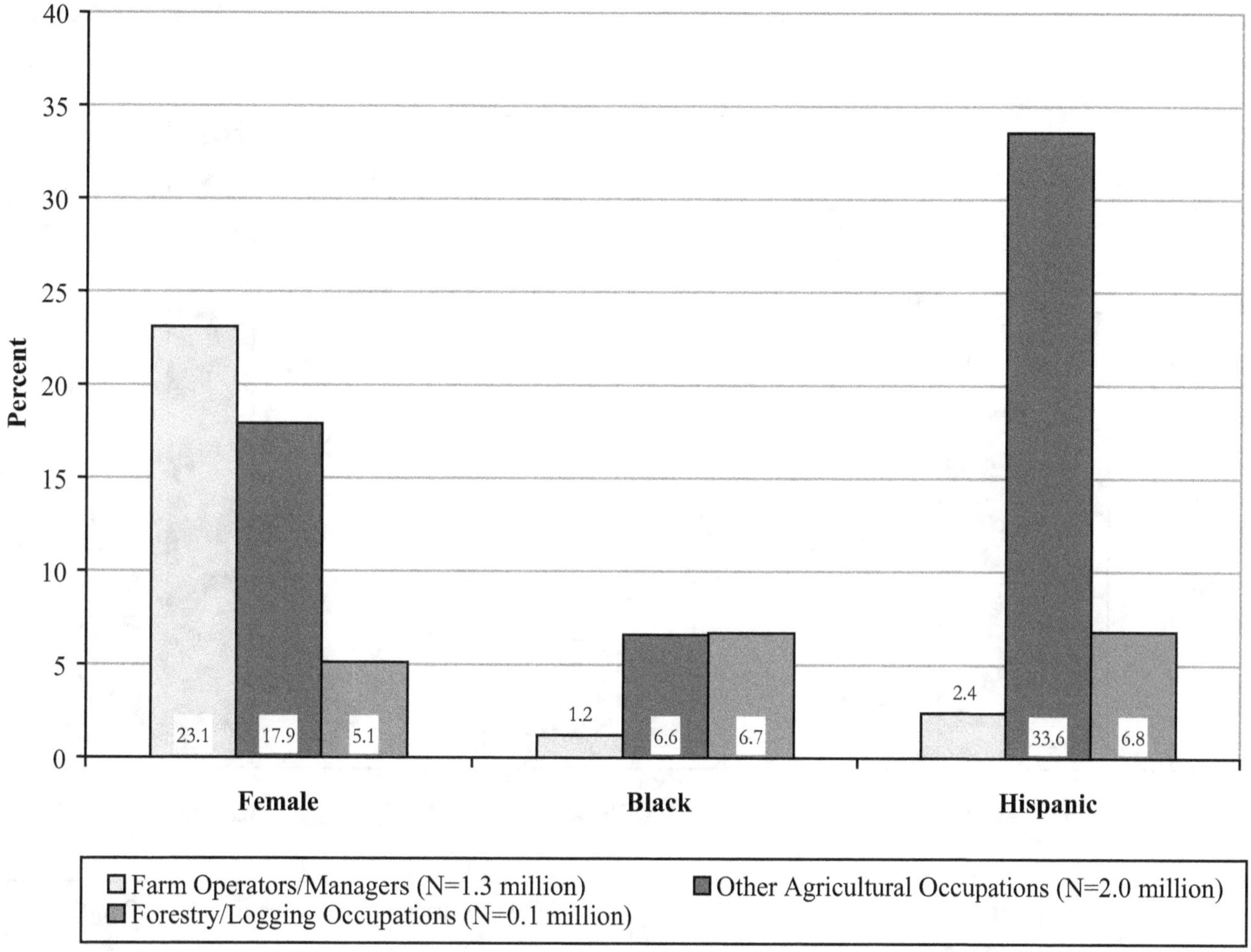

Group	Female	Black	Hispanic
Farm Operators/Managers (N=1.3 million)	23.1	1.2	2.4
Other Agricultural Occupations (N=2.0 million)	17.9	6.6	33.6
Forestry/Logging Occupations (N=0.1 million)	5.1	6.7	6.8

SOURCE: U.S. Bureau of Labor Statistics: *Current Population Survey* (ftp://ftp.bls.gov/pub/special.requests/lf/aa97/aat11.txt)

Demographics

Figure 1-3. Distribution of employed U.S. agricultural groups by sex, race, and ethnicity, 2002

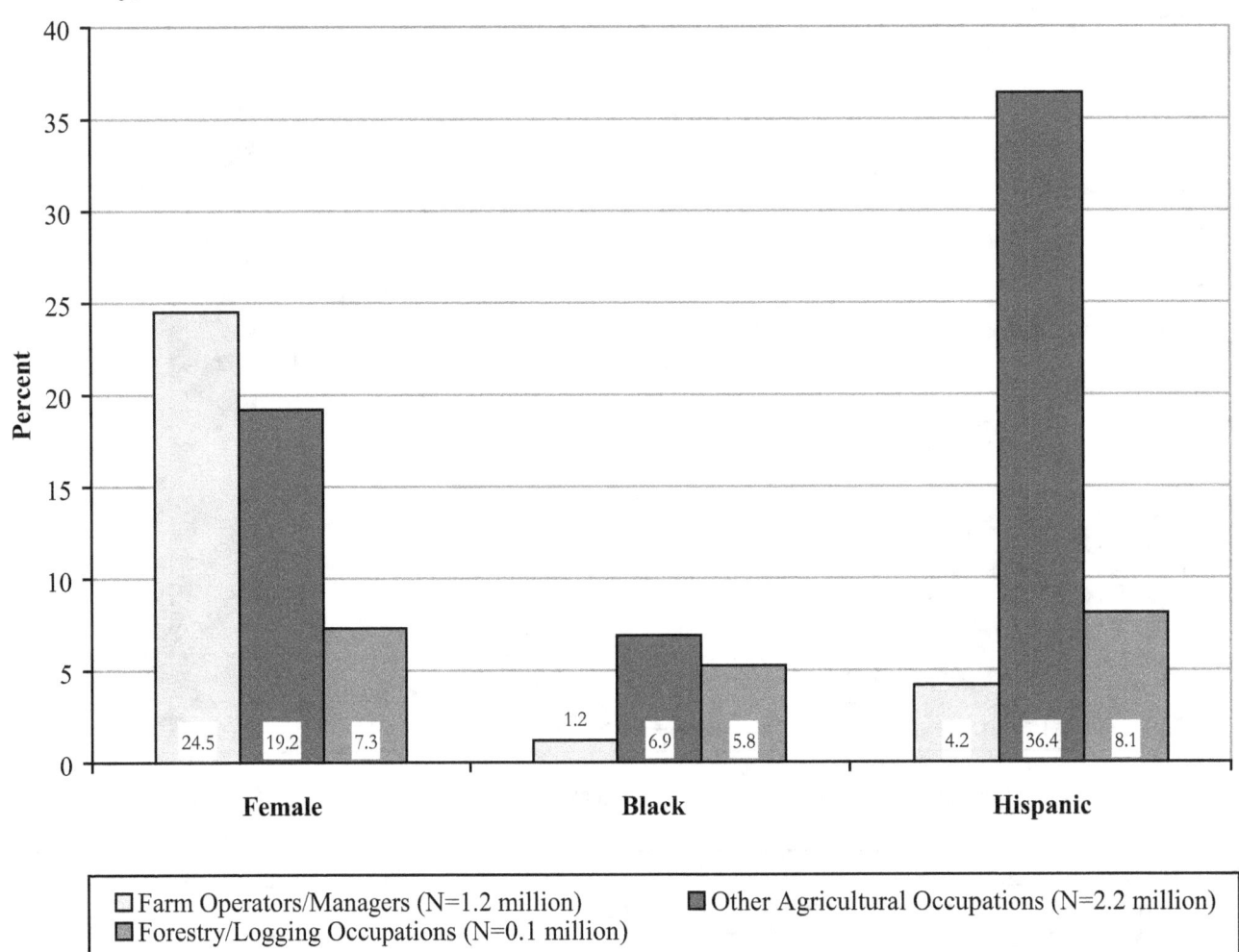

SOURCE: U.S. Bureau of Labor Statistics: *Current Population Survey* (ftp://ftp.bls.gov/pub/special.requests/lf/aa2002/aat11.txt)

Demographics

Table 1-2. Distribution of employed U.S. agricultural workers by state, 2002

States Used in the Mortality Analysis (Section 2)	Number Employed in Agriculture (in thousands)	Percent of all Agricultural Workers	Other States	Number Employed in Agriculture (in thousands)	Percent of all Agricultural Workers
North Carolina	107	3.1	California	513	14.7
Ohio	95	2.7	Texas	297	8.5
Washington	94	2.7	Florida	193	5.5
Wisconsin	91	2.6	New York	122	3.5
Georgia	89	2.5	Illinois	111	3.2
Indiana	76	2.2	Pennsylvania	109	3.1
Tennessee	67	1.9	Minnesota	102	2.9
Kansas	56	1.6	Michigan	90	2.6
Kentucky	56	1.6	Missouri	79	2.3
Colorado	55	1.6	Iowa	71	2.0
Oklahoma	53	1.5	Oregon	64	1.8
New Jersey	46	1.3	Virginia	64	1.8
South Carolina	40	1.1	Arizona	62	1.8
Idaho	37	1.1	Nebraska	62	1.8
New Mexico	27	0.8	Maryland	55	1.6
Utah	26	0.7	Arkansas	53	1.5
Maine	21	0.6	Louisiana	49	1.4
Hawaii	20	0.6	Mississippi	47	1.3
Nevada	18	0.5	Massachusetts	44	1.3
New Hampshire	13	0.4	Alabama	38	1.1
Vermont	13	0.4	Montana	35	1.0
West Virginia	13	0.4	South Dakota	35	1.0
Alaska	7	0.2	North Dakota	28	0.8
Rhode Island	6	0.2	Connecticut	22	0.6
Total	*1,126*	*32.2*	Wyoming	15	0.4
			Delaware	7	0.2
			District of Columbia	3	0.1
			Total, All States	*3,496*	*100.0*

SOURCE: U.S. Bureau of Labor Statistics, *Current Population Survey* (http://www.bls.gov/opub/gp/pdf/gp02_14.pdf and http://www.bls.gov/opub/gp/pdf/gp02_15.pdf)

Section 2

Mortality

Mortality by Disease Category within Agricultural Group

Table 2-1. Crop farm workers: Proportionate mortality ratio (PMR) adjusted for age, sex, and race/ethnicity by disease category, U.S. residents age 15 and over, selected states, 1988–1998

Disease Category (ICD Codes)	Number of Deaths	PMR	95% Confidence Interval LCL	95% Confidence Interval UCL
Tuberculosis (010-018)	522	**148**	136	161
Mycoses (110-118)	376	110	99	122
Sarcoidosis (135)	41	**71**	56	96
Malignant neoplasms of trachea/bronchus/lung/pleura (162-163)	13,099	**80**	78	82
Acute respiratory infections (460-466)	329	**124**	111	138
Other diseases of upper respiratory tract (470-478)	97	90	73	110
Pneumonia and influenza (480-487)	26,114	**109**	107	111
Chronic obstructive pulmonary disease and allied conditions (490-496)	26,186	97	95	99
Pneumoconiosis and other lung diseases - external agents (500-508)	5,224	**84**	82	86
Other diseases of respiratory system (510-519)	7,706	95	93	97

ICD - International Classification of Diseases, 9th Revision   LCL - lower confidence limit   UCL - upper confidence limit
NOTE: PMRs in **bold** are significantly different from 100 (p<0.05). PMRs in *italics* are based on fewer than five observed deaths. PMRs are based on underlying and contributing cause of death. Some values could not be calculated because the number of observed or expected deaths was zero; such values are indicated by ---. See appendices for source description, methods, ICD codes, and a list of selected states.
SOURCE: National Center for Health Statistics multiple-cause-of-death data

Mortality by Disease Category within Agricultural Group

Table 2-2. Livestock farm workers: Proportionate mortality ratio (PMR) adjusted for age, sex, and race/ethnicity by disease category, U.S. residents age 15 and over, selected states, 1988-1998

Disease Category (ICD Codes)	Number of Deaths	PMR	95% Confidence Interval LCL	95% Confidence Interval UCL
Tuberculosis (010-018)	56	**75**	87	97
Mycoses (110-118)	79	94	75	117
Sarcoidosis (135)	9	103	47	195
Malignant neoplasms of trachea/bronchus/lung/pleura (162-163)	2,960	**68**	66	70
Acute respiratory infections (460-466)	59	80	61	103
Other diseases of upper respiratory tract (470-478)	20	72	44	111
Pneumonia and influenza (480-487)	6,391	99	97	101
Chronic obstructive pulmonary disease and allied conditions (490-496)	6,956	**93**	91	95
Pneumoconiosis and other lung diseases - external agents (500-508)	1,381	**83**	79	87
Other diseases of respiratory system (510-519)	1,917	88	84	92

ICD - International Classification of Diseases, 9th Revision   LCL - lower confidence limit   UCL - upper confidence limit
NOTE: PMRs in **bold** are significantly different from 100 (p<0.05). PMRs in *italics* are based on fewer than five observed deaths. PMRs are based on underlying and contributing cause of death. Some values could not be calculated because the number of observed or expected deaths was zero; such values are indicated by ---. See appendices for source description, methods, ICD codes, and a list of selected states.
SOURCE: National Center for Health Statistics multiple-cause-of-death data

Mortality by Disease Category within Agricultural Group

Table 2-3. Farm managers: Proportionate mortality ratio (PMR) adjusted for age, sex, and race/ethnicity by disease category, U.S. residents age 15 and over, selected states, 1988–1998

Disease Category (ICD Codes)	Number of Deaths	PMR	95% Confidence Interval LCL	95% Confidence Interval UCL
Tuberculosis (010-018)	5	86	28	201
Mycoses (110-118)	5	84	27	196
Sarcoidosis (135)	0	*0*	---	---
Malignant neoplasms of trachea/bronchus/lung/pleura (162-163)	251	94	84	106
Acute respiratory infections (460-466)	2	*48*	6	173
Other diseases of upper respiratory tract (470-478)	1	*58*	1	322
Pneumonia and influenza (480-487)	373	104	94	115
Chronic obstructive pulmonary disease and allied conditions (490-496)	407	99	90	109
Pneumoconiosis and other lung diseases - external agents (500-508)	65	**70**	55	89
Other diseases of respiratory system (510-519)	117	88	73	106

ICD - International Classification of Diseases, 9th Revision   LCL - lower confidence limit   UCL - upper confidence limit
NOTE: PMRs in **bold** are significantly different from 100 (p<0.05). PMRs in *italics* are based on fewer than five observed deaths. PMRs are based on underlying and contributing cause of death. Some values could not be calculated because the number of observed or expected deaths was zero; such values are indicated by ---. See appendices for source description, methods, ICD codes, and a list of selected states.
SOURCE: National Center for Health Statistics multiple-cause-of-death data

Mortality by Disease Category within Agricultural Group

Table 2-4. Landscape and horticultural workers: Proportionate mortality ratio (PMR) adjusted for age, sex, and race/ethnicity by disease category, U.S. residents age 15 and over, selected states, 1988–1998

Disease Category (ICD Codes)	Number of Deaths	PMR	95% Confidence Interval LCL	95% Confidence Interval UCL
Tuberculosis (010-018)	16	80	46	130
Mycoses (110-118)	27	106	70	154
Sarcoidosis (135)	2	*34*	4	123
Malignant neoplasms of trachea/bronchus/lung/pleura (162-163)	647	97	90	105
Acute respiratory infections (460-466)	9	108	50	205
Other diseases of upper respiratory tract (470-478)	4	76	21	194
Pneumonia and influenza (480-487)	607	94	87	102
Chronic obstructive pulmonary disease and allied conditions (490-496)	799	**109**	102	117
Pneumoconiosis and other lung diseases - external agents (500-508)	154	98	83	115
Other diseases of respiratory system (510-519)	252	**84**	75	95

ICD - International Classification of Diseases, 9th Revision LCL - lower confidence limit UCL - upper confidence limit
NOTE: PMRs in **bold** are significantly different from 100 (p<0.05). PMRs in *italics* are based on fewer than five observed deaths. PMRs are based on underlying and contributing cause of death. Some values could not be calculated because the number of observed or expected deaths was zero; such values are indicated by ---. See appendices for source description, methods, ICD codes, and a list of selected states.
SOURCE: National Center for Health Statistics multiple-cause-of-death data

Mortality by Disease Category within Agricultural Group

Table 2-5. Forestry workers: Proportionate mortality ratio (PMR) adjusted for age, sex, and race/ethnicity by disease category, U.S. residents age 15 and over, selected states, 1988–1998

Disease Category (ICD Codes)	Number of Deaths	PMR	95% Confidence Interval	
			LCL	UCL
Tuberculosis (010-018)	45	129	94	173
Mycoses (110-118)	23	**60**	38	90
Sarcoidosis (135)	4	*49*	13	125
Malignant neoplasms of trachea/bronchus/lung/pleura (162-163)	1,553	102	97	107
Acute respiratory infections (460-466)	15	83	46	137
Other diseases of upper respiratory tract (470-478)	8	86	37	169
Pneumonia and influenza (480-487)	1,771	**116**	111	122
Chronic obstructive pulmonary disease and allied conditions (490-496)	2,318	**122**	117	127
Pneumoconiosis and other lung diseases - external agents (500-508)	354	90	81	100
Other diseases of respiratory system (510-519)	549	**88**	81	96

ICD - International Classification of Diseases, 9th Revision   LCL - lower confidence limit   UCL - upper confidence limit
NOTE: PMRs in **bold** are significantly different from 100 (p<0.05). PMRs in *italics* are based on fewer than five observed deaths. PMRs are based on underlying and contributing cause of death. Some values could not be calculated because the number of observed or expected deaths was zero; such values are indicated by ---. See appendices for source description, methods, ICD codes, and a list of selected states.
SOURCE: National Center for Health Statistics multiple-cause-of-death data

Mortality by Disease Category within Agricultural Group

Table 2-6. Fishery workers: Proportionate mortality ratio (PMR) adjusted for age, sex, and race/ethnicity by disease category, U.S. residents age 15 and over, selected states, 1988–1998

Disease Category (ICD Codes)	Number of Deaths	PMR	95% Confidence Interval	
			LCL	UCL
Tuberculosis (010-018)	14	175	96	294
Mycoses (110-118)	7	64	26	132
Sarcoidosis (135)	4	256	70	655
Malignant neoplasms of trachea/bronchus/lung/pleura (162-163)	426	108	98	119
Acute respiratory infections (460-466)	4	78	21	199
Other diseases of upper respiratory tract (470-478)	5	193	62	451
Pneumonia and influenza (480-487)	422	103	94	113
Chronic obstructive pulmonary disease and allied conditions (490-496)	568	**113**	104	123
Pneumoconiosis and other lung diseases - external agents (500-508)	98	94	76	115
Other diseases of respiratory system (510-519)	150	88	75	103

ICD - International Classification of Diseases, 9th Revision LCL - lower confidence limit UCL - upper confidence limit
NOTE: PMRs in **bold** are significantly different from 100 (p<0.05). PMRs in *italics* are based on fewer than five observed deaths. PMRs are based on underlying and contributing cause of death. Some values could not be calculated because the number of observed or expected deaths was zero; such values are indicated by ---. See appendices for source description, methods, ICD codes, and a list of selected states.
SOURCE: National Center for Health Statistics multiple-cause-of-death data

Mortality by Agricultural Group within Disease Category

Figure 2-1. Tuberculosis: Proportionate mortality ratio (PMR) adjusted for age, sex, and race/ethnicity by agricultural group, U.S. residents age 15 and over, selected states, 1988–1998

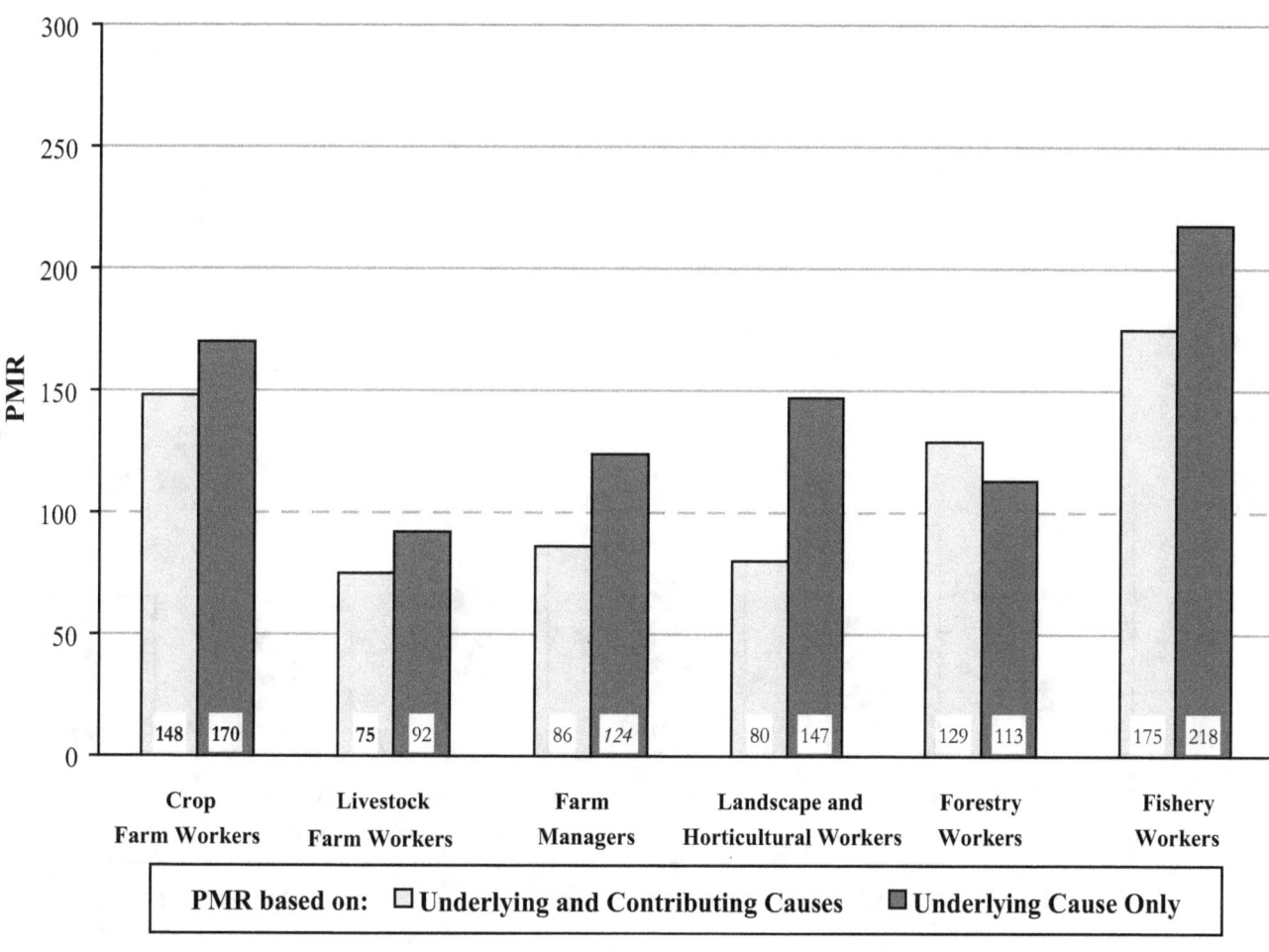

ICD - International Classification of Diseases, 9th Revision
NOTE: Tuberculosis = ICD-9 codes 010-018. PMRs in **bold** are significantly different from 100 ($p<0.05$). PMRs in *italics* are based on fewer than five observed deaths. PMRs are based on underlying and contributing cause of death. See appendices for source description, methods, ICD codes, and a list of selected states.
SOURCE: National Center for Health Statistics multiple-cause-of-death data

Mortality by Agricultural Group within Disease Category

Figure 2-2. Mycoses: Proportionate mortality ratio (PMR) adjusted for age, sex, and race/ethnicity by agricultural group, U.S. residents age 15 and over, selected states, 1988–1998

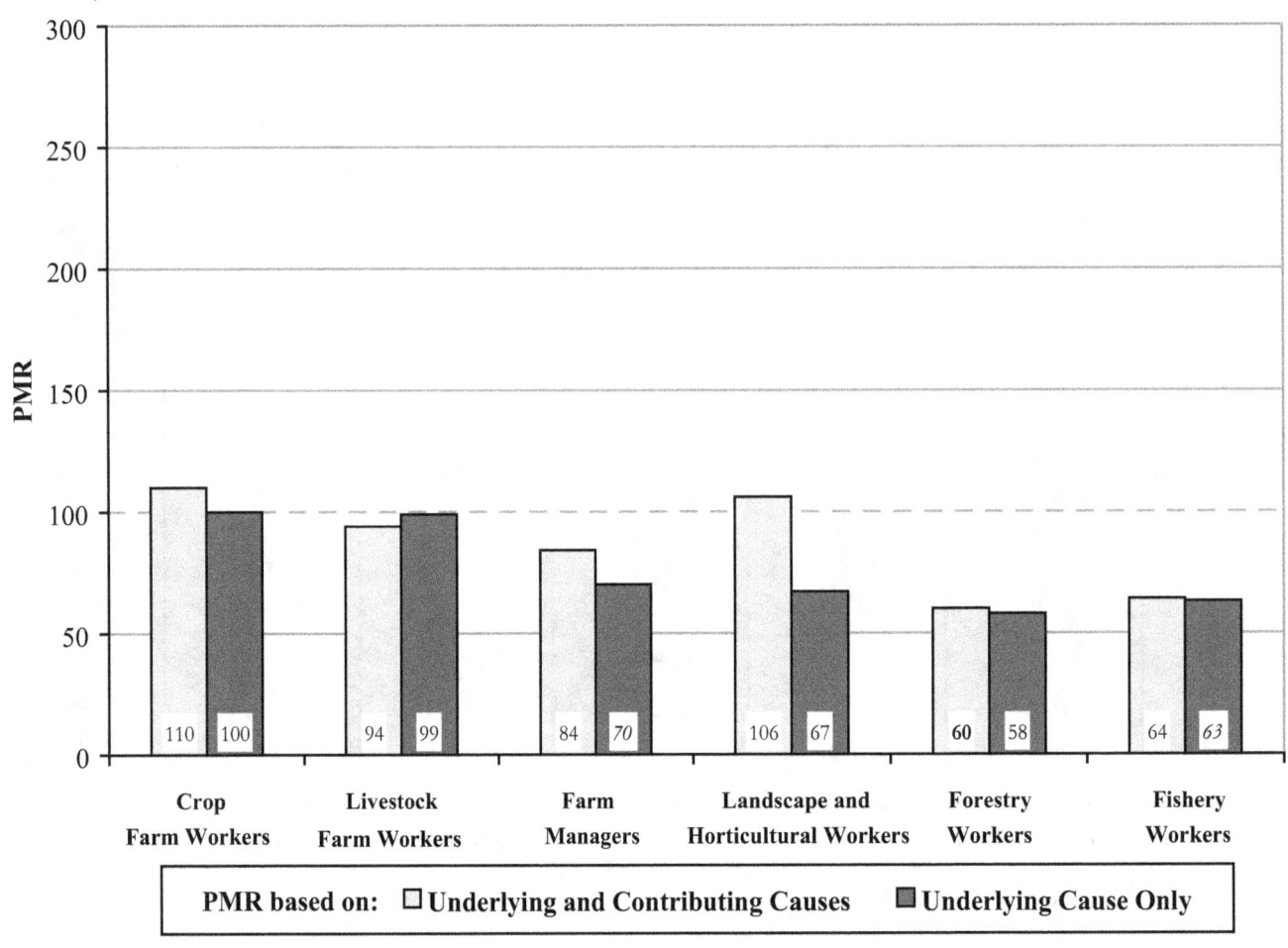

ICD - International Classification of Diseases, 9th Revision
NOTE: Mycoses = ICD-9 codes 110-118. PMRs in **bold** are significantly different from 100 ($p<0.05$). PMRs in *italics* are based on fewer than five observed deaths. PMRs are based on underlying and contributing cause of death. See appendices for source description, methods, ICD codes, and a list of selected states.
SOURCE: National Center for Health Statistics multiple-cause-of-death data

Figure 2-3. Sarcoidosis: Proportionate mortality ratio (PMR) adjusted for age, sex, and race/ethnicity by agricultural group, U.S. residents age 15 and over, selected states, 1988–1998

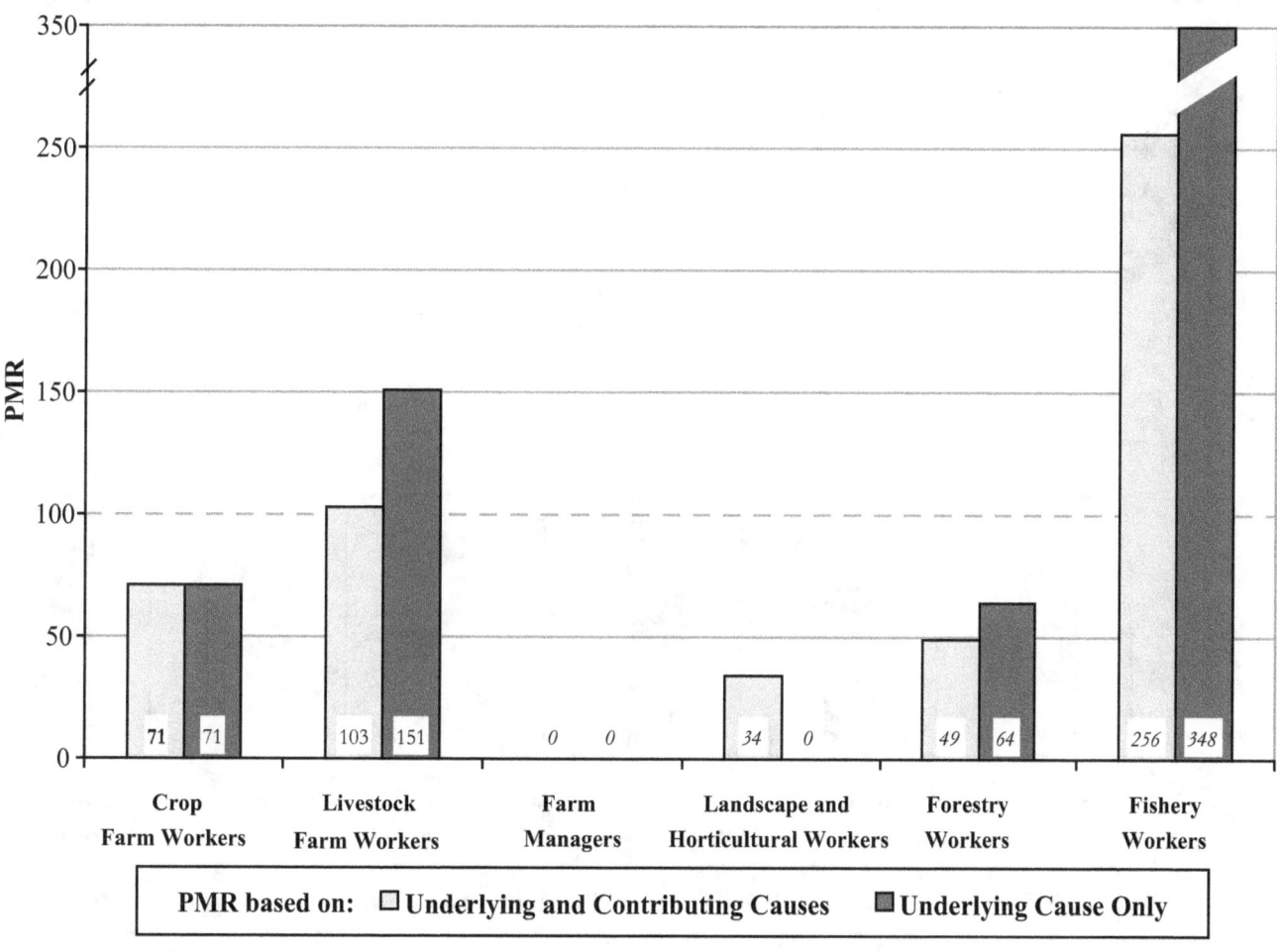

ICD - International Classification of Diseases, 9th Revision
NOTE: Sarcoidosis = ICD-9 code 135. PMRs in **bold** are significantly different from 100 ($p<0.05$). PMRs in *italics* are based on fewer than five observed deaths. PMRs are based on underlying and contributing cause of death. See appendices for source description, methods, ICD codes, and a list of selected states.
SOURCE: National Center for Health Statistics multiple-cause-of-death data

Mortality by Agricultural Group within Disease Category

Figure 2-4. Malignant neoplasms of trachea/bronchus/lung/pleura: Proportionate mortality ratio (PMR) adjusted for age, sex, and race/ethnicity by agricultural group, U.S. residents age 15 and over, selected states, 1988–1998

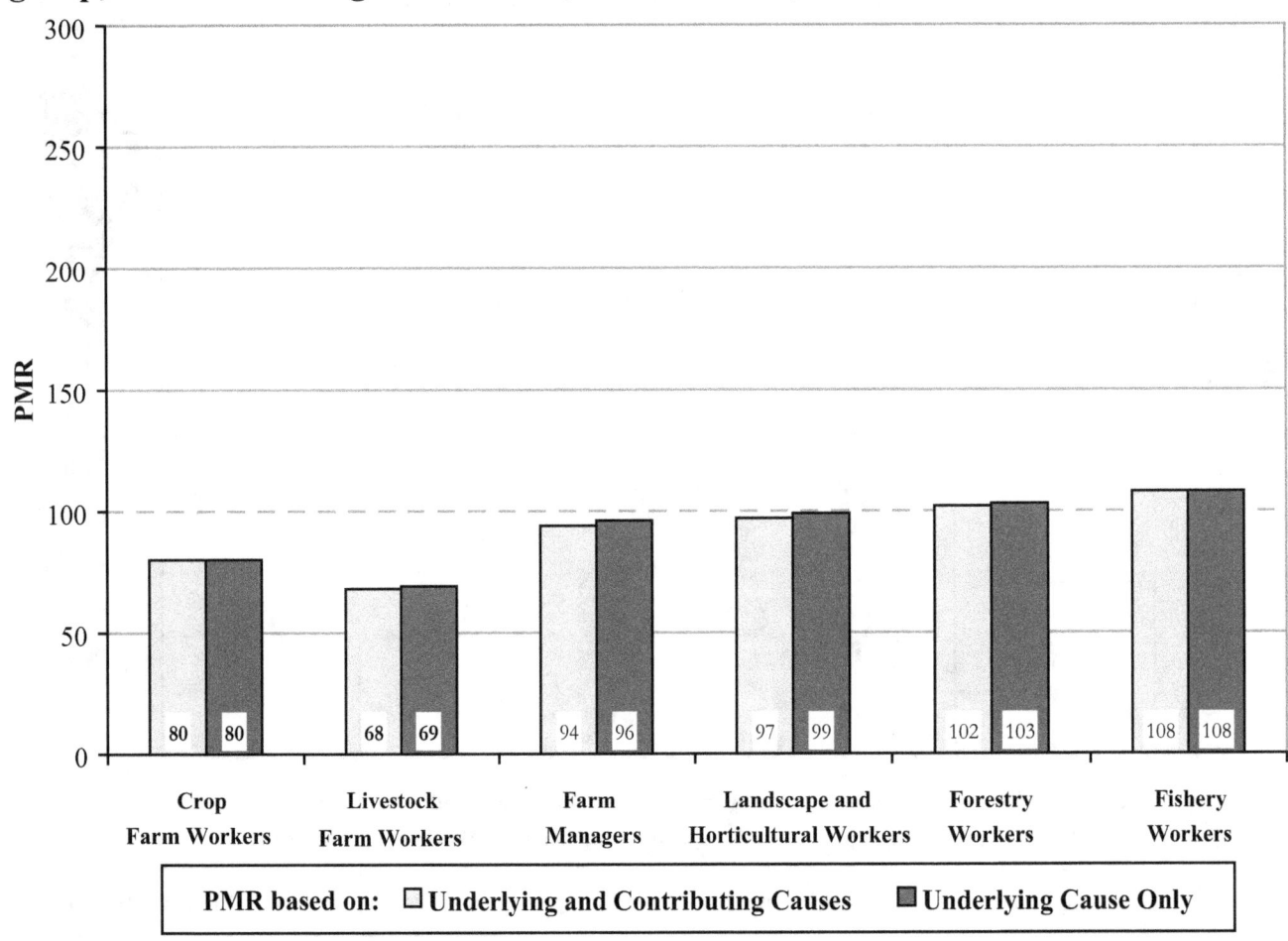

ICD - International Classification of Diseases, 9th Revision
NOTE: Malignant neoplasms of trachea/bronchus/lung/pleura = ICD-9 codes 162-163. PMRs in **bold** are significantly different from 100 (p<0.05). PMRs in *italics* are based on fewer than five observed deaths. PMRs are based on underlying and contributing cause of death. See appendices for source description, methods, ICD codes, and a list of selected states.
SOURCE: National Center for Health Statistics multiple-cause-of-death data

Mortality by Agricultural Group within Disease Category

Figure 2-5. Acute respiratory infections: Proportionate mortality ratio (PMR) adjusted for age, sex, and race/ethnicity by agricultural group, U.S. residents age 15 and over, selected states, 1988–1998

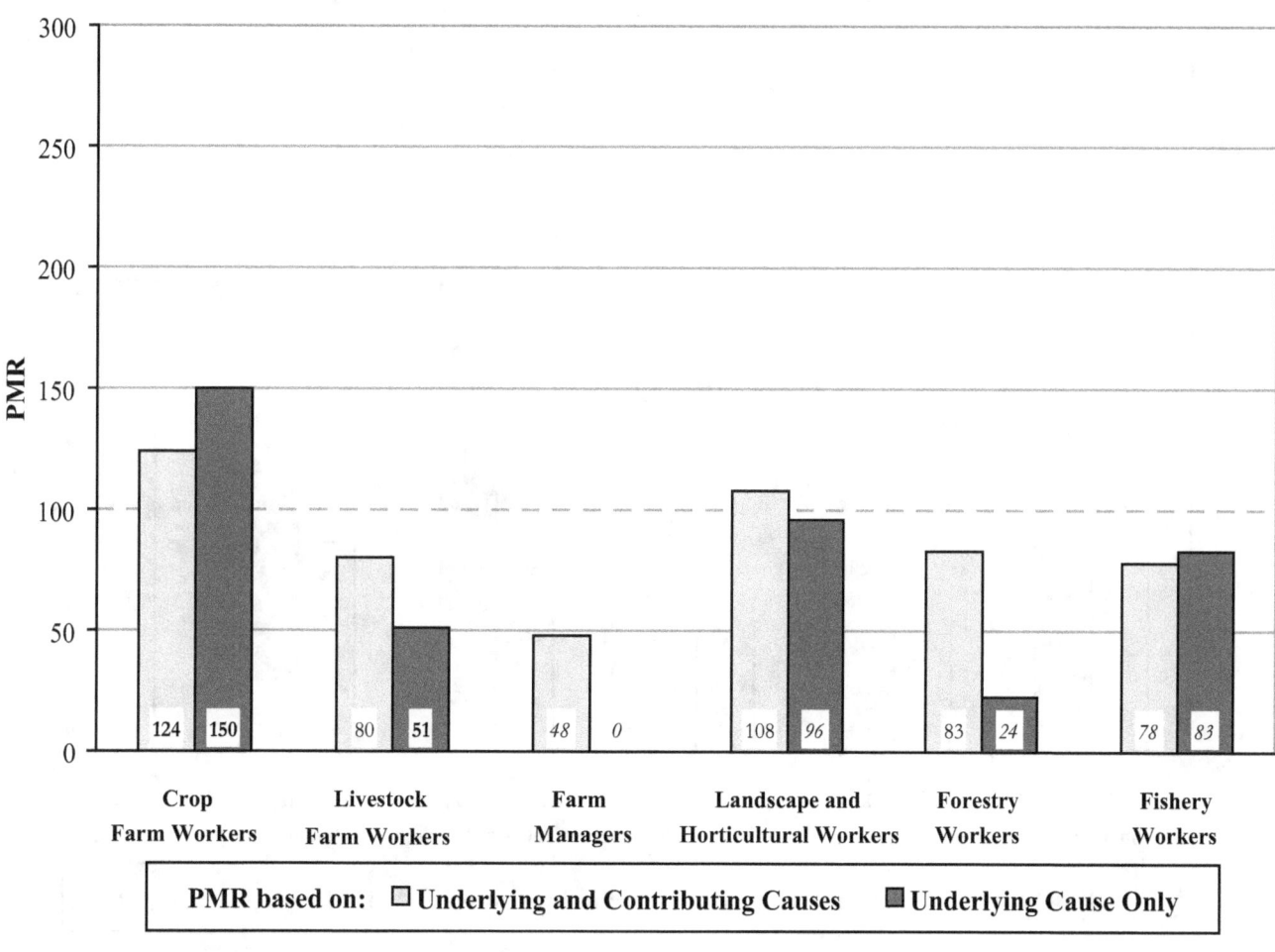

ICD - International Classification of Diseases, 9th Revision
NOTE: Acute respiratory infections = ICD-9 codes 460-466. PMRs in **bold** are significantly different from 100 ($p<0.05$). PMRs in *italics* are based on fewer than five observed deaths. PMRs are based on underlying and contributing cause of death. See appendices for source description, methods, ICD codes, and a list of selected states.
SOURCE: National Center for Health Statistics multiple-cause-of-death data

Mortality by Agricultural Group within Disease Category

Figure 2-6. Other diseases of upper respiratory tract: Proportionate mortality ratio (PMR) adjusted for age, sex, and race/ethnicity by agricultural group, U.S. residents age 15 and over, selected states, 1988–1998

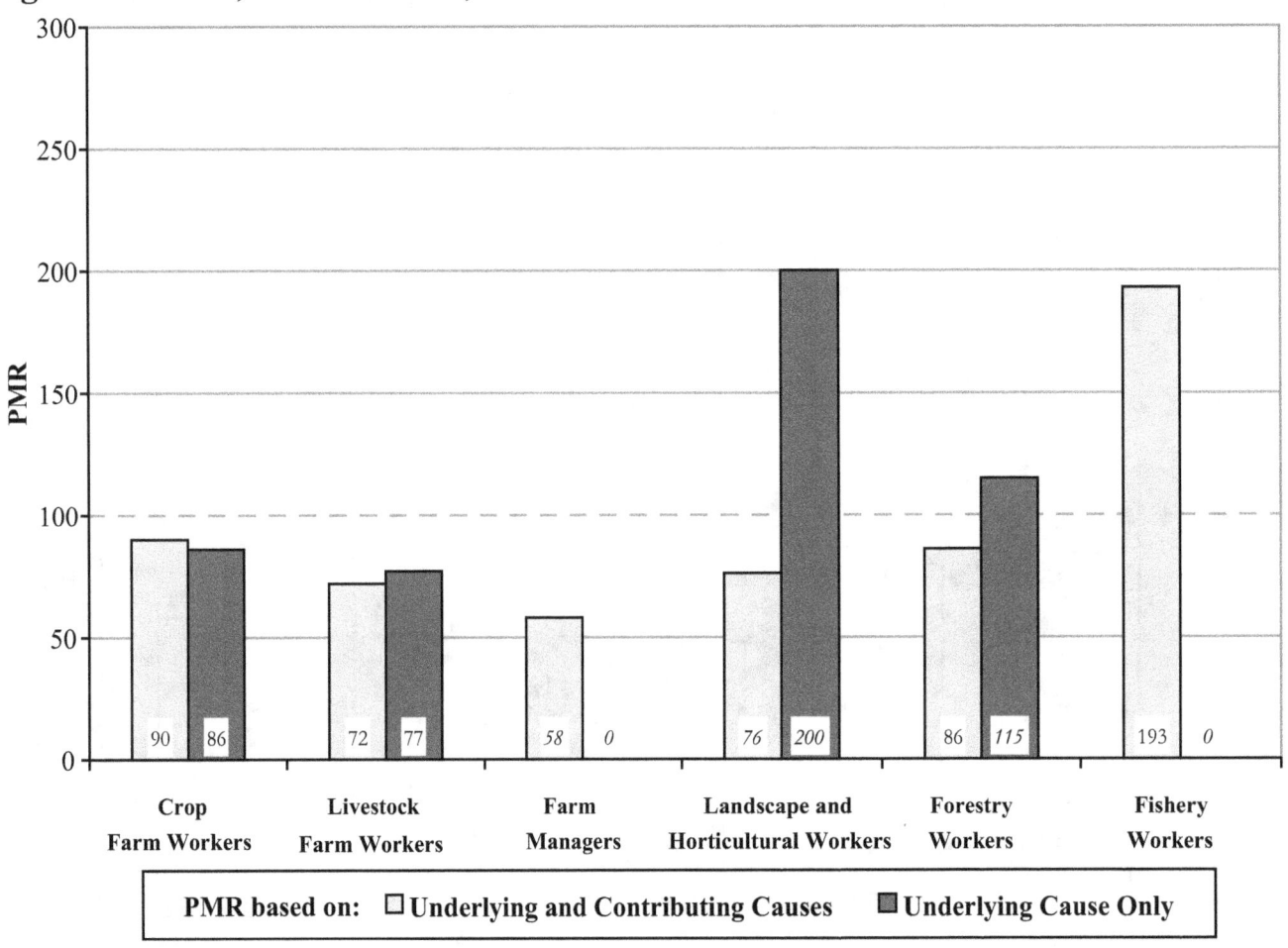

ICD - International Classification of Diseases, 9th Revision
NOTE: Other diseases of upper respiratory tract = ICD-9 codes 470-478. PMRs in **bold** are significantly different from 100 (p<0.05). PMRs in *italics* are based on fewer than five observed deaths. PMRs are based on underlying and contributing cause of death. See appendices for source description, methods, ICD codes, and a list of selected states.
SOURCE: National Center for Health Statistics multiple-cause-of-death data

Mortality by Agricultural Group within Disease Category

Figure 2-7. Pneumonia and influenza: Proportionate mortality ratio (PMR) adjusted for age, sex, and race/ethnicity by agricultural group, U.S. residents age 15 and over, selected states, 1988–1998

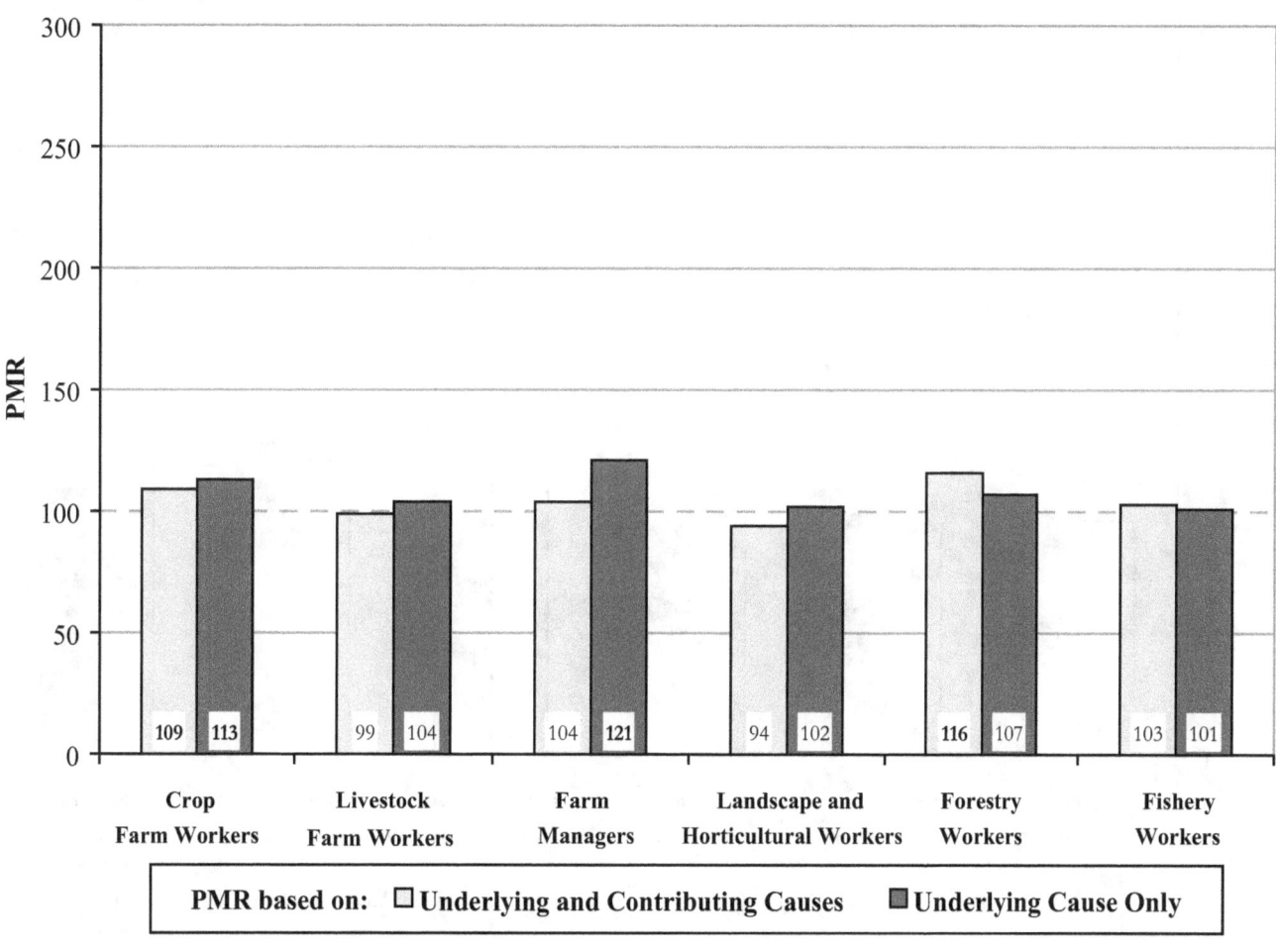

ICD - International Classification of Diseases, 9th Revision
NOTE: Pneumonia and influenza = ICD-9 codes 480-487. PMRs in **bold** are significantly different from 100 ($p<0.05$). PMRs in *italics* are based on fewer than five observed deaths. PMRs are based on underlying and contributing cause of death. See appendices for source description, methods, ICD codes, and a list of selected states.
SOURCE: National Center for Health Statistics multiple-cause-of-death data

Mortality by Agricultural Group within Disease Category

Figure 2-8. Chronic obstructive pulmonary disease and allied conditions: Proportionate mortality ratio (PMR) adjusted for age, sex, and race/ethnicity by agricultural group, U.S. residents age 15 and over, selected states, 1988–1998

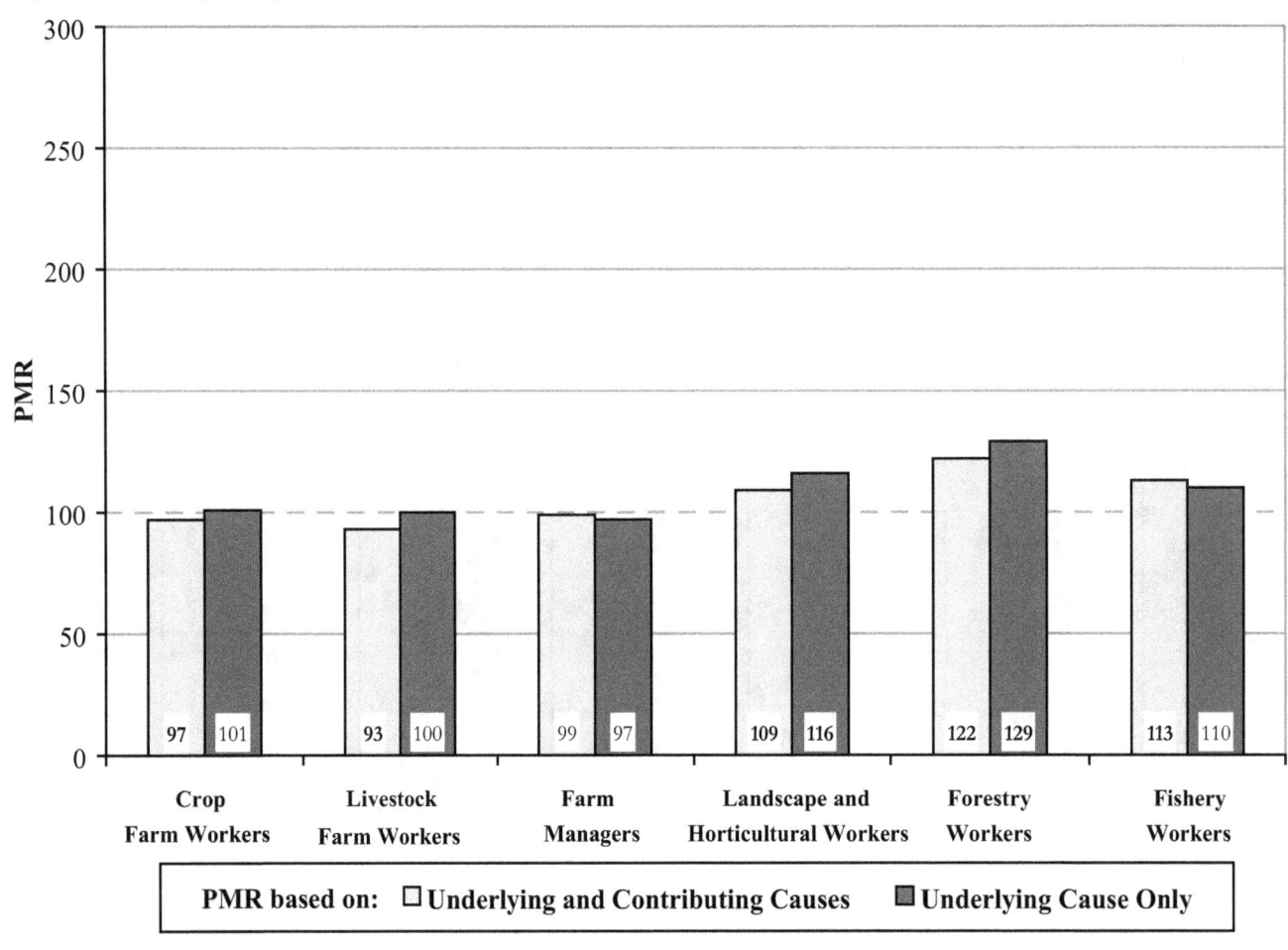

ICD - International Classification of Diseases, 9th Revision
NOTE: Chronic obstructive pulmonary disease and allied conditions = ICD-9 codes 490-496. PMRs in **bold** are significantly different from 100 (p<0.05). PMRs in *italics* are based on fewer than five observed deaths. PMRs are based on underlying and contributing cause of death. See appendices for source description, methods, ICD codes, and a list of selected states.
SOURCE: National Center for Health Statistics multiple-cause-of-death data

Mortality by Agricultural Group within Disease Category

Figure 2-9. Pneumoconioses and other lung diseases–external agents: Proportionate mortality ratio (PMR) adjusted for age, sex, and race/ethnicity by agricultural group, U.S. residents age 15 and over, selected states, 1988–1998

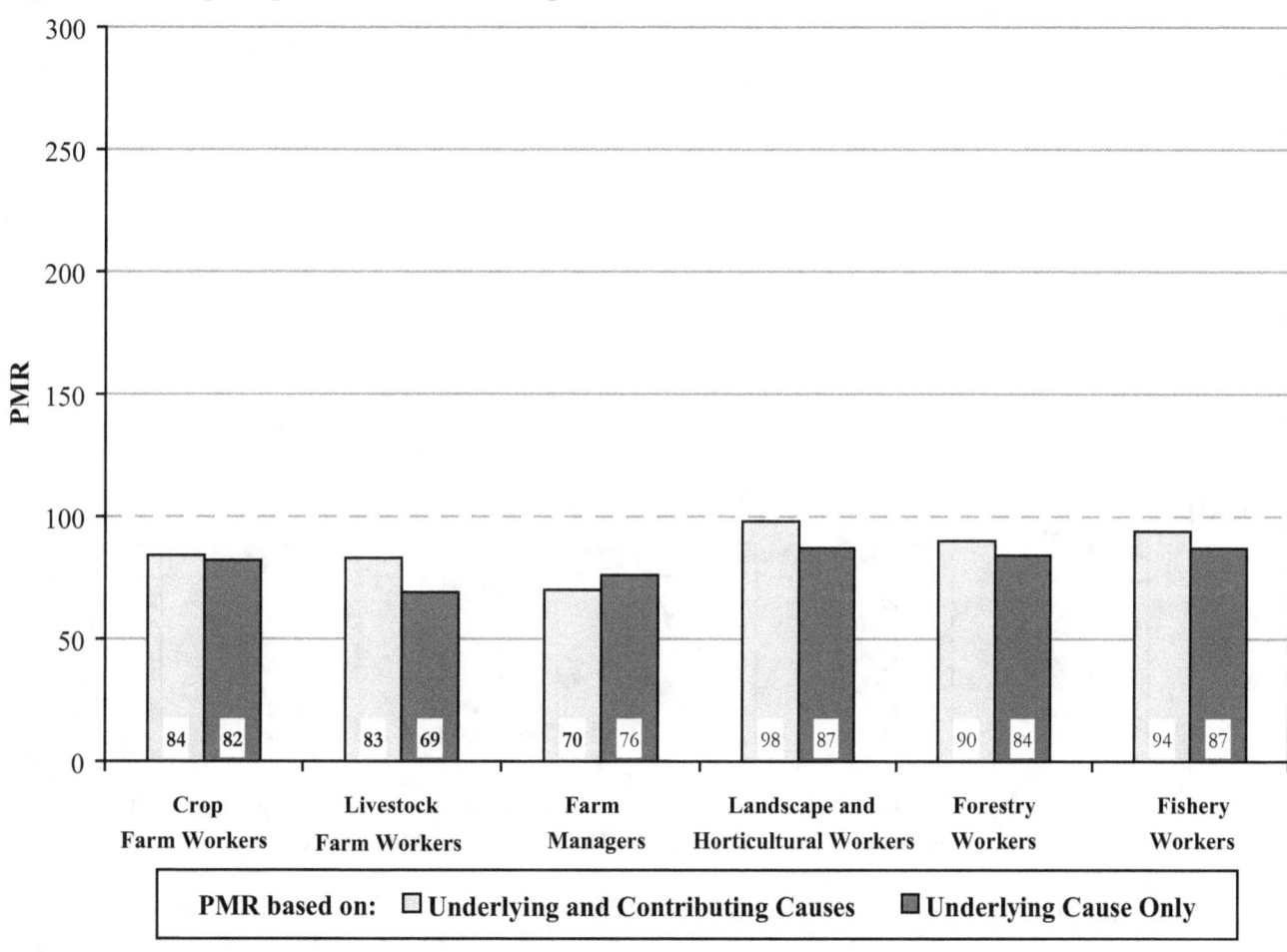

ICD - International Classification of Diseases, 9th Revision
NOTE: Pneumoconioses and other lung diseases - external agents = ICD-9 codes 500-508. PMRs in **bold** are significantly different from 100 ($p<0.05$). PMRs in *italics* are based on fewer than five observed deaths. PMRs are based on underlying and contributing cause of death. See appendices for source description, methods, ICD codes, and a list of selected states.
SOURCE: National Center for Health Statistics multiple-cause-of-death data

Mortality by Agricultural Group within Disease Category

Figure 2-10. Other diseases of respiratory system: Proportionate mortality ratio (PMR) adjusted for age, sex, and race/ethnicity by agricultural group, U.S. residents age 15 and over, selected states, 1988–1998

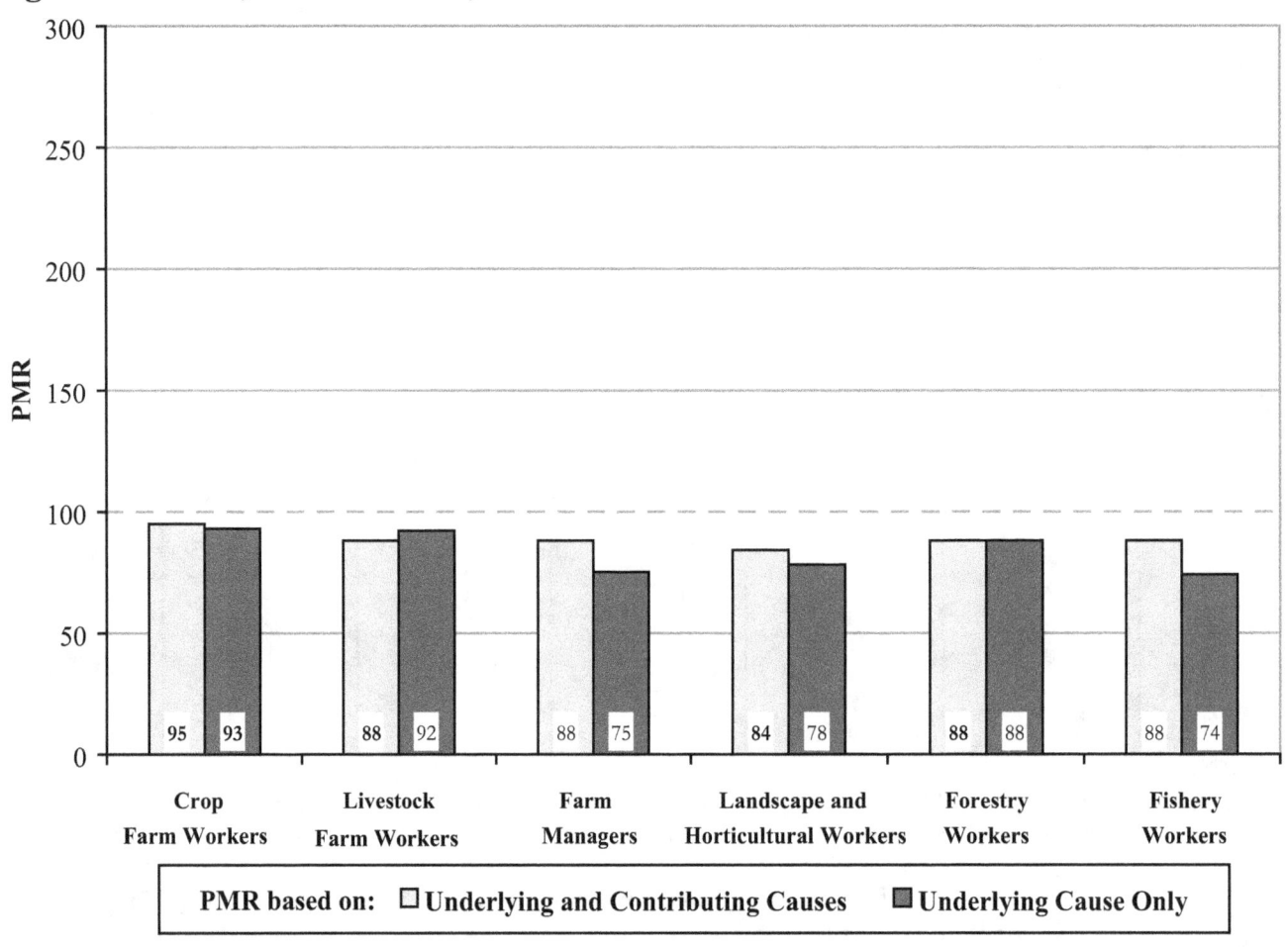

ICD - International Classification of Diseases, 9th Revision
NOTE: Other diseases of respiratory system = ICD-9 codes 510-519. PMRs in **bold** are significantly different from 100 (p<0.05). PMRs in *italics* are based on fewer than five observed deaths. PMRs are based on underlying and contributing cause of death. See appendices for source description, methods, ICD codes, and a list of selected states.
SOURCE: National Center for Health Statistics multiple-cause-of-death data

Mortality by Disease Category within Agricultural and Sex Group

Table 2-7. Crop farm workers, males: Proportionate mortality ratio (PMR) adjusted for age and race/ethnicity by disease category, U.S. residents age 15 and over, selected states, 1988–1998

Disease Category (ICD Codes)	Number of Deaths	PMR	95% Confidence Interval	
			LCL	UCL
Tuberculosis (010-018)	481	**147**	134	161
Mycoses (110-018)	361	**112**	101	124
Sarcoidosis (135)	34	76	52	106
Malignant neoplasms of trachea/bronchus/lung/pleura (162-163)	12,765	**80**	78	82
Acute respiratory infections (460-466)	312	**125**	112	140
Other diseases of upper respiratory tract (470-478)	93	92	75	113
Pneumonia and influenza (480-487)	24,848	**110**	108	112
Chronic obstructive pulmonary disease and allied conditions (490-496)	25,521	97	95	99
Pneumoconiosis and other lung diseases - external agents (500-508)	4,970	**83**	81	85
Other diseases of respiratory system (510-519)	7,282	**95**	93	97

ICD - International Classification of Diseases, 9th Revision   LCL - lower confidence limit   UCL - upper confidence limit
NOTE: PMRs in **bold** are significantly different from 100 (p<0.05). PMRs in *italics* are based on fewer than five observed deaths. PMRs are based on underlying and contributing cause of death. Some values could not be calculated because the number of observed or expected deaths was zero; such values are indicated by ---. See appendices for source description, methods, ICD codes, and a list of selected states.
SOURCE: National Center for Health Statistics multiple-cause-of-death data

Mortality by Disease Category within Agricultural and Sex Group

Table 2-8. Livestock farm workers, males: Proportionate mortality ratio (PMR) adjusted for age and race/ethnicity by disease category, U.S. residents age 15 and over, selected states, 1988–1998

Disease Category (ICD Codes)	Number of Deaths	PMR	95% Confidence Interval LCL	95% Confidence Interval UCL
Tuberculosis (010-018)	51	**72**	55	95
Mycoses (110-018)	73	92	73	116
Sarcoidosis (135)	9	119	55	226
Malignant neoplasms of trachea/bronchus/lung/pleura (162-163)	2,835	**68**	66	71
Acute respiratory infections (460-466)	55	80	61	104
Other diseases of upper respiratory tract (470-478)	17	65	38	104
Pneumonia and influenza (480-487)	6,060	99	97	102
Chronic obstructive pulmonary disease and allied conditions (490-496)	6,712	**93**	91	95
Pneumoconiosis and other lung diseases - external agents (500-508)	1,331	**83**	79	88
Other diseases of respiratory system (510-519)	1,773	**87**	83	91

ICD - International Classification of Diseases, 9th Revision LCL - lower confidence limit UCL - upper confidence limit
NOTE: PMRs in **bold** are significantly different from 100 (p<0.05). PMRs in *italics* are based on fewer than five observed deaths. PMRs are based on underlying and contributing cause of death. Some values could not be calculated because the number of observed or expected deaths was zero; such values are indicated by ---. See appendices for source description, methods, ICD codes, and a list of selected states.
SOURCE: National Center for Health Statistics multiple-cause-of-death data

Mortality by Disease Category within Agricultural and Sex Group

Table 2-9. Farm managers, males: Proportionate mortality ratio (PMR) adjusted for age and race/ethnicity by disease category, U.S. residents age 15 and over, selected states, 1988–1998

Disease Category (ICD Codes)	Number of Deaths	PMR	95% Confidence Interval LCL	95% Confidence Interval UCL
Tuberculosis (010-018)	4	77	21	197
Mycoses (110-018)	5	94	30	220
Sarcoidosis (135)	0	*0*	---	---
Malignant neoplasms of trachea/bronchus/lung/pleura (162-163)	231	93	82	106
Acute respiratory infections (460-466)	2	55	7	199
Other diseases of upper respiratory tract (470-478)	0	*0*	---	---
Pneumonia and influenza (480-487)	330	104	93	116
Chronic obstructive pulmonary disease and allied conditions (490-496)	373	99	90	110
Pneumoconiosis and other lung diseases - external agents (500-508)	61	**72**	56	92
Other diseases of respiratory system (510-519)	102	88	73	107

ICD - International Classification of Diseases, 9th Revision LCL - lower confidence limit UCL - upper confidence limit
NOTE: PMRs in **bold** are significantly different from 100 (p<0.05). PMRs in *italics* are based on fewer than five observed deaths. PMRs are based on underlying and contributing cause of death. Some values could not be calculated because the number of observed or expected deaths was zero; such values are indicated by ---. See appendices for source description, methods, ICD codes, and a list of selected states.
SOURCE: National Center for Health Statistics multiple-cause-of-death data

Mortality by Disease Category within Agricultural and Sex Group

Table 2-10. Landscape and horticultural workers, males: Proportionate mortality ratio (PMR) adjusted for age and race/ethnicity by disease category, U.S. residents age 15 and over, selected states, 1988–1998

Disease Category (ICD Codes)	Number of Deaths	PMR	95% Confidence Interval LCL	95% Confidence Interval UCL
Tuberculosis (010-018)	16	84	48	136
Mycoses (110-018)	26	108	71	158
Sarcoidosis (135)	1	*18*	0	100
Malignant neoplasms of trachea/bronchus/lung/pleura (162-163)	615	98	91	106
Acute respiratory infections (460-466)	7	92	37	190
Other diseases of upper respiratory tract (470-478)	4	*80*	22	205
Pneumonia and influenza (480-487)	563	94	87	102
Chronic obstructive pulmonary disease and allied conditions (490-496)	741	**108**	101	116
Pneumoconiosis and other lung diseases - external agents (500-508)	149	100	85	117
Other diseases of respiratory system (510-519)	228	**82**	72	93

ICD - International Classification of Diseases, 9th Revision   LCL - lower confidence limit   UCL - upper confidence limit
NOTE: PMRs in **bold** are significantly different from 100 (p<0.05). PMRs in *italics* are based on fewer than five observed deaths. PMRs are based on underlying and contributing cause of death. Some values could not be calculated because the number of observed or expected deaths was zero; such values are indicated by ---. See appendices for source description, methods, ICD codes, and a list of selected states.
SOURCE: National Center for Health Statistics multiple-cause-of-death data

Mortality by Disease Category within Agricultural and Sex Group

Table 2-11. Forestry workers, males: Proportionate mortality ratio (PMR) adjusted for age and race/ethnicity by disease category, U.S. residents age 15 and over, selected states, 1988–1998

Disease Category (ICD Codes)	Number of Deaths	PMR	95% Confidence Interval LCL	95% Confidence Interval UCL
Tuberculosis (010-018)	44	127	93	171
Mycoses (110-018)	23	**60**	38	90
Sarcoidosis (135)	4	*50*	14	128
Malignant neoplasms of trachea/bronchus/lung/pleura (162-163)	1,540	102	97	107
Acute respiratory infections (460-466)	15	84	47	139
Other diseases of upper respiratory tract (470-478)	8	87	38	171
Pneumonia and influenza (480-487)	1,758	**116**	111	122
Chronic obstructive pulmonary disease and allied conditions (490-496)	2,308	**123**	118	128
Pneumoconiosis and other lung diseases - external agents (500-508)	353	90	81	100
Other diseases of respiratory system (510-519)	546	**88**	81	96

ICD - International Classification of Diseases, 9th Revision   LCL - lower confidence limit   UCL - upper confidence limit
NOTE: PMRs in **bold** are significantly different from 100 (p<0.05). PMRs in *italics* are based on fewer than five observed deaths. PMRs are based on underlying and contributing cause of death. Some values could not be calculated because the number of observed or expected deaths was zero; such values are indicated by —. See appendices for source description, methods, ICD codes, and a list of selected states.
SOURCE: National Center for Health Statistics multiple-cause-of-death data

Mortality by Disease Category within Agricultural and Sex Group

Table 2-12. Fishery workers, males: Proportionate mortality ratio (PMR) adjusted for age and race/ethnicity by disease category, U.S. residents age 15 and over, selected states, 1988–1998

Disease Category (ICD Codes)	Number of Deaths	PMR	95% Confidence Interval LCL	UCL
Tuberculosis (010-018)	14	178	97	299
Mycoses (110-018)	7	65	26	134
Sarcoidosis (135)	4	*265*	*72*	*678*
Malignant neoplasms of trachea/bronchus/lung/pleura (162-163)	415	107	97	118
Acute respiratory infections (460-466)	4	*80*	*22*	*205*
Other diseases of upper respiratory tract (470-478)	5	197	64	460
Pneumonia and influenza (480-487)	416	104	95	114
Chronic obstructive pulmonary disease and allied conditions (490-496)	556	**112**	103	122
Pneumoconiosis and other lung diseases - external agents (500-508)	97	95	77	116
Other diseases of respiratory system (510-519)	148	89	76	105

ICD - International Classification of Diseases, 9th Revision   LCL - lower confidence limit   UCL - upper confidence limit
NOTE: PMRs in **bold** are significantly different from 100 (p<0.05). PMRs in *italics* are based on fewer than five observed deaths. PMRs are based on underlying and contributing cause of death. Some values could not be calculated because the number of observed or expected deaths was zero; such values are indicated by ---. See appendices for source description, methods, ICD codes, and a list of selected states.
SOURCE: National Center for Health Statistics multiple-cause-of-death data

Mortality by Disease Category within Agricultural and Sex Group

Table 2-13. Crop farm workers, females: Proportionate mortality ratio (PMR) adjusted for age and race/ethnicity by disease category, U.S. residents age 15 and over, selected states, 1988–1998

Disease Category (ICD Codes)	Number of Deaths	PMR	95% Confidence Interval LCL	95% Confidence Interval UCL
Tuberculosis (010-018)	41	**168**	123	228
Mycoses (110-018)	15	71	40	117
Sarcoidosis (135)	7	53	21	109
Malignant neoplasms of trachea/bronchus/lung/pleura (162-163)	334	**67**	60	75
Acute respiratory infections (460-466)	17	122	71	195
Other diseases of upper respiratory tract (470-478)	4	63	17	161
Pneumonia and influenza (480-487)	1,266	106	100	112
Chronic obstructive pulmonary disease and allied conditions (490-496)	665	**80**	74	86
Pneumoconiosis and other lung diseases - external agents (500-508)	254	97	86	110
Other diseases of respiratory system (510-519)	424	91	83	100

ICD - International Classification of Diseases, 9th Revision   LCL - lower confidence limit   UCL - upper confidence limit
NOTE: PMRs in **bold** are significantly different from 100 (p<0.05). PMRs in *italics* are based on fewer than five observed deaths. PMRs are based on underlying and contributing cause of death. Some values could not be calculated because the number of observed or expected deaths was zero; such values are indicated by ---. See appendices for source description, methods, ICD codes, and a list of selected states.
SOURCE: National Center for Health Statistics multiple-cause-of-death data

Mortality by Disease Category within Agricultural and Sex Group

Table 2-14. Livestock farm workers, females: Proportionate mortality ratio (PMR) adjusted for age and race/ethnicity by disease category, U.S. residents age 15 and over, selected states, 1988–1998

Disease Category (ICD Codes)	Number of Deaths	PMR	95% Confidence Interval LCL	95% Confidence Interval UCL
Tuberculosis (010-018)	5	133	43	311
Mycoses (110-018)	6	129	47	281
Sarcoidosis (135)	0	0	---	---
Malignant neoplasms of trachea/bronchus/lung/pleura (162-163)	125	**78**	66	93
Acute respiratory infections (460-466)	4	79	22	202
Other diseases of upper respiratory tract (470-478)	3	*191*	39	558
Pneumonia and influenza (480-487)	331	91	82	101
Chronic obstructive pulmonary disease and allied conditions (490-496)	244	**81**	71	92
Pneumoconiosis and other lung diseases - external agents (500-508)	50	**71**	53	94
Other diseases of respiratory system (510-519)	144	108	92	127

ICD - International Classification of Diseases, 9th Revision   LCL - lower confidence limit   UCL - upper confidence limit
NOTE: PMRs in **bold** are significantly different from 100 (p<0.05). PMRs in *italics* are based on fewer than five observed deaths. PMRs are based on underlying and contributing cause of death. Some values could not be calculated because the number of observed or expected deaths was zero; such values are indicated by ---. See appendices for source description, methods, ICD codes, and a list of selected states.
SOURCE: National Center for Health Statistics multiple-cause-of-death data

Mortality by Disease Category within Agricultural and Sex Group

Table 2-15. Farm managers, females: Proportionate mortality ratio (PMR) adjusted for age and race/ethnicity by disease category, U.S. residents age 15 and over, selected states, 1988–1998

Disease Category (ICD Codes)	Number of Deaths	PMR	95% Confidence Interval LCL	95% Confidence Interval UCL
Tuberculosis (010-018)	1	*168*	4	933
Mycoses (110-018)	0	*0*	---	---
Sarcoidosis (135)	0	*0*	---	---
Malignant neoplasms of trachea/bronchus/lung/pleura (162-163)	20	100	61	155
Acute respiratory infections (460-466)	0	*0*	---	---
Other diseases of upper respiratory tract (470-478)	1	*496*	13	2,756
Pneumonia and influenza (480-487)	43	105	77	142
Chronic obstructive pulmonary disease and allied conditions (490-496)	34	99	68	138
Pneumoconiosis and other lung diseases - external agents (500-508)	4	*49*	13	125
Other diseases of respiratory system (510-519)	15	93	52	153

ICD - International Classification of Diseases, 9th Revision   LCL - lower confidence limit   UCL - upper confidence limit
NOTE: PMRs in **bold** are significantly different from 100 (p<0.05). PMRs in *italics* are based on fewer than five observed deaths. PMRs are based on underlying and contributing cause of death. Some values could not be calculated because the number of observed or expected deaths was zero; such values are indicated by ---. See appendices for source description, methods, ICD codes, and a list of selected states.
SOURCE: National Center for Health Statistics multiple-cause-of-death data

Mortality by Disease Category within Agricultural and Sex Group

Table 2-16. Landscape and horticultural workers, females: Proportionate mortality ratio (PMR) adjusted for age and race/ethnicity by disease category, U.S. residents age 15 and over, selected states, 1988–1998

Disease Category (ICD Codes)	Number of Deaths	PMR	95% Confidence Interval LCL	95% Confidence Interval UCL
Tuberculosis (010-018)	0	0	---	---
Mycoses (110-018)	1	*77*	2	428
Sarcoidosis (135)	1	*219*	6	1,217
Malignant neoplasms of trachea/bronchus/lung/pleura (162-163)	32	89	61	126
Acute respiratory infections (460-466)	2	*253*	31	913
Other diseases of upper respiratory tract (470-478)	0	0	---	---
Pneumonia and influenza (480-487)	44	88	64	118
Chronic obstructive pulmonary disease and allied conditions (490-496)	58	116	89	150
Pneumoconiosis and other lung diseases - external agents (500-508)	5	53	17	124
Other diseases of respiratory system (510-519)	24	101	65	150

ICD - International Classification of Diseases, 9th Revision   LCL - lower confidence limit   UCL - upper confidence limit
NOTE: PMRs in **bold** are significantly different from 100 (p<0.05). PMRs in *italics* are based on fewer than five observed deaths. PMRs are based on underlying and contributing cause of death. Some values could not be calculated because the number of observed or expected deaths was zero; such values are indicated by ---. See appendices for source description, methods, ICD codes, and a list of selected states.
SOURCE: National Center for Health Statistics multiple-cause-of-death data

Mortality by Disease Category within Agricultural and Sex Group

Table 2-17. Forestry workers, females: Proportionate mortality ratio (PMR) adjusted for age and race/ethnicity by disease category, U.S. residents age 15 and over, selected states, 1988–1998

Disease Category (ICD Codes)	Number of Deaths	PMR	95% Confidence Interval LCL	95% Confidence Interval UCL
Tuberculosis (010-018)	1	*387*	10	2,150
Mycoses (110-018)	0	*0*	--	--
Sarcoidosis (135)	0	*0*	--	--
Malignant neoplasms of trachea/bronchus/lung/pleura (162-163)	13	122	65	209
Acute respiratory infections (460-466)	0	*0*	--	--
Other diseases of upper respiratory tract (470-478)	0	*0*	--	--
Pneumonia and influenza (480-487)	13	94	50	161
Chronic obstructive pulmonary disease and allied conditions (490-496)	10	68	33	125
Pneumoconiosis and other lung diseases - external agents (500-508)	1	*38*	1	211
Other diseases of respiratory system (510-519)	3	*45*	9	132

ICD - International Classification of Diseases, 9th Revision   LCL - lower confidence limit   UCL - upper confidence limit
NOTE: PMRs in **bold** are significantly different from 100 (p<0.05). PMRs in *italics* are based on fewer than five observed deaths. PMRs are based on underlying and contributing cause of death. Some values could not be calculated because the number of observed or expected deaths was zero; such values are indicated by ---.   See appendices for source description, methods, ICD codes, and a list of selected states.
SOURCE: National Center for Health Statistics multiple-cause-of-death data

Mortality by Disease Category within Agricultural and Sex Group

Table 2-18. Fishery workers, females: Proportionate mortality ratio (PMR) adjusted for age and race/ethnicity by disease category, U.S. residents age 15 and over, selected states, 1988–1998

Disease Category (ICD Codes)	Number of Deaths	PMR	95% Confidence Interval	
			LCL	UCL
Tuberculosis (010-018)	0	*0*	---	---
Mycoses (110-018)	0	*0*	---	---
Sarcoidosis (135)	0	*0*	---	---
Malignant neoplasms of trachea/bronchus/lung/pleura (162-163)	11	*177*	89	317
Acute respiratory infections (460-466)	0	*0*	---	---
Other diseases of upper respiratory tract (470-478)	0	*0*	---	---
Pneumonia and influenza (480-487)	6	*72*	26	157
Chronic obstructive pulmonary disease and allied conditions (490-496)	12	*136*	70	237
Pneumoconiosis and other lung diseases - external agents (500-508)	1	*62*	2	344
Other diseases of respiratory system (510-519)	2	*52*	6	188

ICD - International Classification of Diseases, 9th Revision   LCL - lower confidence limit   UCL - upper confidence limit
NOTE: PMRs in **bold** are significantly different from 100 (p<0.05). PMRs in *italics* are based on fewer than five observed deaths. PMRs are based on underlying and contributing cause of death. Some values could not be calculated because the number of observed or expected deaths was zero; such values are indicated by ---. See appendices for source description, methods, ICD codes, and a list of selected states.
SOURCE: National Center for Health Statistics multiple-cause-of-death data

Mortality by Agricultural Group and Sex within Disease Category

Figure 2-11. Tuberculosis: Proportionate mortality ratio (PMR) adjusted for age and race/ethnicity by agricultural group and sex, U.S. residents age 15 and over, selected states, 1988–1998

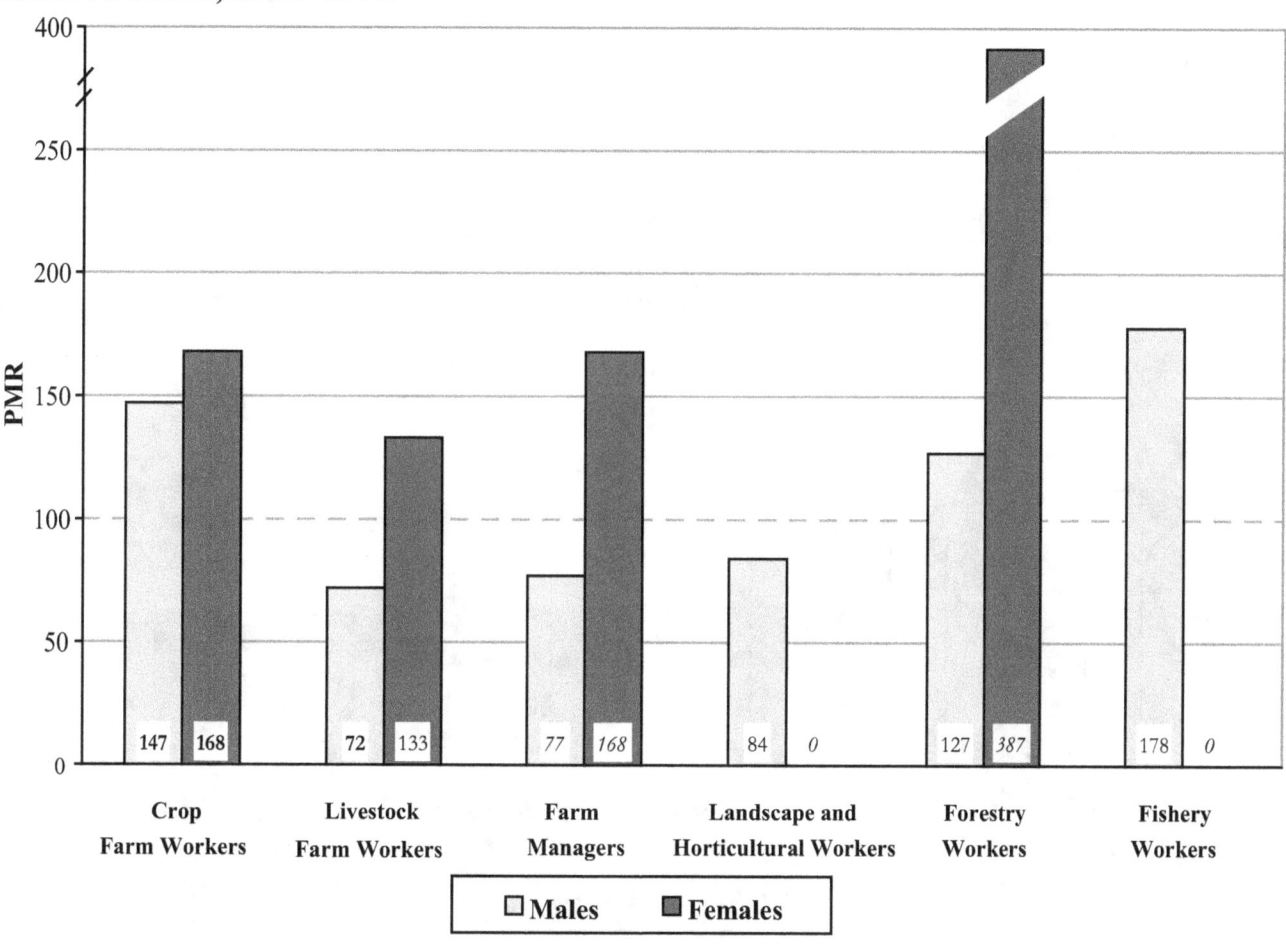

ICD – International Classification of Diseases, 9th Revision
NOTE: Tuberculosis = ICD-9 codes 010-018. PMRs in **bold** are significantly different from 100 (p<0.05). PMRs in *italics* are based on fewer than five observed deaths. PMRs are based on underlying and contributing cause of death. See appendices for source description, methods, ICD codes, and a list of selected states.
SOURCE: National Center for Health Statistics multiple-cause-of-death data

Mortality by Agricultural Group and Sex within Disease Category

Figure 2-12. Mycoses: Proportionate mortality ratio (PMR) adjusted for age and race/ethnicity by agricultural group and sex, U.S. residents age 15 and over, selected states, 1988–1998

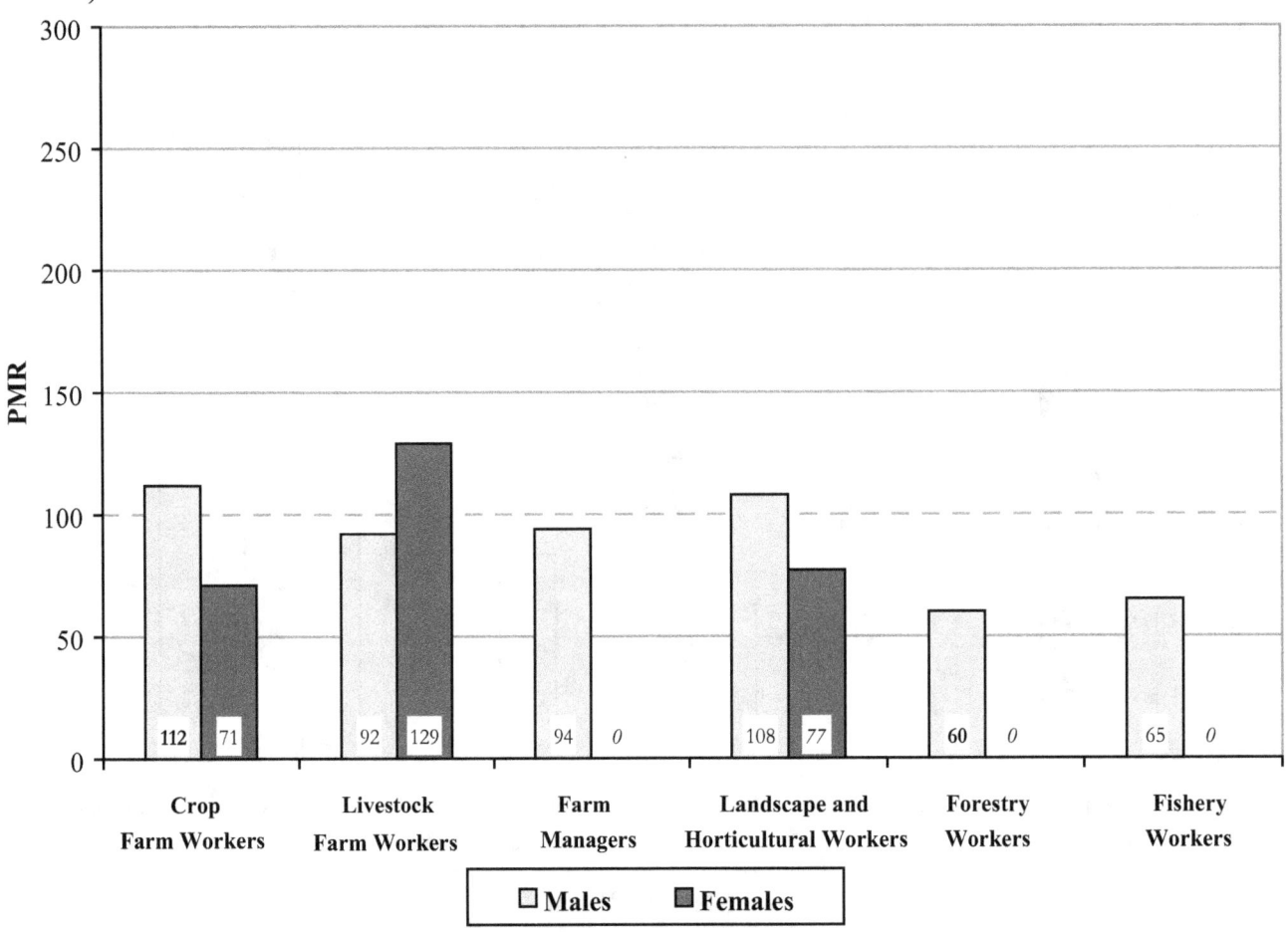

ICD – International Classification of Diseases, 9th Revision
NOTE: Mycoses = ICD-9 codes 110-118. PMRs in **bold** are significantly different from 100 (p<0.05). PMRs in *italics* are based on fewer than five observed deaths. PMRs are based on underlying and contributing cause of death. See appendices for source description, methods, ICD codes, and a list of selected states.
SOURCE: National Center for Health Statistics multiple-cause-of-death data

Mortality by Agricultural Group and Sex within Disease Category

Figure 2-13. Sarcoidosis: Proportionate mortality ratio (PMR) adjusted for age and race/ethnicity by agricultural group and sex, U.S. residents age 15 and over, selected states, 1988–1998

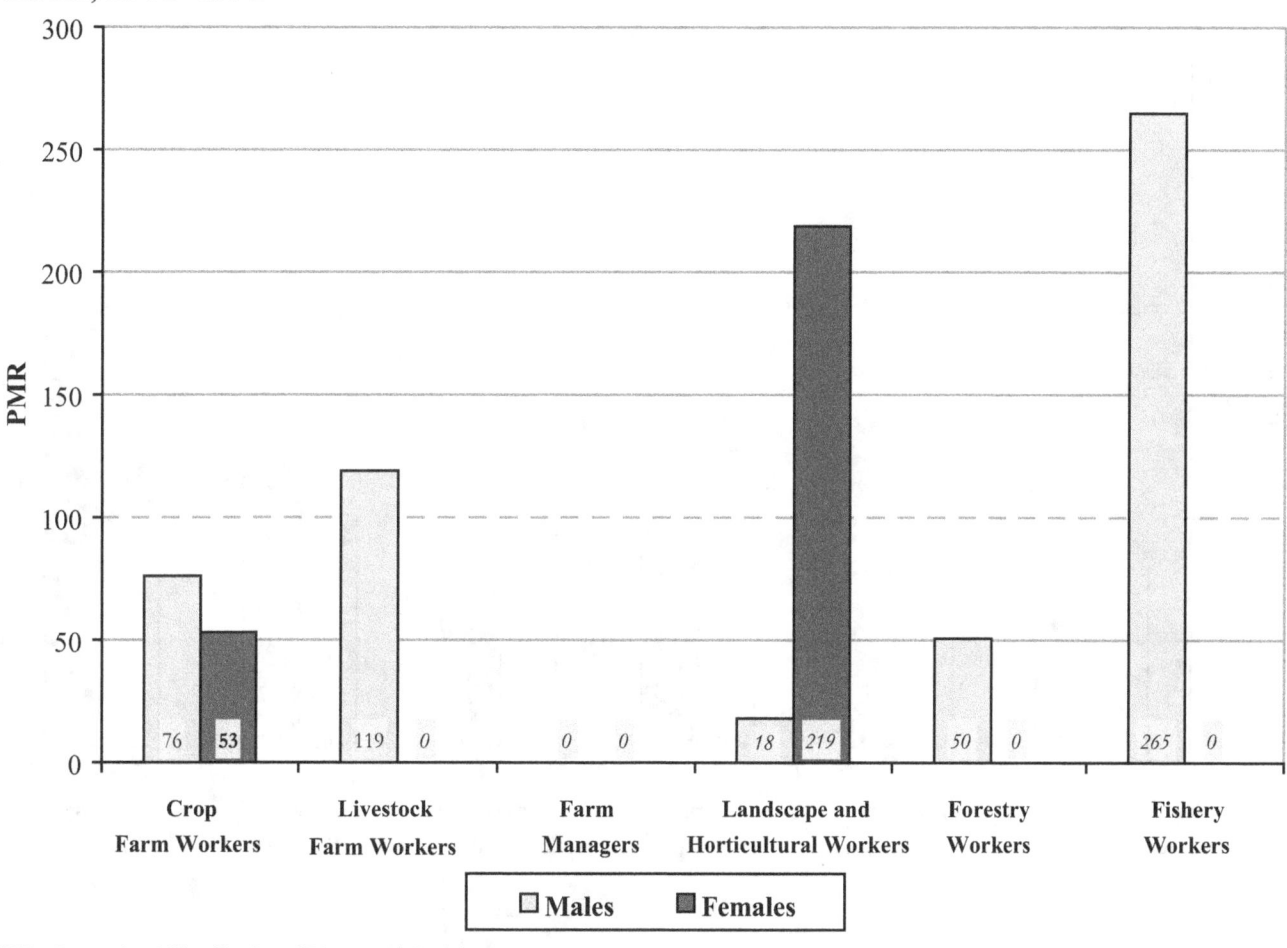

ICD – International Classification of Diseases, 9th Revision
NOTE: Sarcoidosis = ICD-9 code 135. PMRs in **bold** are significantly different from 100 (p<0.05). PMRs in *italics* are based on fewer than five observed deaths. PMRs are based on underlying and contributing cause of death. See appendices for source description, methods, ICD codes, and a list of selected states.
SOURCE: National Center for Health Statistics multiple-cause-of-death data

Mortality by Agricultural Group and Sex within Disease Category

Figure 2-14. Malignant neoplasms of trachea/bronchus/lung/pleura: Proportionate mortality ratio (PMR) adjusted for age and race/ethnicity by agricultural group and sex, U.S. residents age 15 and over, selected states, 1988–1998

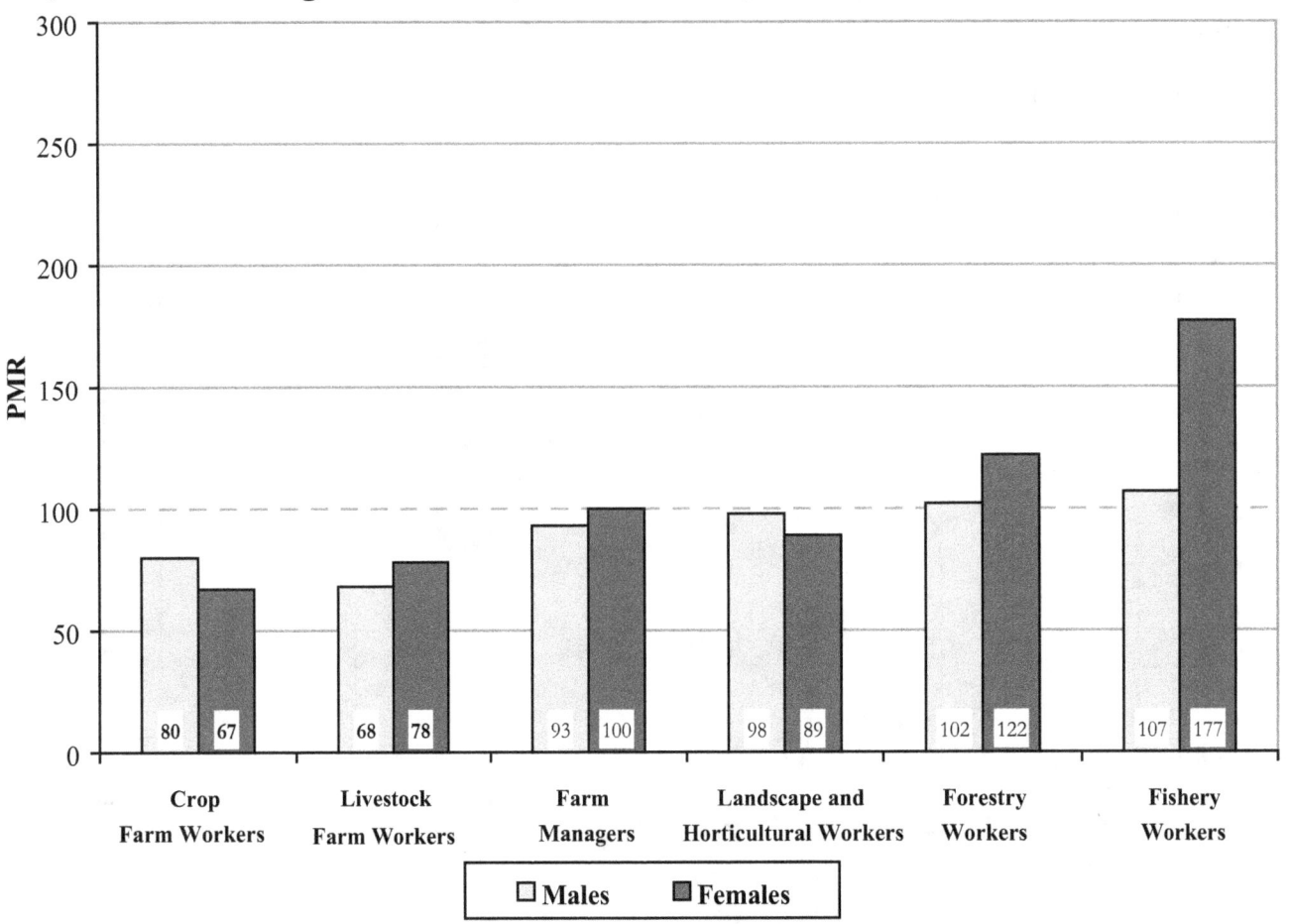

ICD – International Classification of Diseases, 9th Revision
NOTE: Malignant neoplasms of trachea/bronchus/lung/pleura = ICD-9 codes 162-163. PMRs in **bold** are significantly different from 100 (p<0.05). PMRs in *italics* are based on fewer than five observed deaths. PMRs are based on underlying and contributing cause of death. See appendices for source description, methods, ICD codes, and a list of selected states.
SOURCE: National Center for Health Statistics multiple-cause-of-death data

Figure 2-15. Acute respiratory infections: Proportionate mortality ratio (PMR) adjusted for age and race/ethnicity by agricultural group and sex, U.S. residents age 15 and over, selected states, 1988–1998

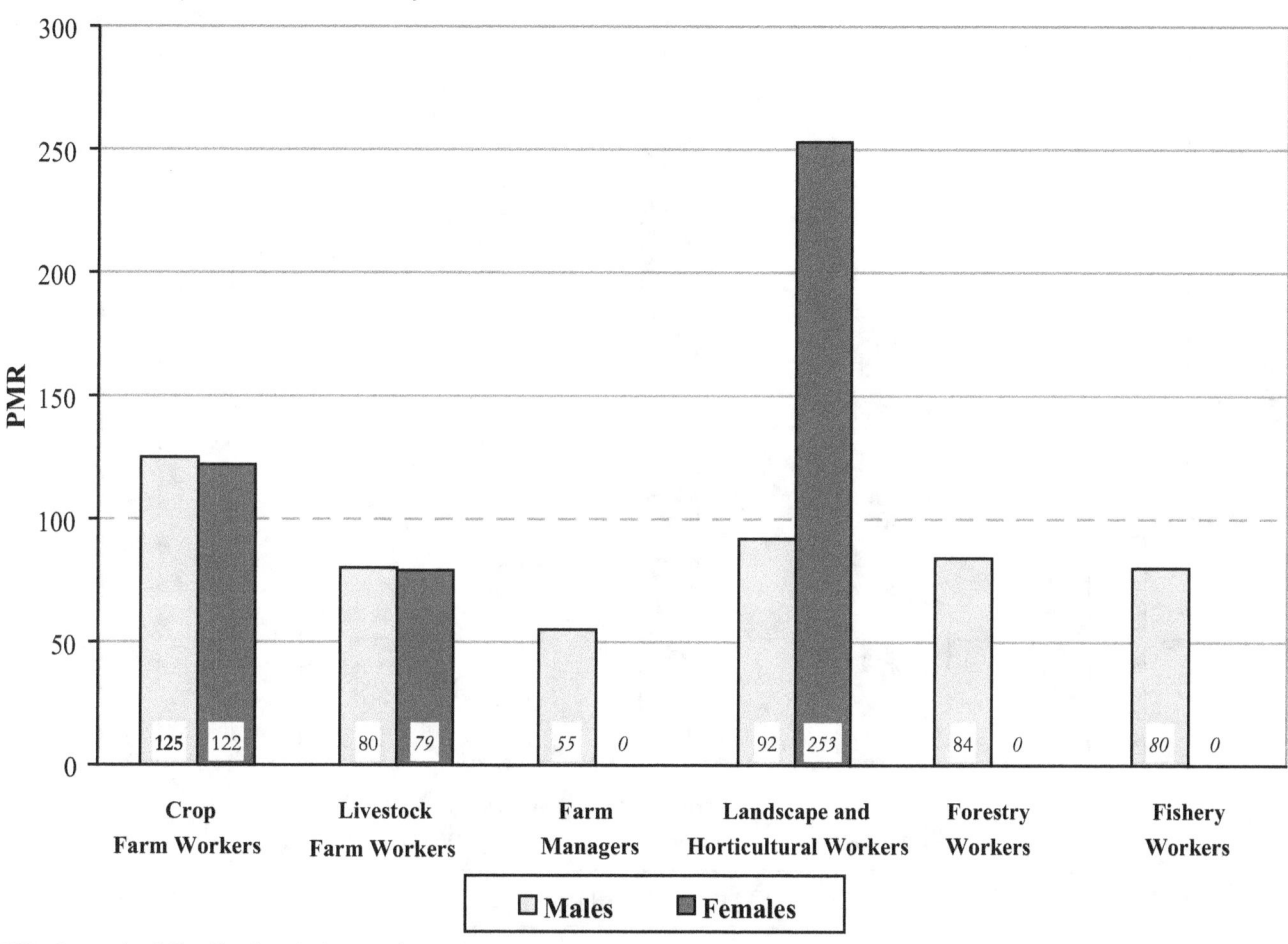

ICD – International Classification of Diseases, 9th Revision
NOTE: Acute respiratory infections = ICD-9 codes 460-466. PMRs in **bold** are significantly different from 100 (p<0.05). PMRs in *italics* are based on fewer than five observed deaths. PMRs are based on underlying and contributing cause of death. See appendices for source description, methods, ICD codes, and a list of selected states.
SOURCE: National Center for Health Statistics multiple-cause-of-death data

Mortality by Agricultural Group and Sex within Disease Category

Figure 2-16. Other diseases of upper respiratory tract: Proportionate mortality ratio (PMR) adjusted for age and race/ethnicity by agricultural group and sex, U.S. residents age 15 and over, selected states, 1988–1998

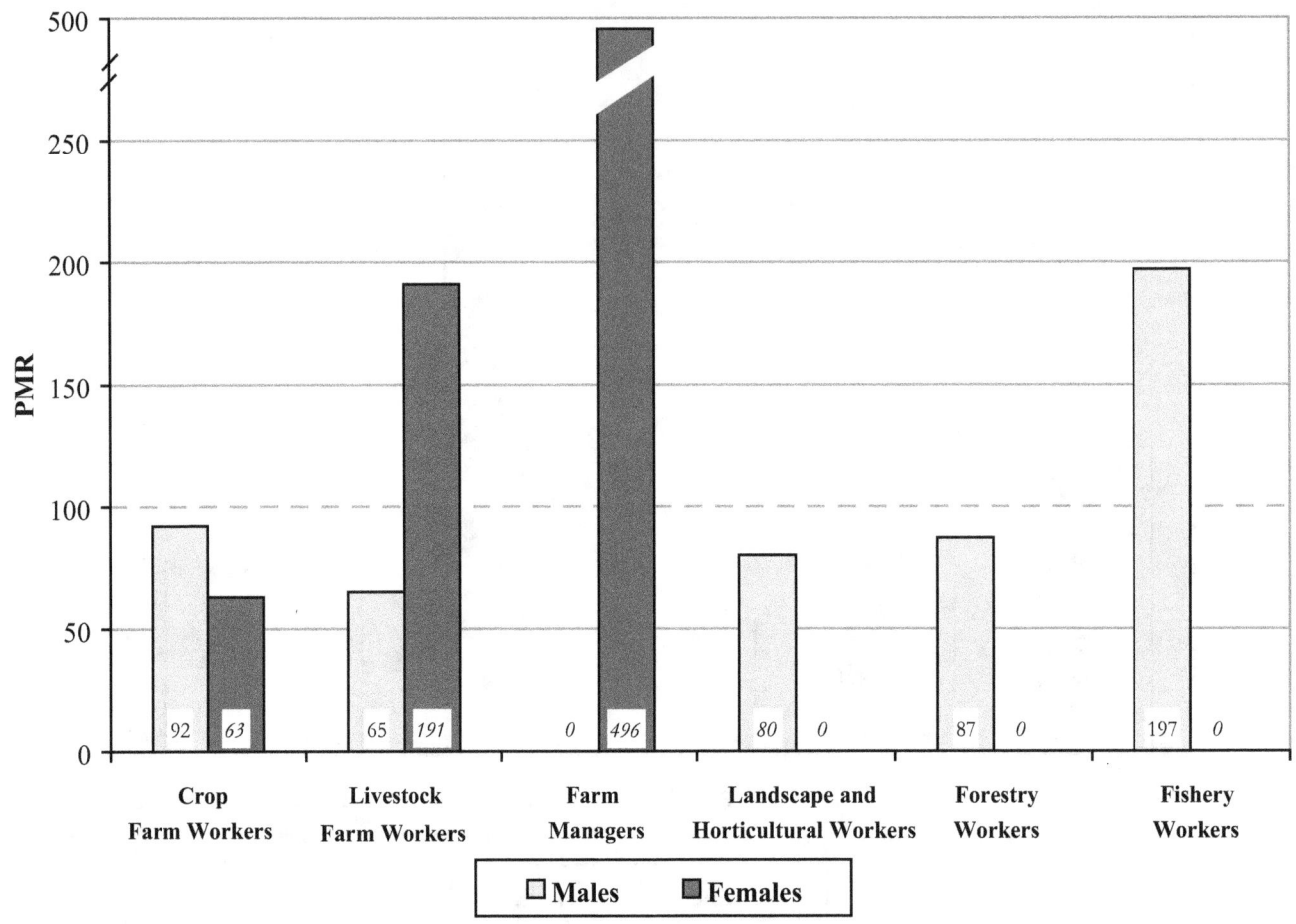

ICD – International Classification of Diseases, 9th Revision
NOTE: Other diseases of upper respiratory tract = ICD-9 codes 470-478. PMRs in **bold** are significantly different from 100 ($p<0.05$). PMRs in *italics* are based on fewer than five observed deaths. PMRs are based on underlying and contributing cause of death. See appendices for source description, methods, ICD codes, and a list of selected states.
SOURCE: National Center for Health Statistics multiple-cause-of-death data

Mortality by Agricultural Group and Sex within Disease Category

Figure 2-17. Pneumonia and influenza: Proportionate mortality ratio (PMR) adjusted for age and race/ethnicity by agricultural group and sex, U.S. residents age 15 and over, selected states, 1988–1998

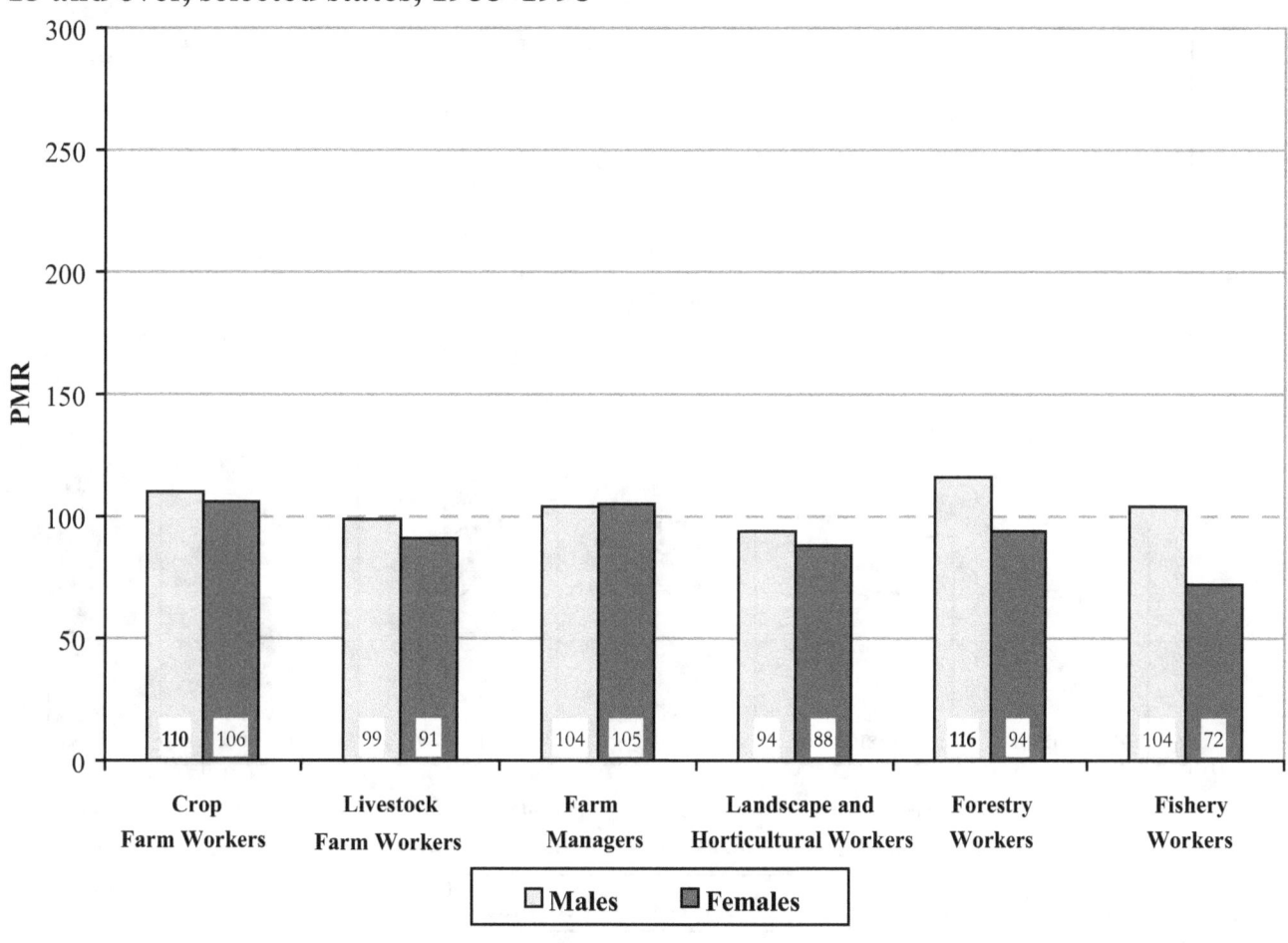

ICD – International Classification of Diseases, 9th Revision
NOTE: Pneumonia and influenza = ICD-9 codes 480-487. PMRs in **bold** are significantly different from 100 (p<0.05). PMRs in *italics* are based on fewer than five observed deaths. PMRs are based on underlying and contributing cause of death. See appendices for source description, methods, ICD codes, and a list of selected states.
SOURCE: National Center for Health Statistics multiple-cause-of-death data

Mortality by Agricultural Group and Sex within Disease Category

Figure 2-18. Chronic obstructive pulmonary disease and allied conditions: Proportionate mortality ratio (PMR) adjusted for age and race/ethnicity by agricultural group and sex, U.S. residents age 15 and over, selected states, 1988–1998

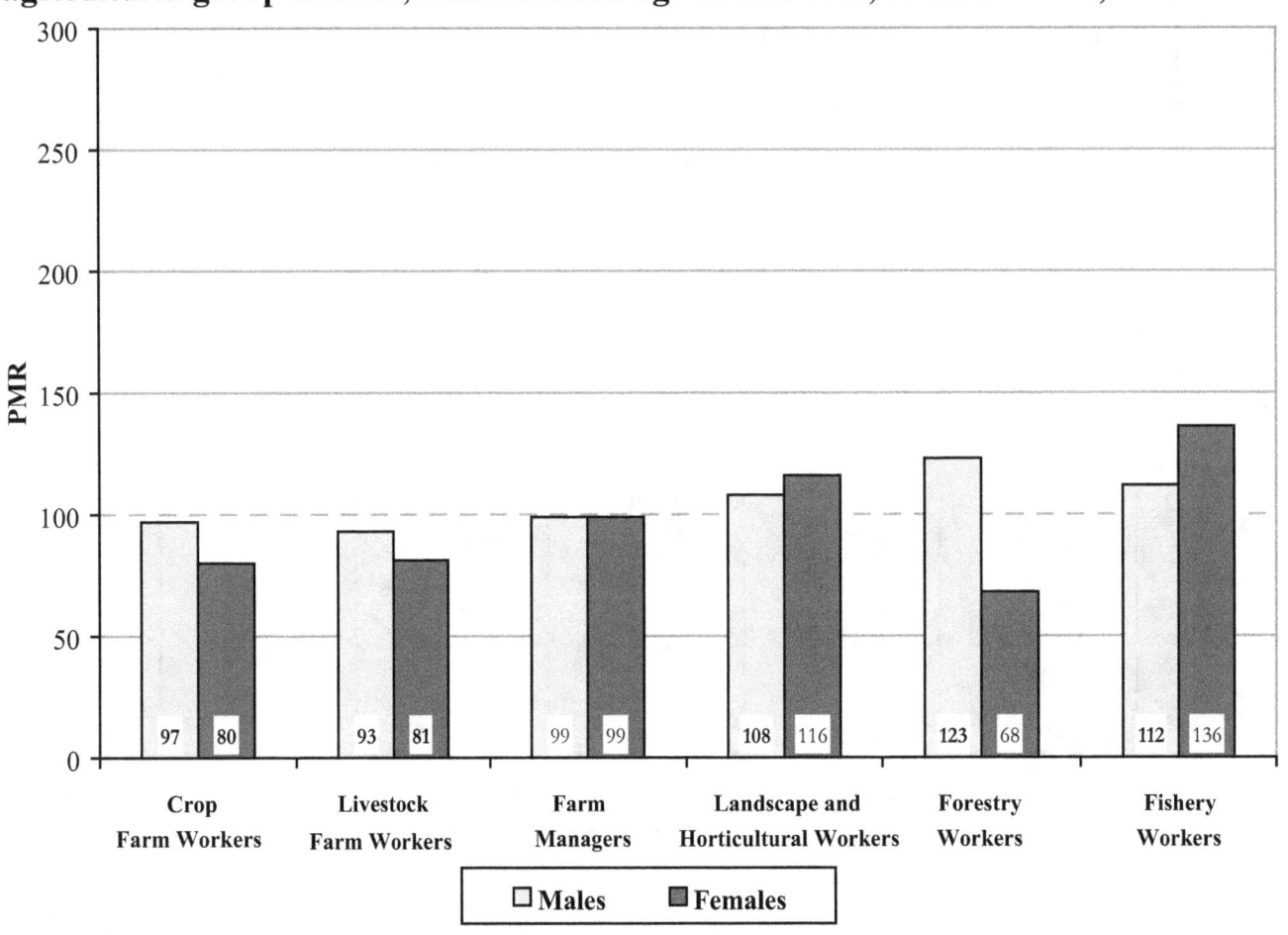

ICD – International Classification of Diseases, 9th Revision
NOTE: Chronic obstructive pulmonary disease and allied conditions = ICD-9 codes 490-496. PMRs in **bold** are significantly different from 100 ($p<0.05$). PMRs in *italics* are based on fewer than five observed deaths. PMRs are based on underlying and contributing cause of death. See appendices for source description, methods, ICD codes, and a list of selected states.
SOURCE: National Center for Health Statistics multiple-cause-of-death data

Mortality by Agricultural Group and Sex within Disease Category

Figure 2-19. Pneumoconioses and other lung diseases–external agents: Proportionate mortality ratio (PMR) adjusted for age and race/ethnicity by agricultural group and sex, U.S. residents age 15 and over, selected states, 1988–1998

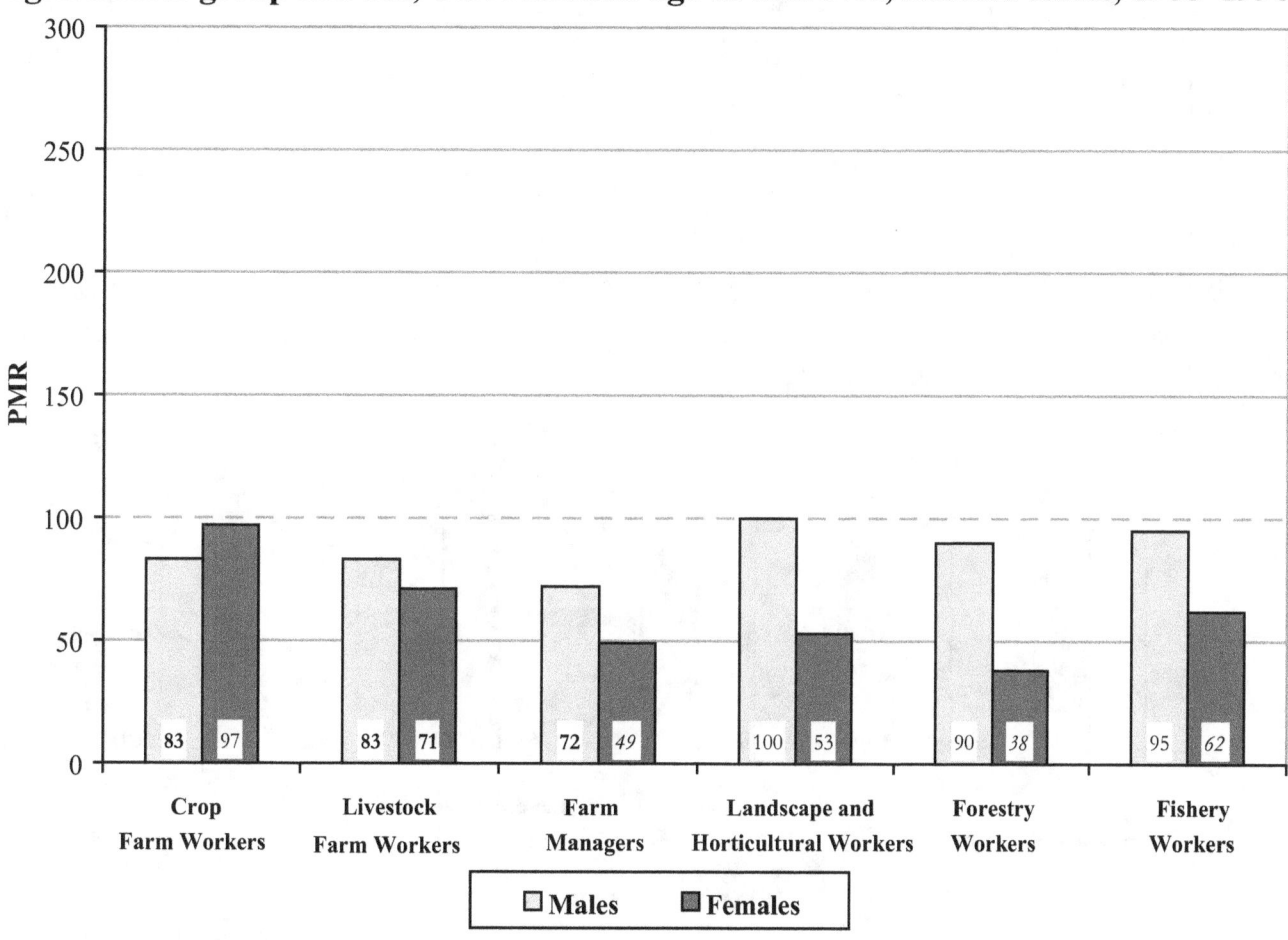

ICD – International Classification of Diseases, 9th Revision
NOTE: Pneumoconioses and other lung diseases - external agents = ICD-9 codes 500-508. PMRs in **bold** are significantly different from 100 (p<0.05). PMRs in *italics* are based on fewer than five observed deaths. PMRs are based on underlying and contributing cause of death. See appendices for source description, methods, ICD codes, and a list of selected states.
SOURCE: National Center for Health Statistics multiple-cause-of-death data

Mortality by Agricultural Group and Sex within Disease Category

Figure 2-20. Other diseases of respiratory system: Proportionate mortality ratio (PMR) adjusted for age and race/ethnicity by agricultural group and sex, U.S. residents age 15 and over, selected states, 1988–1998

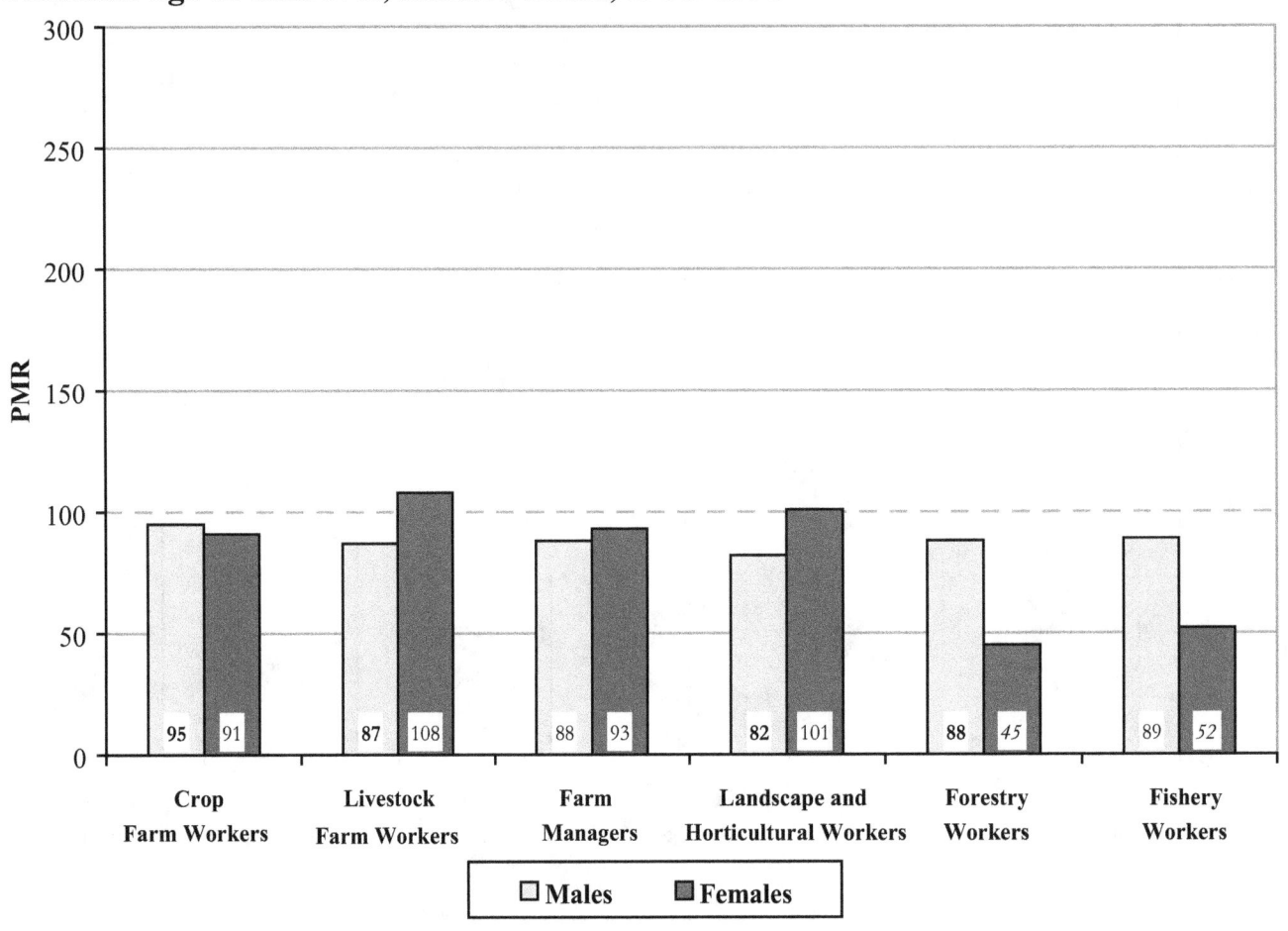

ICD – International Classification of Diseases, 9th Revision
NOTE: Other diseases of respiratory system = ICD-9 codes 510-519. PMRs in **bold** are significantly different from 100 (p<0.05). PMRs in *italics* are based on fewer than five observed deaths. PMRs are based on underlying and contributing cause of death. See appendices for source description, methods, ICD codes, and a list of selected states.
SOURCE: National Center for Health Statistics multiple-cause-of-death data

Mortality by Disease Category within Agricultural and Race/Ethnicity Group

Table 2-19. Crop farm workers, white, non-Hispanic: Proportionate mortality ratio (PMR) adjusted for age and sex by disease category, U.S. residents age 15 and over, selected states, 1988–1998

Disease Category (ICD Codes)	Number of Deaths	PMR	95% Confidence Interval LCL	95% Confidence Interval UCL
Tuberculosis (010-018)	223	**134**	118	153
Mycoses (110-018)	268	**118**	105	133
Sarcoidosis (135)	18	90	53	142
Malignant neoplasms of trachea/bronchus/lung/pleura (162-163)	9,869	**79**	77	81
Acute respiratory infections (460-466)	272	**126**	112	142
Other diseases of upper respiratory tract (470-478)	69	89	70	113
Pneumonia and influenza (480-487)	19,980	**109**	107	111
Chronic obstructive pulmonary disease and allied conditions (490-496)	21,215	**97**	95	99
Pneumoconiosis and other lung diseases - external agents (500-508)	3,944	**82**	79	85
Other diseases of respiratory system (510-519)	5,832	**95**	93	97

ICD - International Classification of Diseases, 9th Revision LCL - lower confidence limit UCL - upper confidence limit
NOTE: PMRs in **bold** are significantly different from 100 (p≤0.05). PMRs in *italics* are based on fewer than five observed deaths. PMRs are based on underlying and contributing cause of death. Some values could not be calculated because the number of observed or expected deaths was zero; such values are indicated by ---. See appendices for source description, methods, ICD codes, and a list of selected states.
SOURCE: National Center for Health Statistics multiple-cause-of-death data

Mortality by Disease Category within Agricultural and Race/Ethnicity Group

Table 2-20. Livestock farm workers, white, non-Hispanic: Proportionate mortality ratio (PMR) adjusted for age and sex by disease category, U.S. residents age 15 and over, selected states, 1988–1998

Disease Category (ICD Codes)	Number of Deaths	PMR	95% Confidence Interval LCL	UCL
Tuberculosis (010-018)	22	**46**	29	70
Mycoses (110-018)	69	104	81	132
Sarcoidosis (135)	8	134	58	264
Malignant neoplasms of trachea/bronchus/lung/pleura (162-163)	2,324	**65**	63	68
Acute respiratory infections (460-466)	52	83	63	109
Other diseases of upper respiratory tract (470-478)	17	75	44	120
Pneumonia and influenza (480-487)	5,103	97	95	100
Chronic obstructive pulmonary disease and allied conditions (490-496)	5,791	**93**	91	95
Pneumoconiosis and other lung diseases - external agents (500-508)	1,129	**82**	77	87
Other diseases of respiratory system (510-519)	1,532	**86**	82	90

ICD - International Classification of Diseases, 9th Revision   LCL - lower confidence limit   UCL - upper confidence limit
NOTE: PMRs in **bold** are significantly different from 100 (p<0.05). PMRs in *italics* are based on fewer than five observed deaths. PMRs are based on underlying and contributing cause of death. Some values could not be calculated because the number of observed or expected deaths was zero; such values are indicated by ---. See appendices for source description, methods, ICD codes, and a list of selected states.
SOURCE: National Center for Health Statistics multiple-cause-of-death data

Mortality by Disease Category within Agricultural and Race/Ethnicity Group

Table 2-21. Farm managers, white, non-Hispanic: Proportionate mortality ratio (PMR) adjusted for age and sex by disease category, U.S. residents age 15 and over, selected states, 1988–1998

Disease Category (ICD Codes)	Number of Deaths	PMR	95% Confidence Interval LCL	95% Confidence Interval UCL
Tuberculosis (010-018)	3	*108*	22	316
Mycoses (110-018)	4	*91*	25	233
Sarcoidosis (135)	0	0	---	---
Malignant neoplasms of trachea/bronchus/lung/pleura (162-163)	202	92	81	106
Acute respiratory infections (460-466)	2	*57*	7	206
Other diseases of upper respiratory tract (470-478)	0	0	---	---
Pneumonia and influenza (480-487)	280	100	89	112
Chronic obstructive pulmonary disease and allied conditions (490-496)	350	102	92	113
Pneumoconiosis and other lung diseases - external agents (500-508)	49	**69**	51	91
Other diseases of respiratory system (510-519)	92	90	73	110

ICD - International Classification of Diseases, 9th Revision LCL - lower confidence limit UCL - upper confidence limit
NOTE: PMRs in **bold** are significantly different from 100 (p<0.05). PMRs in *italics* are based on fewer than five observed deaths. PMRs are based on underlying and contributing cause of death. Some values could not be calculated because the number of observed or expected deaths was zero; such values are indicated by ---. See appendices for source description, methods, ICD codes, and a list of selected states.
SOURCE: National Center for Health Statistics multiple-cause-of-death data

Mortality by Disease Category within Agricultural and Race/Ethnicity Group

Table 2-22. Landscape and horticultural workers, white, non-Hispanic: Proportionate mortality ratio (PMR) adjusted for age and sex by disease category, U.S. residents age 15 and over, selected states, 1988–1998

Disease Category (ICD Codes)	Number of Deaths	PMR	95% Confidence Interval LCL	95% Confidence Interval UCL
Tuberculosis (010-018)	3	*50*	10	146
Mycoses (110-018)	16	110	63	179
Sarcoidosis (135)	0	*0*	---	---
Malignant neoplasms of trachea/bronchus/lung/pleura (162-163)	444	98	89	108
Acute respiratory infections (460-466)	4	*65*	18	166
Other diseases of upper respiratory tract (470-478)	2	*61*	7	220
Pneumonia and influenza (480-487)	385	92	83	102
Chronic obstructive pulmonary disease and allied conditions (490-496)	604	**113**	105	122
Pneumoconiosis and other lung diseases - external agents (500-508)	99	97	79	118
Other diseases of respiratory system (510-519)	164	**84**	72	98

ICD - International Classification of Diseases, 9th Revision   LCL - lower confidence limit   UCL - upper confidence limit
NOTE: PMRs in **bold** are significantly different from 100 (p<0.05). PMRs in *italics* are based on fewer than five observed deaths. PMRs are based on underlying and contributing cause of death. Some values could not be calculated because the number of observed or expected deaths was zero; such values are indicated by ---. See appendices for source description, methods, ICD codes, and a list of selected states.
SOURCE: National Center for Health Statistics multiple-cause-of-death data

Mortality by Disease Category within Agricultural and Race/Ethnicity Group

Table 2-23. Forestry workers, white, non-Hispanic: Proportionate mortality ratio (PMR) adjusted for age and sex by disease category, U.S. residents age 15 and over, selected states, 1988–1998

Disease Category (ICD Codes)	Number of Deaths	PMR	95% Confidence Interval LCL	95% Confidence Interval UCL
Tuberculosis (010-018)	14	112	61	188
Mycoses (110-018)	16	70	40	114
Sarcoidosis (135)	1	*48*	1	267
Malignant neoplasms of trachea/bronchus/lung/pleura (162-163)	1,061	101	95	107
Acute respiratory infections (460-466)	12	87	45	152
Other diseases of upper respiratory tract (470-478)	4	*65*	18	166
Pneumonia and influenza (480-487)	1,231	**115**	109	122
Chronic obstructive pulmonary disease and allied conditions (490-496)	1,835	**127**	122	133
Pneumoconiosis and other lung diseases - external agents (500-508)	255	92	82	104
Other diseases of respiratory system (510-519)	362	**84**	76	93

ICD - International Classification of Diseases, 9th Revision   LCL - lower confidence limit   UCL - upper confidence limit
NOTE: PMRs in **bold** are significantly different from 100 (p<0.05). PMRs in *italics* are based on fewer than five observed deaths. PMRs are based on underlying and contributing cause of death. Some values could not be calculated because the number of observed or expected deaths was zero; such values are indicated by ---. See appendices for source description, methods, ICD codes, and a list of selected states.
SOURCE: National Center for Health Statistics multiple-cause-of-death data

Mortality by Disease Category within Agricultural and Race/Ethnicity Group

Table 2-24. Fishery workers, white, non–Hispanic: Proportionate mortality ratio (PMR) adjusted for age and sex by disease category, U.S. residents age 15 and over, selected states, 1988–1998

Disease Category (ICD Codes)	Number of Deaths	PMR	95% Confidence Interval LCL	95% Confidence Interval UCL
Tuberculosis (010-018)	6	154	56	336
Mycoses (110-018)	4	*51*	14	130
Sarcoidosis (135)	1	*143*	4	794
Malignant neoplasms of trachea/bronchus/lung/pleura (162-163)	333	107	96	119
Acute respiratory infections (460-466)	3	70	14	205
Other diseases of upper respiratory tract (470-478)	3	*150*	31	439
Pneumonia and influenza (480-487)	321	99	89	110
Chronic obstructive pulmonary disease and allied conditions (490-496)	502	**119**	110	130
Pneumoconiosis and other lung diseases - external agents (500-508)	78	94	75	117
Other diseases of respiratory system (510-519)	119	90	75	108

ICD - International Classification of Diseases, 9th Revision   LCL - lower confidence limit   UCL - upper confidence limit
NOTE: PMRs in **bold** are significantly different from 100 (p<0.05). PMRs in *italics* are based on fewer than five observed deaths. PMRs are based on underlying and contributing cause of death. Some values could not be calculated because the number of observed or expected deaths was zero; such values are indicated by ---. See appendices for source description, methods, ICD codes, and a list of selected states.
SOURCE: National Center for Health Statistics multiple-cause-of-death data

Mortality by Disease Category within Agricultural and Race/Ethnicity Group

Table 2-25. Crop farm workers, black, non-Hispanic: Proportionate mortality ratio (PMR) adjusted for age and sex by disease category, U.S. residents age 15 and over, selected states, 1988–1998

Disease Category (ICD Codes)	Number of Deaths	PMR	95% Confidence Interval LCL	95% Confidence Interval UCL
Tuberculosis (010-018)	231	**188**	165	214
Mycoses (110-018)	69	94	73	119
Sarcoidosis (135)	22	**63**	39	95
Malignant neoplasms of trachea/bronchus/lung/pleura (162-163)	1,916	**85**	81	89
Acute respiratory infections (460-466)	31	138	94	196
Other diseases of upper respiratory tract (470-478)	20	115	70	178
Pneumonia and influenza (480-487)	3,215	**114**	110	118
Chronic obstructive pulmonary disease and allied conditions (490-496)	2,322	96	92	100
Pneumoconiosis and other lung diseases - external agents (500-508)	734	95	89	102
Other diseases of respiratory system (510-519)	1,011	94	89	100

ICD - International Classification of Diseases, 9th Revision LCL - lower confidence limit UCL - upper confidence limit
NOTE: PMRs in **bold** are significantly different from 100 (p<0.05). PMRs in *italics* are based on fewer than five observed deaths. PMRs are based on underlying and contributing cause of death. Some values could not be calculated because the number of observed or expected deaths was zero; such values are indicated by ---. See appendices for source description, methods, ICD codes, and a list of selected states.
SOURCE: National Center for Health Statistics multiple-cause-of-death data

Mortality by Disease Category within Agricultural and Race/Ethnicity Group

Table 2-26. Livestock farm workers, black, non-Hispanic: Proportionate mortality ratio (PMR) adjusted for age and sex by disease category, U.S. residents age 15 and over, selected states, 1988–1998

Disease Category (ICD Codes)	Number of Deaths	PMR	95% Confidence Interval LCL	95% Confidence Interval UCL
Tuberculosis (010-018)	4	*114*	31	292
Mycoses (110-018)	0	*0*	---	---
Sarcoidosis (135)	0	*0*	---	---
Malignant neoplasms of trachea/bronchus/lung/pleura (162-163)	57	103	79	134
Acute respiratory infections (460-466)	1	*183*	5	1,017
Other diseases of upper respiratory tract (470-478)	0	*0*	---	---
Pneumonia and influenza (480-487)	69	111	87	141
Chronic obstructive pulmonary disease and allied conditions (490-496)	49	92	68	122
Pneumoconiosis and other lung diseases - external agents (500-508)	15	93	52	153
Other diseases of respiratory system (510-519)	20	75	46	116

ICD - International Classification of Diseases, 9th Revision   LCL - lower confidence limit   UCL - upper confidence limit
NOTE: PMRs in **bold** are significantly different from 100 (p<0.05). PMRs in *italics* are based on fewer than five observed deaths. PMRs are based on underlying and contributing cause of death. Some values could not be calculated because the number of observed or expected deaths was zero; such values are indicated by ---. See appendices for source description, methods, ICD codes, and a list of selected states.
SOURCE: National Center for Health Statistics multiple-cause-of-death data

Mortality by Disease Category within Agricultural and Race/Ethnicity Group

Table 2-27. Farm managers, black, non-Hispanic: Proportionate mortality ratio (PMR) adjusted for age and sex by disease category, U.S. residents age 15 and over, selected states, 1988–1998

Disease Category (ICD Codes)	Number of Deaths	PMR	95% Confidence Interval	
			LCL	UCL
Tuberculosis (010-018)	0	0	---	---
Mycoses (110-018)	0	0	---	---
Sarcoidosis (135)	0	0	---	---
Malignant neoplasms of trachea/bronchus/lung/pleura (162-163)	8	93	40	183
Acute respiratory infections (460-466)	0	0	---	---
Other diseases of upper respiratory tract (470-478)	0	0	---	---
Pneumonia and influenza (480-487)	9	92	42	175
Chronic obstructive pulmonary disease and allied conditions (490-496)	4	47	13	120
Pneumoconiosis and other lung diseases - external agents (500-508)	4	*152*	41	389
Other diseases of respiratory system (510-519)	2	*52*	6	188

ICD - International Classification of Diseases, 9th Revision   LCL - lower confidence limit   UCL - upper confidence limit
NOTE: PMRs in **bold** are significantly different from 100 (p<0.05). PMRs in *italics* are based on fewer than five observed deaths. PMRs are based on underlying and contributing cause of death. Some values could not be calculated because the number of observed or expected deaths was zero; such values are indicated by ---. See appendices for source description, methods, ICD codes, and a list of selected states.
SOURCE: National Center for Health Statistics multiple-cause-of-death data

Mortality by Disease Category within Agricultural and Race/Ethnicity Group

Table 2-28. Landscape and horticultural workers, black, non-Hispanic: Proportionate mortality ratio (PMR) adjusted for age and sex by disease category, U.S. residents age 15 and over; selected states, 1988–1998

Disease Category (ICD Codes)	Number of Deaths	PMR	95% Confidence Interval LCL	95% Confidence Interval UCL
Tuberculosis (010-018)	9	86	39	163
Mycoses (110-018)	8	102	44	201
Sarcoidosis (135)	1	23	1	128
Malignant neoplasms of trachea/bronchus/lung/pleura (162-163)	147	100	85	118
Acute respiratory infections (460-466)	4	281	77	719
Other diseases of upper respiratory tract (470-478)	2	*149*	18	538
Pneumonia and influenza (480-487)	141	96	82	113
Chronic obstructive pulmonary disease and allied conditions (490-496)	122	99	83	118
Pneumoconiosis and other lung diseases - external agents (500-508)	32	89	61	126
Other diseases of respiratory system (510-519)	61	88	69	113

ICD - International Classification of Diseases, 9th Revision LCL - lower confidence limit UCL - upper confidence limit
NOTE: PMRs in **bold** are significantly different from 100 (p<0.05). PMRs in *italics* are based on fewer than five observed deaths. PMRs are based on underlying and contributing cause of death. Some values could not be calculated because the number of observed or expected deaths was zero; such values are indicated by ---. See appendices for source description, methods, ICD codes, and a list of selected states.
SOURCE: National Center for Health Statistics multiple-cause-of-death data

Mortality by Disease Category within Agricultural and Race/Ethnicity Group

Table 2-29. Forestry workers, black, non-Hispanic: Proportionate mortality ratio (PMR) adjusted for age and sex by disease category, U.S. residents age 15 and over, selected states, 1988–1998

Disease Category (ICD Codes)	Number of Deaths	PMR	95% Confidence Interval LCL	95% Confidence Interval UCL
Tuberculosis (010-018)	22	121	76	183
Mycoses (110-018)	5	**42**	14	98
Sarcoidosis (135)	3	*52*	11	152
Malignant neoplasms of trachea/bronchus/lung/pleura (162-163)	357	101	91	112
Acute respiratory infections (460-466)	3	*109*	22	319
Other diseases of upper respiratory tract (470-478)	4	*174*	47	445
Pneumonia and influenza (480-487)	381	**124**	112	137
Chronic obstructive pulmonary disease and allied conditions (490-496)	269	91	81	103
Pneumoconiosis and other lung diseases - external agents (500-508)	77	95	75	119
Other diseases of respiratory system (510-519)	138	101	85	119

ICD - International Classification of Diseases, 9th Revision   LCL - lower confidence limit   UCL - upper confidence limit
NOTE: PMRs in **bold** are significantly different from 100 (p<0.05). PMRs in *italics* are based on fewer than five observed deaths. PMRs are based on underlying and contributing cause of death. Some values could not be calculated because the number of observed or expected deaths was zero; such values are indicated by ---. See appendices for source description, methods, ICD codes, and a list of selected states.
SOURCE: National Center for Health Statistics multiple-cause-of-death data

Mortality by Disease Category within Agricultural and Race/Ethnicity Group

Table 2-30. Fishery workers, black, non-Hispanic: Proportionate mortality ratio (PMR) adjusted for age and sex by disease category, U.S. residents age 15 and over, selected states, 1988–1998

Disease Category (ICD Codes)	Number of Deaths	PMR	95% Confidence Interval LCL	UCL
Tuberculosis (010-018)	2	*191*	23	690
Mycoses (110-018)	1	*71*	2	394
Sarcoidosis (135)	2	*291*	35	1,051
Malignant neoplasms of trachea/bronchus/lung/pleura (162-163)	46	123	91	164
Acute respiratory infections (460-466)	1	*317*	8	1,761
Other diseases of upper respiratory tract (470-478)	0	0	---	---
Pneumonia and influenza (480-487)	46	130	96	174
Chronic obstructive pulmonary disease and allied conditions (490-496)	21	**64**	40	98
Pneumoconiosis and other lung diseases - external agents (500-508)	11	119	60	213
Other diseases of respiratory system (510-519)	13	84	45	144

ICD - International Classification of Diseases, 9th Revision   LCL - lower confidence limit   UCL - upper confidence limit
NOTE: PMRs in **bold** are significantly different from 100 (p<0.05). PMRs in *italics* are based on fewer than five observed deaths. PMRs are based on underlying and contributing cause of death. Some values could not be calculated because the number of observed or expected deaths was zero; such values are indicated by ---. See appendices for source description, methods, ICD codes, and a list of selected states.
SOURCE: National Center for Health Statistics multiple-cause-of-death data

Mortality by Disease Category within Agricultural and Race/Ethnicity Group

Table 2-31. Crop farm workers, Hispanic: Proportionate mortality ratio (PMR) adjusted for age and sex by disease category, U.S. residents age 15 and over, selected states, 1988–1998

Disease Category (ICD Codes)	Number of Deaths	PMR	95% Confidence Interval LCL	95% Confidence Interval UCL
Tuberculosis (010-018)	12	76	39	133
Mycoses (110-018)	8	67	29	132
Sarcoidosis (135)	0	0	---	---
Malignant neoplasms of trachea/bronchus/lung/pleura (162-163)	155	**81**	69	95
Acute respiratory infections (460-466)	9	**349**	160	662
Other diseases of upper respiratory tract (470-478)	3	*108*	22	316
Pneumonia and influenza (480-487)	447	109	99	120
Chronic obstructive pulmonary disease and allied conditions (490-496)	385	109	99	120
Pneumoconiosis and other lung diseases - external agents (500-508)	78	**72**	57	90
Other diseases of respiratory system (510-519)	160	**85**	72	99

ICD - International Classification of Diseases, 9th Revision   LCL - lower confidence limit   UCL - upper confidence limit
NOTE: PMRs in **bold** are significantly different from 100 (p<0.05). PMRs in *italics* are based on fewer than five observed deaths. PMRs are based on underlying and contributing cause of death. Some values could not be calculated because the number of observed or expected deaths was zero; such values are indicated by ---.   See appendices for source description, methods, ICD codes, and a list of selected states.
SOURCE: National Center for Health Statistics multiple-cause-of-death data

Mortality by Disease Category within Agricultural and Race/Ethnicity Group

Table 2-32. Livestock farm workers, Hispanic: Proportionate mortality ratio (PMR) adjusted for age and sex by disease category, U.S. residents age 15 and over, selected states, 1988–1998

Disease Category (ICD Codes)	Number of Deaths	PMR	95% Confidence Interval LCL	95% Confidence Interval UCL
Tuberculosis (010-018)	9	165	76	313
Mycoses (110-018)	1	36	1	200
Sarcoidosis (135)	0	0	---	---
Malignant neoplasms of trachea/bronchus/lung/pleura (162-163)	73	91	72	115
Acute respiratory infections (460-466)	0	0	---	---
Other diseases of upper respiratory tract (470-478)	1	*131*	3	728
Pneumonia and influenza (480-487)	223	114	100	130
Chronic obstructive pulmonary disease and allied conditions (490-496)	194	110	95	127
Pneumoconiosis and other lung diseases - external agents (500-508)	55	100	76	130
Other diseases of respiratory system (510-519)	53	**69**	53	90

ICD - International Classification of Diseases, 9th Revision   LCL - lower confidence limit   UCL - upper confidence limit
NOTE: PMRs in **bold** are significantly different from 100 (p<0.05). PMRs in *italics* are based on fewer than five observed deaths. PMRs are based on underlying and contributing cause of death. Some values could not be calculated because the number of observed or expected deaths was zero; such values are indicated by ---. See appendices for source description, methods, ICD codes, and a list of selected states.
SOURCE: National Center for Health Statistics multiple-cause-of-death data

Mortality by Disease Category within Agricultural and Race/Ethnicity Group

Table 2-33. Farm managers, Hispanic: Proportionate mortality ratio (PMR) adjusted for age and sex by disease category, U.S. residents age 15 and over, selected states, 1988–1998

Disease Category (ICD Codes)	Number of Deaths	PMR	95% Confidence Interval	
			LCL	UCL
Tuberculosis (010-018)	1	*165*	4	917
Mycoses (110-018)	1	*264*	7	1,467
Sarcoidosis (135)	0	*0*	---	---
Malignant neoplasms of trachea/bronchus/lung/pleura (162-163)	4	*50*	14	128
Acute respiratory infections (460-466)	0	*0*	---	---
Other diseases of upper respiratory tract (470-478)	0	*0*	---	---
Pneumonia and influenza (480-487)	22	118	74	179
Chronic obstructive pulmonary disease and allied conditions (490-496)	19	116	70	181
Pneumoconiosis and other lung diseases - external agents (500-508)	1	*20*	1	111
Other diseases of respiratory system (510-519)	8	102	44	201

ICD - International Classification of Diseases, 9th Revision   LCL - lower confidence limit   UCL - upper confidence limit
NOTE: PMRs in **bold** are significantly different from 100 ($p<0.05$). PMRs in *italics* are based on fewer than five observed deaths. PMRs are based on underlying and contributing cause of death. Some values could not be calculated because the number of observed or expected deaths was zero; such values are indicated by ---. See appendices for source description, methods, ICD codes, and a list of selected states.
SOURCE: National Center for Health Statistics multiple-cause-of-death data

Table 2-34. Landscape and horticultural workers, Hispanic: Proportionate mortality ratio (PMR) adjusted for age and sex by disease category, U.S. residents age 15 and over, selected states, 1988–1998

Disease Category (ICD Codes)	Number of Deaths	PMR	95% Confidence Interval LCL	95% Confidence Interval UCL
Tuberculosis (010-018)	2	*124*	15	448
Mycoses (110-018)	1	*62*	2	344
Sarcoidosis (135)	0	0	---	---
Malignant neoplasms of trachea/bronchus/lung/pleura (162-163)	10	71	34	131
Acute respiratory infections (460-466)	0	0	---	---
Other diseases of upper respiratory tract (470-478)	0	0	---	---
Pneumonia and influenza (480-487)	23	87	55	131
Chronic obstructive pulmonary disease and allied conditions (490-496)	17	88	51	141
Pneumoconiosis and other lung diseases - external agents (500-508)	6	98	36	214
Other diseases of respiratory system (510-519)	13	83	44	142

ICD - International Classification of Diseases, 9th Revision   LCL - lower confidence limit   UCL - upper confidence limit
NOTE: PMRs in **bold** are significantly different from 100 (p<0.05). PMRs in *italics* are based on fewer than five observed deaths. PMRs are based on underlying and contributing cause of death. Some values could not be calculated because the number of observed or expected deaths was zero; such values are indicated by ---. See appendices for source description, methods, ICD codes, and a list of selected states.
SOURCE: National Center for Health Statistics multiple-cause-of-death data

Mortality by Disease Category within Agricultural and Race/Ethnicity Group

Table 2-35. Forestry workers, Hispanic: Proportionate mortality ratio (PMR) adjusted for age and sex by disease category, U.S. residents age 15 and over, selected states, 1988–1998

Disease Category (ICD Codes)	Number of Deaths	PMR	95% Confidence Interval LCL	UCL
Tuberculosis (010-018)	1	*142*	4	789
Mycoses (110-018)	0	*0*	---	---
Sarcoidosis (135)	0	*0*	---	---
Malignant neoplasms of trachea/bronchus/lung/pleura (162-163)	8	83	36	163
Acute respiratory infections (460-466)	0	*0*	---	---
Other diseases of upper respiratory tract (470-478)	0	*0*	---	---
Pneumonia and influenza (480-487)	18	85	50	134
Chronic obstructive pulmonary disease and allied conditions (490-496)	22	116	73	176
Pneumoconiosis and other lung diseases - external agents (500-508)	3	52	11	152
Other diseases of respiratory system (510-519)	7	77	31	159

ICD - International Classification of Diseases, 9th Revision   LCL - lower confidence limit   UCL - upper confidence limit
NOTE: PMRs in **bold** are significantly different from 100 (p<0.05). PMRs in *italics* are based on fewer than five observed deaths. PMRs are based on underlying and contributing cause of death. Some values could not be calculated because the number of observed or expected deaths was zero; such values are indicated by ---. See appendices for source description, methods, ICD codes, and a list of selected states.
SOURCE: National Center for Health Statistics multiple-cause-of-death data

Mortality by Disease Category within Agricultural and Race/Ethnicity Group

Table 2-36. Fishery workers, Hispanic: Proportionate mortality ratio (PMR) adjusted for age and sex by disease category, U.S. residents age 15 and over, selected states, 1988–1998

Disease Category (ICD Codes)	Number of Deaths	PMR	95% Confidence Interval LCL	UCL
Tuberculosis (010-018)	1	888	22	4,933
Mycoses (110-018)	0	0	---	---
Sarcoidosis (135)	0	0	---	---
Malignant neoplasms of trachea/bronchus/lung/pleura (162-163)	0	0	---	---
Acute respiratory infections (460-466)	0	0	---	---
Other diseases of upper respiratory tract (470-478)	0	0	---	---
Pneumonia and influenza (480-487)	2	*81*	10	292
Chronic obstructive pulmonary disease and allied conditions (490-496)	3	*149*	31	436
Pneumoconiosis and other lung diseases - external agents (500-508)	0	0	---	---
Other diseases of respiratory system (510-519)	0	0	---	---

ICD - International Classification of Diseases, 9th Revision   LCL - lower confidence limit   UCL - upper confidence limit
NOTE: PMRs in **bold** are significantly different from 100 (p<0.05). PMRs in *italics* are based on fewer than five observed deaths. PMRs are based on underlying and contributing cause of death. Some values could not be calculated because the number of observed or expected deaths was zero; such values are indicated by ---. See appendices for source description, methods, ICD codes, and a list of selected states.
SOURCE: National Center for Health Statistics multiple-cause-of-death data

Mortality by Agricultural Group and Race/Ethnicity within Disease Category

Figure 2-21. Tuberculosis: Proportionate mortality ratio (PMR) adjusted for age and sex by agricultural group and race/ethnicity, U.S. residents age 15 and over, selected states, 1988–1998

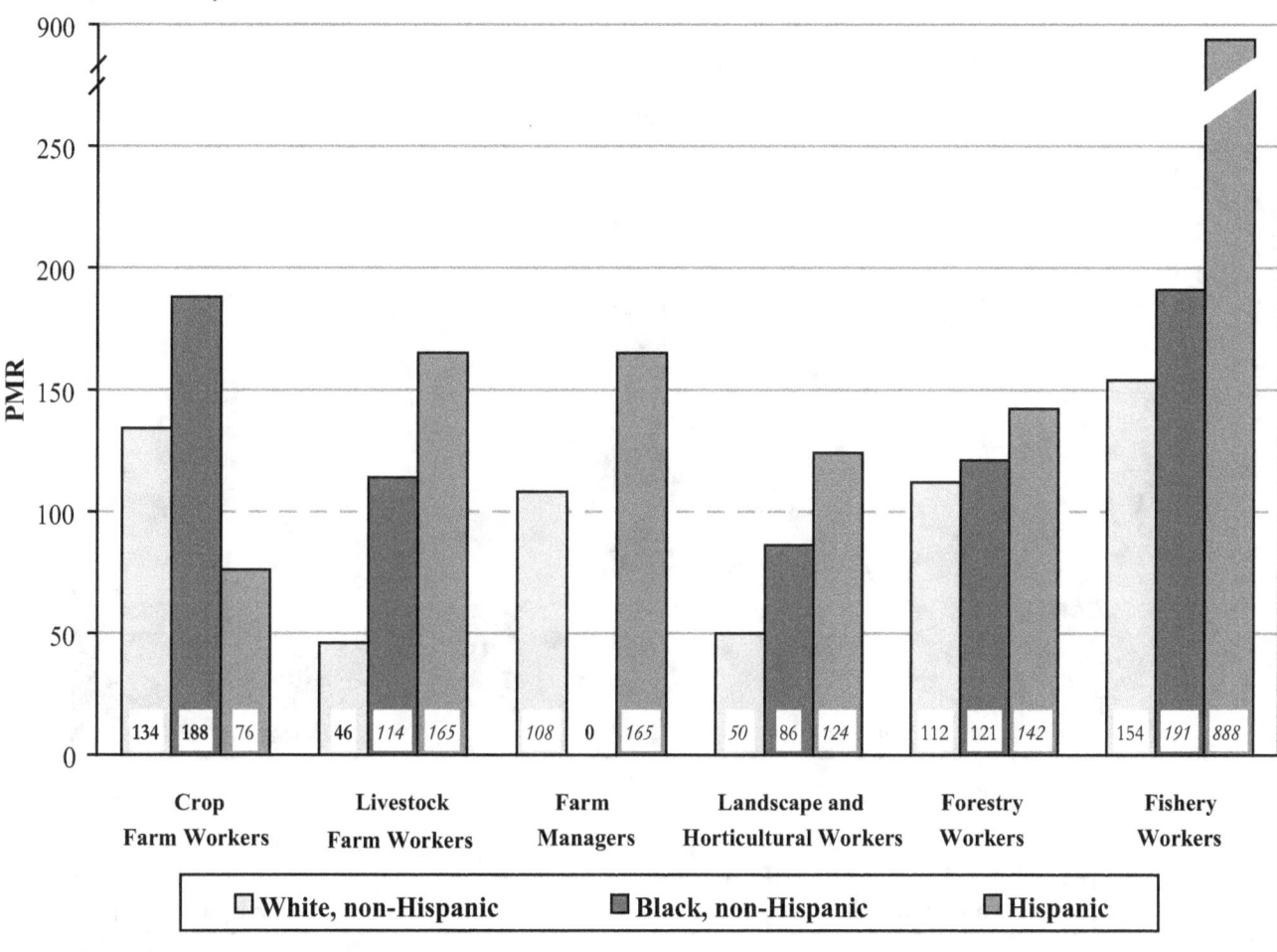

ICD – International Classification of Diseases, 9th Revision
NOTE: Tuberculosis = ICD-9 codes 010-018. PMRs in **bold** are significantly different from 100 (p<0.05). PMRs in *italics* are based on fewer than five observed deaths. PMRs are based on underlying and contributing cause of death. See appendices for source description, methods, ICD codes, and a list of selected states.
SOURCE: National Center for Health Statistics multiple-cause-of-death data

Mortality by Agricultural Group and Race/Ethnicity within Disease Category

Figure 2-22. Mycoses: Proportionate mortality ratio (PMR) adjusted for age and sex by agricultural group and race/ethnicity, U.S. residents age 15 and over, selected states, 1988–1998

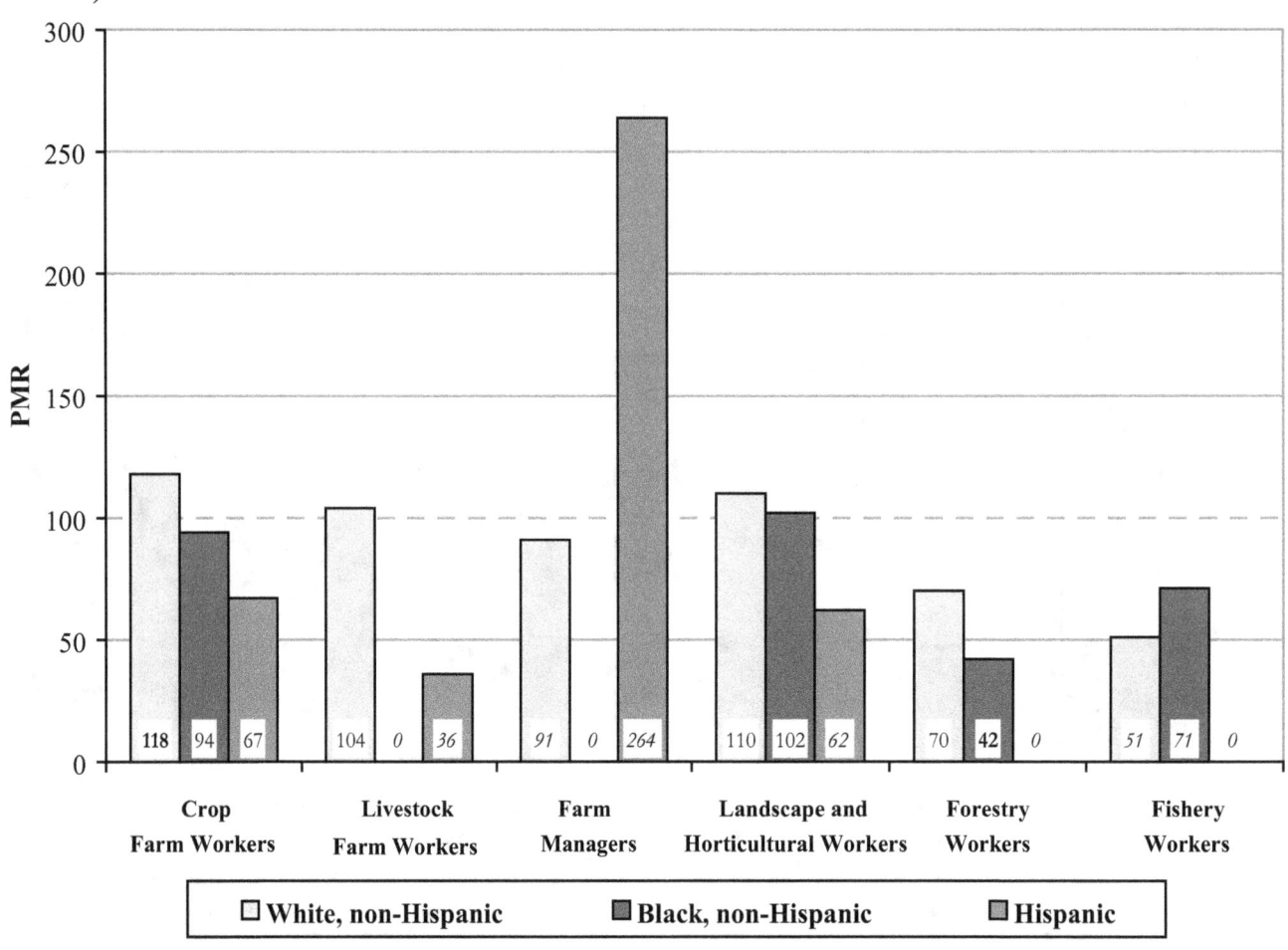

ICD – International Classification of Diseases, 9th Revision
NOTE: Mycoses = ICD-9 codes 110-118. PMRs in **bold** are significantly different from 100 (p<0.05). PMRs in *italics* are based on fewer than five observed deaths. PMRs are based on underlying and contributing cause of death. See appendices for source description, methods, ICD codes, and a list of selected states.
SOURCE: National Center for Health Statistics multiple-cause-of-death data

Mortality by Agricultural Group and Race/Ethnicity within Disease Category

Figure 2-23. Sarcoidosis: Proportionate mortality ratio (PMR) adjusted for age and sex by agricultural group and race/ethnicity, U.S. residents age 15 and over, selected states, 1988–1998

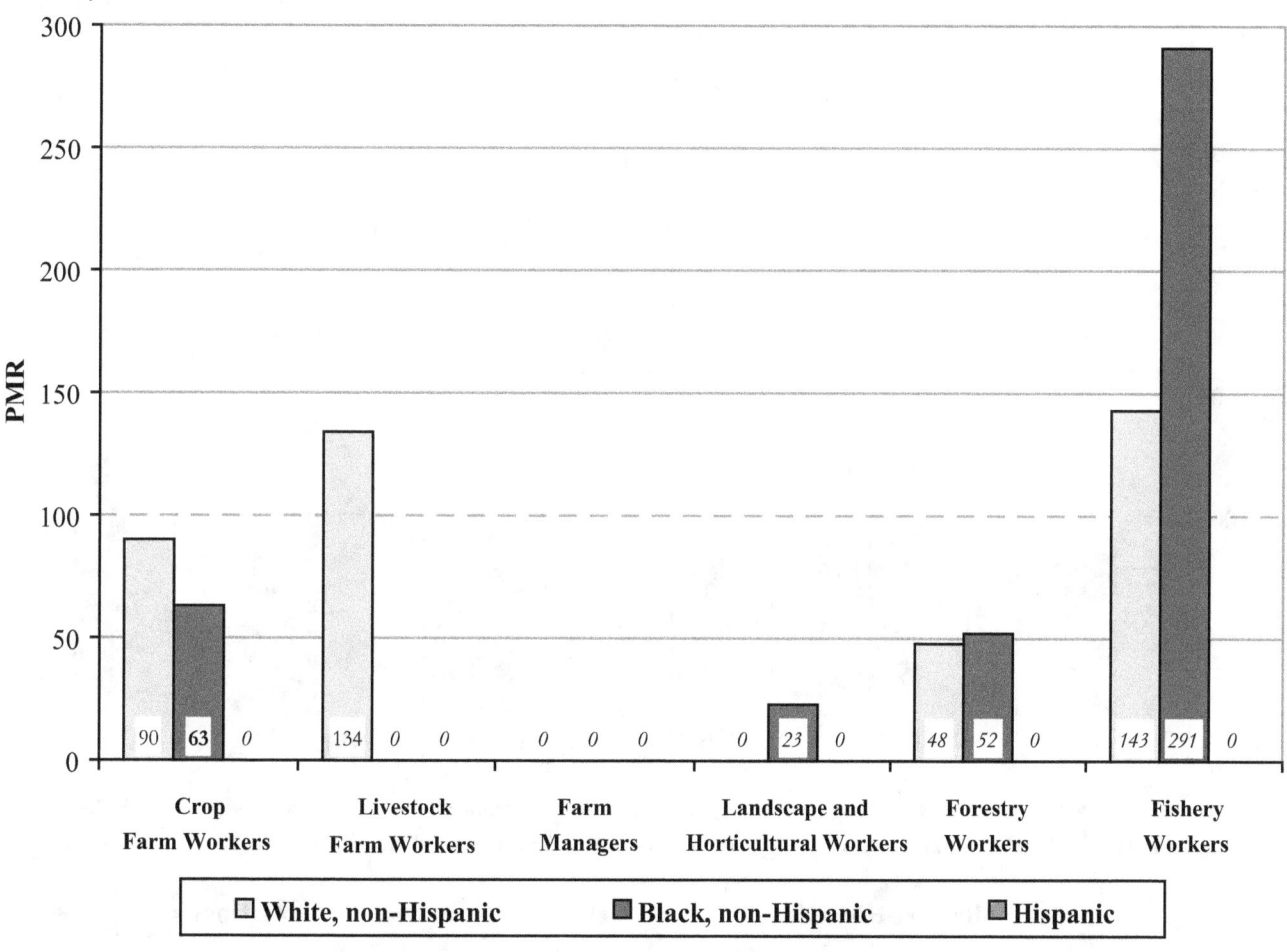

ICD – International Classification of Diseases, 9th Revision
NOTE: Sarcoidosis = ICD-9 code 135. PMRs in **bold** are significantly different from 100 (p<0.05). PMRs in *italics* are based on fewer than five observed deaths. PMRs are based on underlying and contributing cause of death. See appendices for source description, methods, ICD codes, and a list of selected states.
SOURCE: National Center for Health Statistics multiple-cause-of-death data

Mortality by Agricultural Group and Race/Ethnicity within Disease Category

Figure 2-24. Malignant neoplasms of trachea/bronchus/lung/pleura: Proportionate mortality ratio (PMR) adjusted for age and sex by agricultural group and race/ethnicity, U.S. residents age 15 and over, selected states, 1988–1998

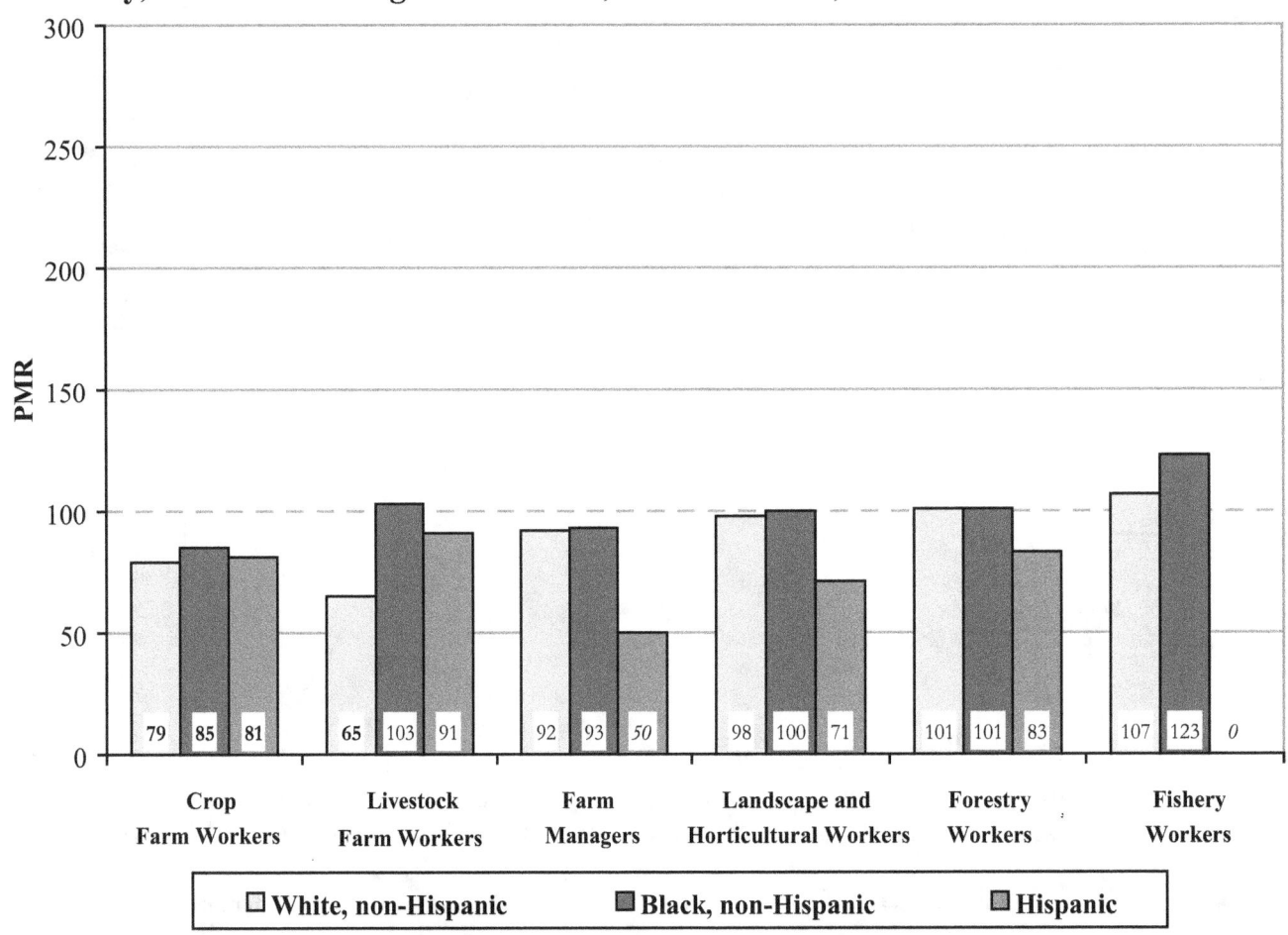

ICD – International Classification of Diseases, 9th Revision
NOTE: Malignant neoplasms of trachea/bronchus/lung/pleura = ICD-9 codes 162-163. PMRs in **bold** are significantly different from 100 (p<0.05). PMRs in *italics* are based on fewer than five observed deaths. PMRs are based on underlying and contributing cause of death. See appendices for source description, methods, ICD codes, and a list of selected states.
SOURCE: National Center for Health Statistics multiple-cause-of-death data

Mortality by Agricultural Group and Race/Ethnicity within Disease Category

Figure 2-25. Acute respiratory infections: Proportionate mortality ratio (PMR) adjusted for age and sex by agricultural group and race/ethnicity, U.S. residents age 15 and over, selected states, 1988–1998

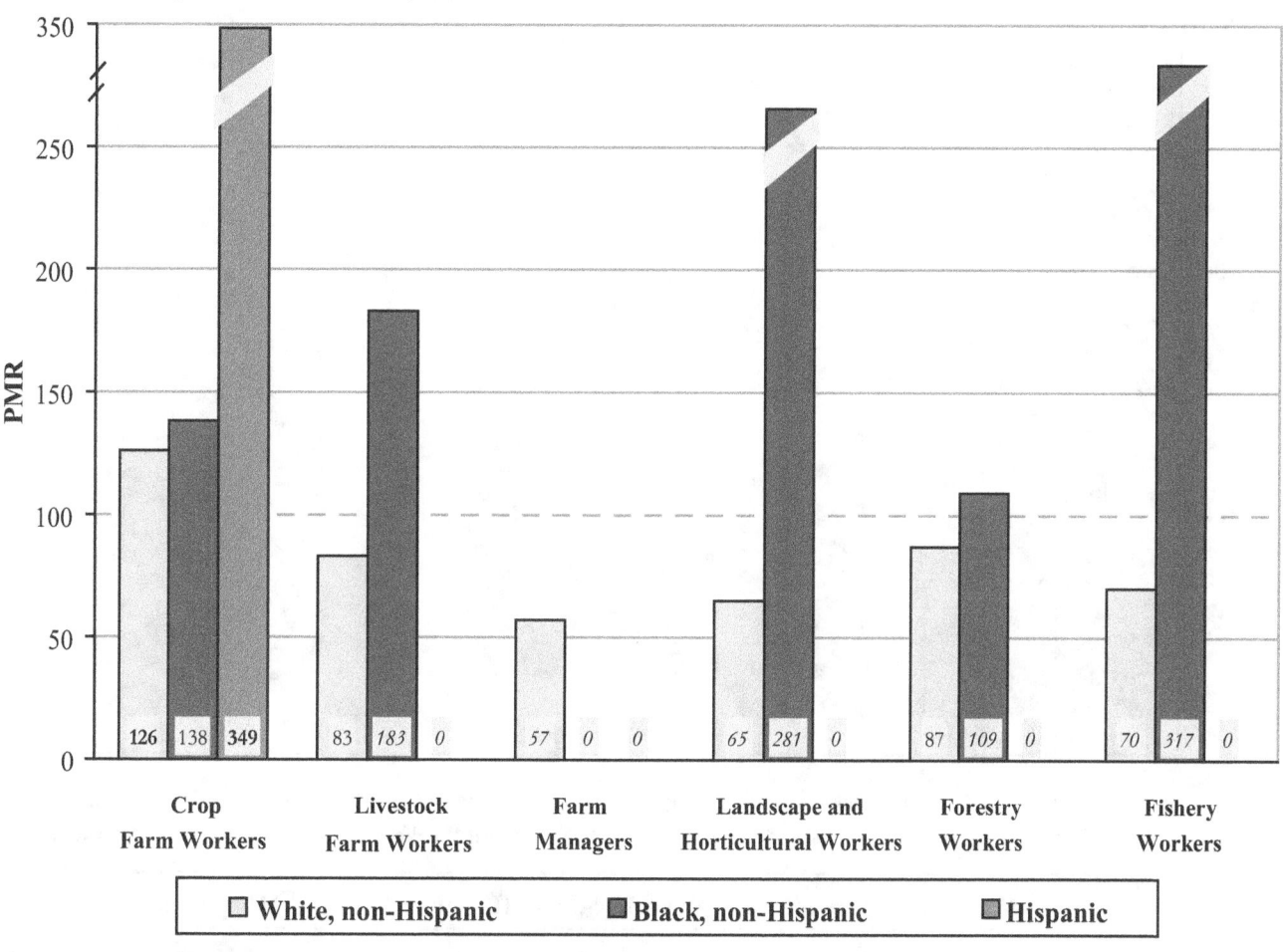

ICD – International Classification of Diseases, 9th Revision
NOTE: Acute respiratory infections = ICD-9 codes 460-466. PMRs in **bold** are significantly different from 100 (p<0.05). PMRs in *italics* are based on fewer than five observed deaths. PMRs are based on underlying and contributing cause of death. See appendices for source description, methods, ICD codes, and a list of selected states.
SOURCE: National Center for Health Statistics multiple-cause-of-death data

Mortality by Agricultural Group and Race/Ethnicity within Disease Category

Figure 2-26. Other diseases of upper respiratory tract: Proportionate mortality ratio (PMR) adjusted for age and sex by agricultural group and race/ethnicity, U.S. residents age 15 and over, selected states, 1988–1998

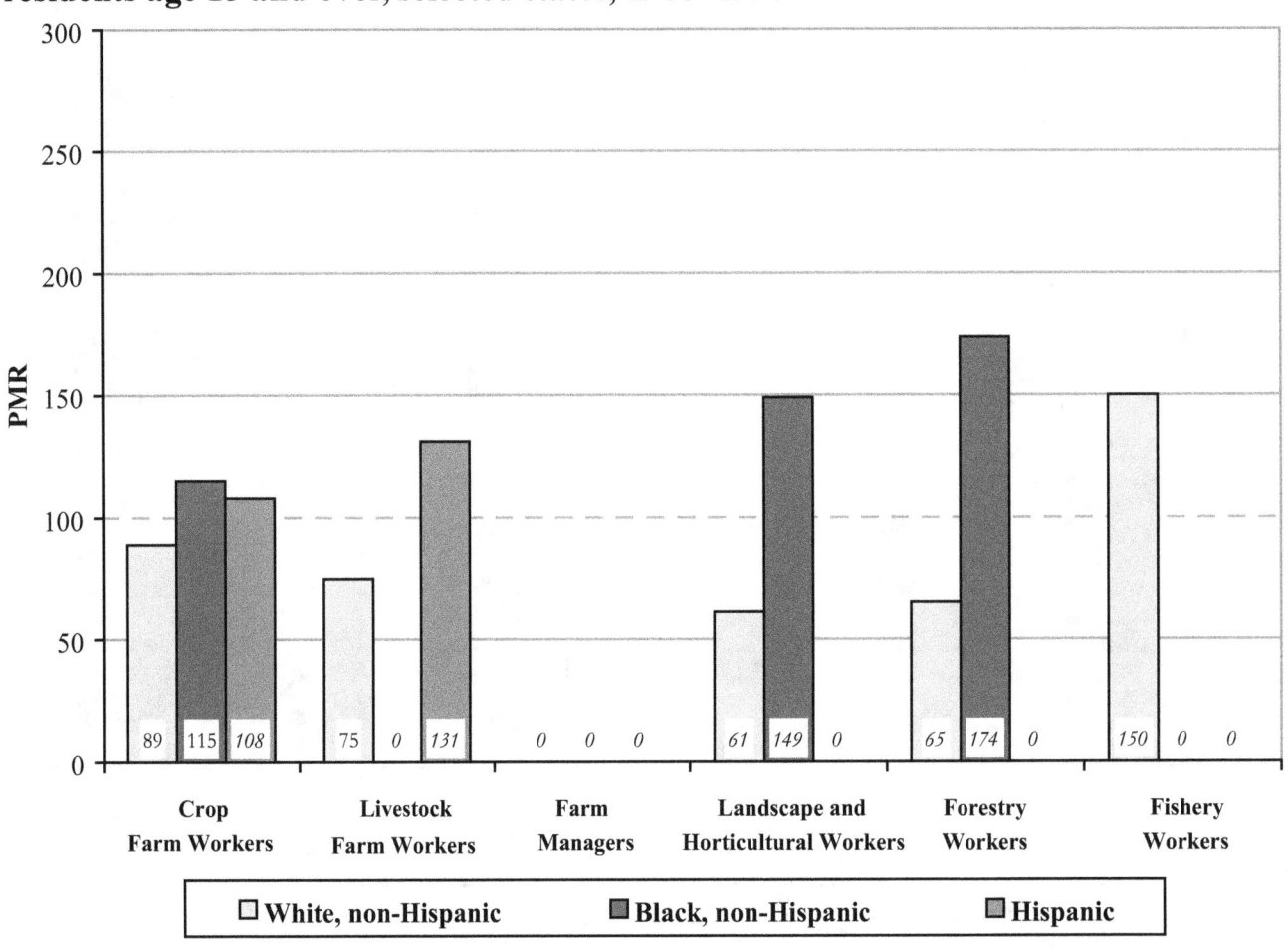

ICD – International Classification of Diseases, 9th Revision
NOTE: Other diseases of upper respiratory tract = ICD-9 codes 470-478. PMRs in **bold** are significantly different from 100 (p<0.05). PMRs in *italics* are based on fewer than five observed deaths. PMRs are based on underlying and contributing cause of death. See appendices for source description, methods, ICD codes, and a list of selected states.
SOURCE: National Center for Health Statistics multiple-cause-of-death data

Mortality by Agricultural Group and Race/Ethnicity within Disease Category

Figure 2-27. Pneumonia and influenza: Proportionate mortality ratio (PMR) adjusted for age and sex by agricultural group and race/ethnicity, U.S. residents age 15 and over, selected states, 1988–1998

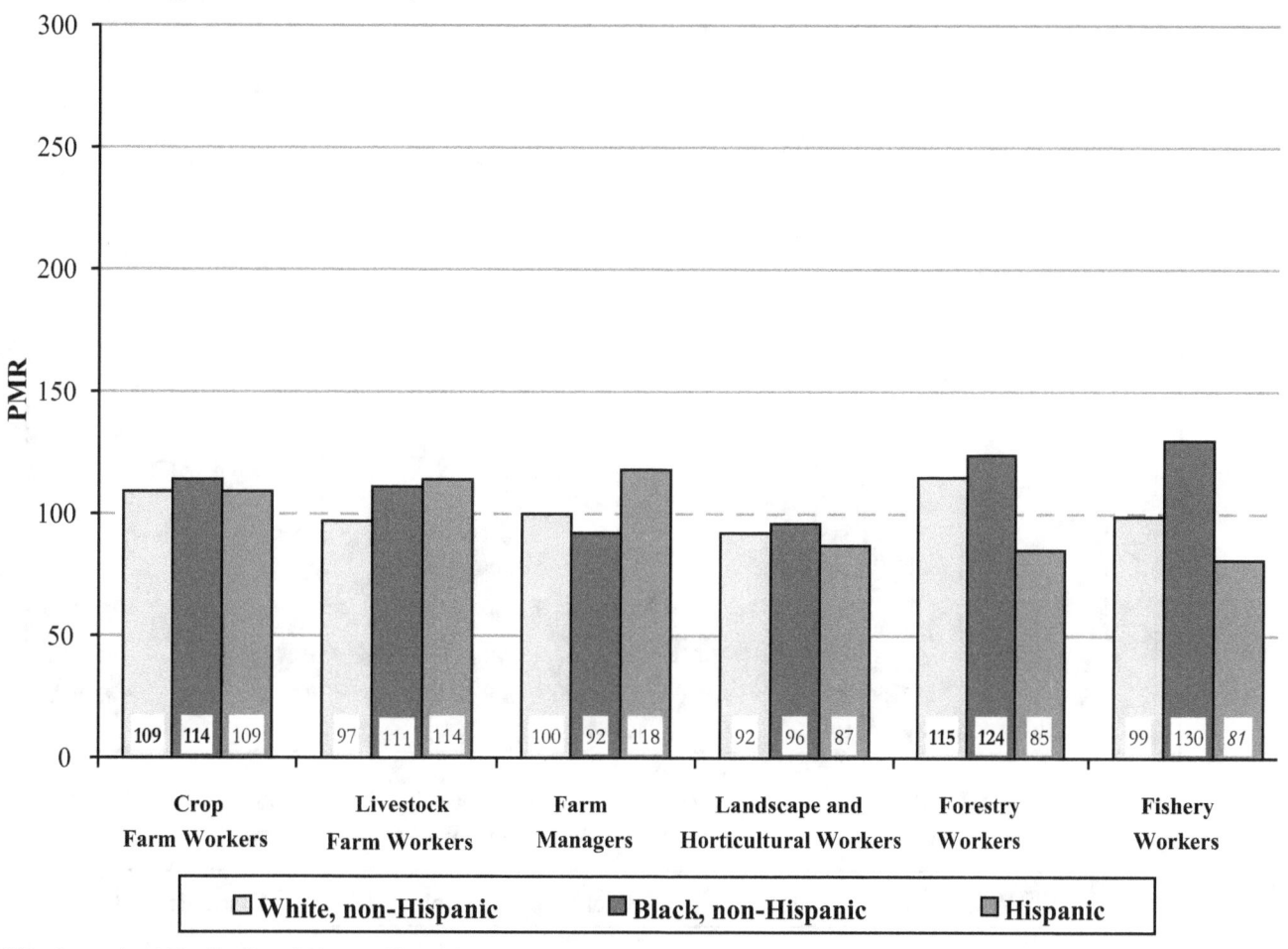

ICD – International Classification of Diseases, 9th Revision
NOTE: Pneumonia and influenza = ICD-9 codes 480-487. PMRs in **bold** are significantly different from 100 (p<0.05). PMRs in *italics* are based on fewer than five observed deaths. PMRs are based on underlying and contributing cause of death. See appendices for source description, methods, ICD codes, and a list of selected states.
SOURCE: National Center for Health Statistics multiple-cause-of-death data

Mortality by Agricultural Group and Race/Ethnicity within Disease Category

Figure 2-28. Chronic obstructive pulmonary disease and allied conditions: Proportionate mortality ratio (PMR) adjusted for age and sex by agricultural group and race/ethnicity, U.S. residents age 15 and over, selected states, 1988–1998

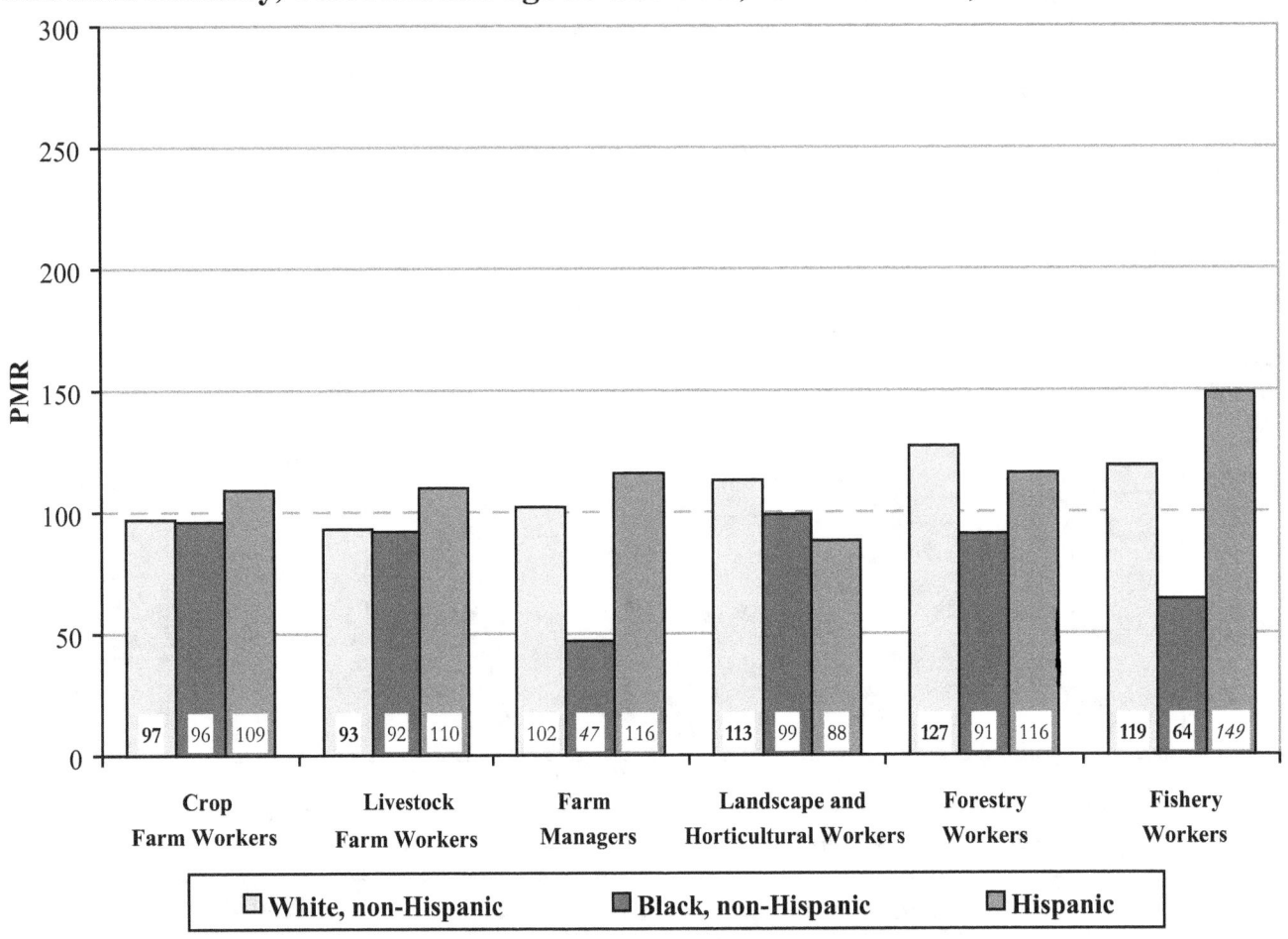

ICD – International Classification of Diseases, 9th Revision
NOTE: Chronic obstructive pulmonary disease and allied conditions = ICD-9 codes 490-496. PMRs in **bold** are significantly different from 100 (p<0.05). PMRs in *italics* are based on fewer than five observed deaths. PMRs are based on underlying and contributing cause of death. See appendices for source description, methods, ICD codes, and a list of selected states.
SOURCE: National Center for Health Statistics multiple-cause-of-death data

Mortality by Agricultural Group and Race/Ethnicity within Disease Category

Figure 2-29. Pneumoconioses and other lung diseases–external agents: Proportionate mortality ratio (PMR) adjusted for age and sex by agricultural group and race/ethnicity, U.S. residents age 15 and over, selected states, 1988–1998

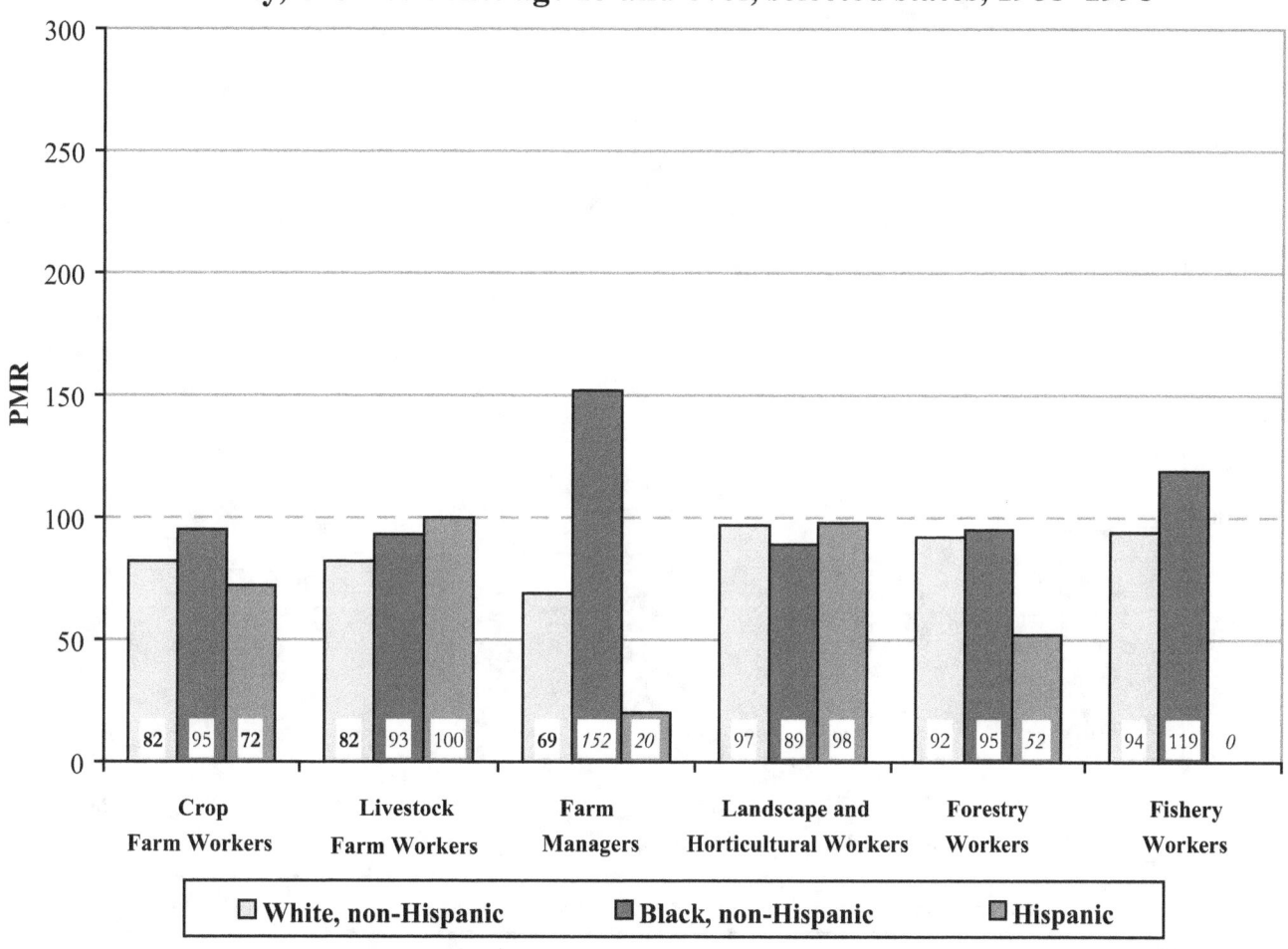

ICD – International Classification of Diseases, 9th Revision
NOTE: Pneumoconioses and other lung diseases - external agents = ICD-9 codes 500-508. PMRs in **bold** are significantly different from 100 (p<0.05). PMRs in *italics* are based on fewer than five observed deaths. PMRs are based on underlying and contributing cause of death. See appendices for source description, methods, ICD codes, and a list of selected states.
SOURCE: National Center for Health Statistics multiple-cause-of-death data

Mortality by Agricultural Group and Race/Ethnicity within Disease Category

Figure 2-30. Other diseases of respiratory system: Proportionate mortality ratio (PMR) adjusted for age and sex by agricultural group and race/ethnicity, U.S. residents age 15 and over, selected states, 1988–1998

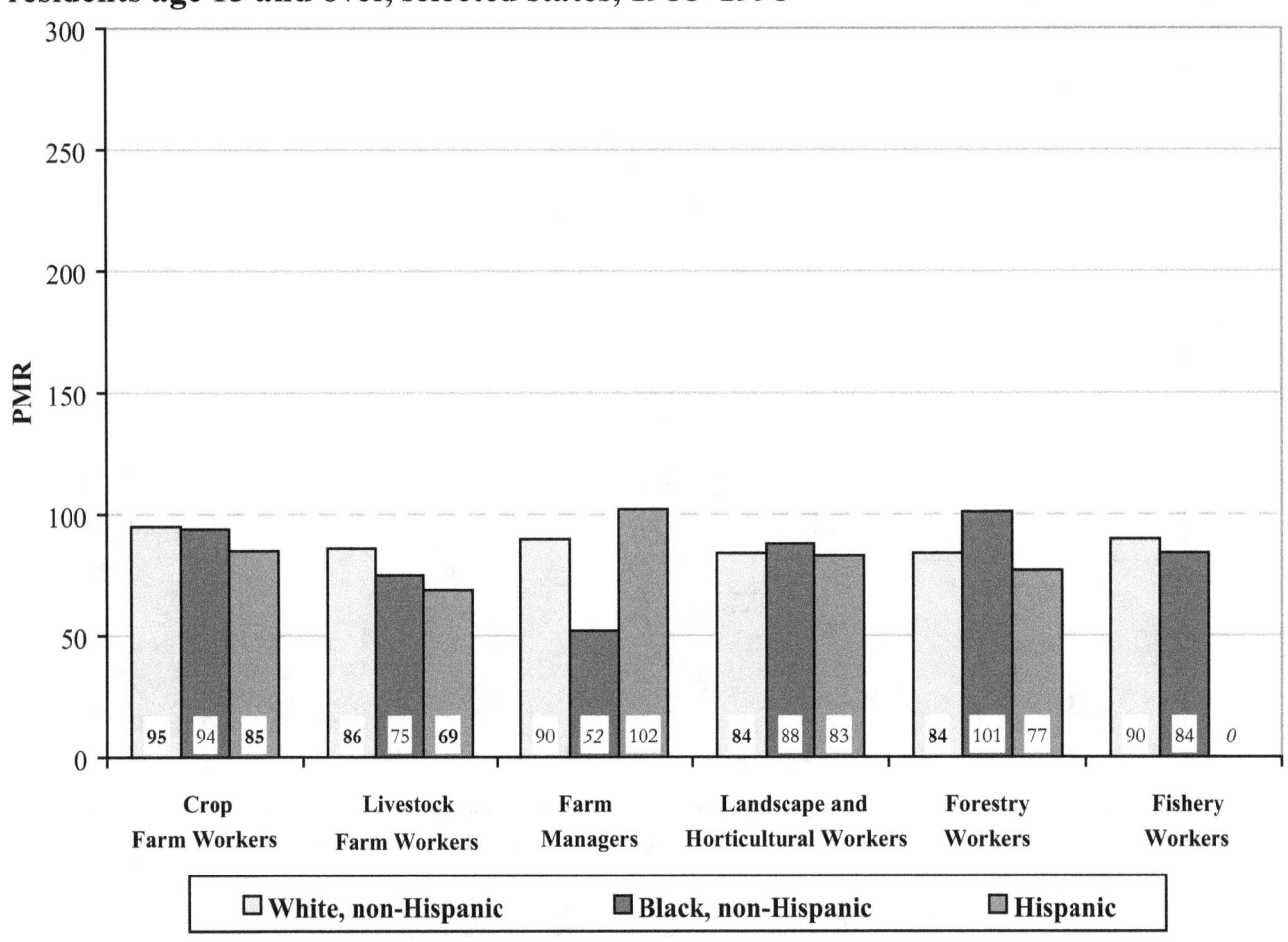

ICD – International Classification of Diseases, 9th Revision
NOTE: Other diseases of respiratory system = ICD-9 codes 510-519. PMRs in **bold** are significantly different from 100 (p<0.05). PMRs in *italics* are based on fewer than five observed deaths. PMRs are based on underlying and contributing cause of death. See appendices for source description, methods, ICD codes, and a list of selected states.
SOURCE: National Center for Health Statistics multiple-cause-of-death data

Table 2-37. Crop farm workers: Proportionate mortality ratio (PMR) adjusted for age, sex, and race/ethnicity for tuberculosis, U.S. residents age 15 and over, selected states, 1988–1998

Disease Category (ICD Code)	Number of Deaths	PMR	95% Confidence Interval	
			LCL	UCL
Pulmonary tuberculosis (011)	437	**152**	138	167
Other respiratory tuberculosis (012)	8	210	91	413
Tuberculosis of meninges and central nervous system (013)	7	134	54	276
Tuberculosis of intestines, peritoneum and mesenteric glands (014)	2	*113*	14	408
Tuberculosis of bones and joints (015)	15	115	64	190
Tuberculosis of genitourinary system (016)	2	*60*	7	217
Tuberculosis of other organs (017)	16	85	49	138
Miliary tuberculosis (018)	35	**196**	136	273

ICD - International Classification of Diseases, 9th Revision   LCL - lower confidence limit   UCL - upper confidence limit
NOTE: PMRs in **bold** are significantly different from 100 (p<0.05). PMRs in *italics* are based on fewer than five observed deaths. PMRs are based on underlying and contributing cause of death. Some values could not be calculated because the number of observed or expected deaths was zero; such values are indicated by ---. See appendices for source description, methods, ICD codes, and a list of selected states.
SOURCE: National Center for Health Statistics multiple-cause-of-death data

Tuberculosis Mortality within and by Agricultural Group

Table 2-38. Livestock farm workers: Proportionate mortality ratio (PMR) adjusted for age, sex, and race/ethnicity for tuberculosis, U.S. residents age 15 and over, selected states, 1988–1998

Disease Category (ICD Code)	Number of Deaths	PMR	95% Confidence Interval LCL	95% Confidence Interval UCL
Pulmonary tuberculosis (011)	37	60	43	83
Other respiratory tuberculosis (012)	5	**675**	218	1,577
Tuberculosis of meninges and central nervous system (013)	5	**546**	177	1,246
Tuberculosis of intestines, peritoneum and mesenteric glands (014)	0	*0*	---	---
Tuberculosis of bones and joints (015)	2	65	8	235
Tuberculosis of genitourinary system (016)	1	130	3	722
Tuberculosis of other organs (017)	2	44	5	159
Miliary tuberculosis (018)	4	*143*	39	366

ICD - International Classification of Diseases, 9th Revision   LCL - lower confidence limit   UCL - upper confidence limit
NOTE: PMRs in **bold** are significantly different from 100 ($p<0.05$). PMRs in *italics* are based on fewer than five observed deaths. PMRs are based on underlying and contributing cause of death. Some values could not be calculated because the number of observed or expected deaths was zero; such values are indicated by ---. See appendices for source description, methods, ICD codes, and a list of selected states.
SOURCE: National Center for Health Statistics multiple-cause-of-death data

Table 2-39. Farm managers: Proportionate mortality ratio (PMR) adjusted for age, sex, and race/ethnicity for tuberculosis, U.S. residents age 15 and over, selected states, 1988–1998

Disease Category (ICD Code)	Number of Deaths	PMR	95% Confidence Interval LCL	95% Confidence Interval UCL
Pulmonary tuberculosis (011)	4	84	23	215
Other respiratory tuberculosis (012)	0	0	---	---
Tuberculosis of meninges and central nervous system (013)	0	0	---	---
Tuberculosis of intestines, peritoneum and mesenteric glands (014)	0	0	---	---
Tuberculosis of bones and joints (015)	0	0	---	---
Tuberculosis of genitourinary system (016)	1	*2,053*	52	11,406
Tuberculosis of other organs (017)	0	0	---	---
Miliary tuberculosis (018)	0	0	---	---

ICD - International Classification of Diseases, 9th Revision   LCL - lower confidence limit   UCL - upper confidence limit
NOTE: PMRs in **bold** are significantly different from 100 (p<0.05). PMRs in *italics* are based on fewer than five observed deaths. PMRs are based on underlying and contributing cause of death. Some values could not be calculated because the number of observed or expected deaths was zero; such values are indicated by ---. See appendices for source description, methods, ICD codes, and a list of selected states.
SOURCE: National Center for Health Statistics multiple-cause-of-death data

Tuberculosis Mortality within and by Agricultural Group

Table 2-40. Landscape and horticultural workers: Proportionate mortality ratio (PMR) adjusted for age, sex, and race/ethnicity for tuberculosis, U.S. residents age 15 and over, selected states, 1988–1998

Disease Category (ICD Code)	Number of Deaths	PMR	95% Confidence Interval LCL	95% Confidence Interval UCL
Pulmonary tuberculosis (011)	14	88	48	148
Other respiratory tuberculosis (012)	0	*0*	---	---
Tuberculosis of meninges and central nervous system (013)	1	*209*	5	1,161
Tuberculosis of intestines, peritoneum and mesenteric glands (014)	0	*0*	---	---
Tuberculosis of bones and joints (015)	0	*0*	---	---
Tuberculosis of genitourinary system (016)	0	*0*	---	---
Tuberculosis of other organs (017)	1	88	2	489
Miliary tuberculosis (018)	0	*0*	---	---

ICD - International Classification of Diseases, 9th Revision   LCL - lower confidence limit   UCL - upper confidence limit
NOTE: PMRs in **bold** are significantly different from 100 (p<0.05). PMRs in *italics* are based on fewer than five observed deaths. PMRs are based on underlying and contributing cause of death. Some values could not be calculated because the number of observed or expected deaths was zero; such values are indicated by ---. See appendices for source description, methods, ICD codes, and a list of selected states.
SOURCE: National Center for Health Statistics multiple-cause-of-death data

Table 2-41. Forestry workers: Proportionate mortality ratio (PMR) adjusted for age, sex, and race/ethnicity for tuberculosis, U.S. residents age 15 and over, selected states, 1988–1998

Disease Category (ICD Code)	Number of Deaths	PMR	95% Confidence Interval LCL	95% Confidence Interval UCL
Pulmonary tuberculosis (011)	41	**143**	104	194
Other respiratory tuberculosis (012)	0	*0*	---	---
Tuberculosis of meninges and central nervous system (013)	0	*0*	---	---
Tuberculosis of intestines, peritoneum and mesenteric glands (014)	2	*914*	111	3,300
Tuberculosis of bones and joints (015)	0	*0*	---	---
Tuberculosis of genitourinary system (016)	0	*0*	---	---
Tuberculosis of other organs (017)	0	*0*	---	---
Miliary tuberculosis (018)	2	*94*	11	339

ICD - International Classification of Diseases, 9th Revision   LCL - lower confidence limit   UCL - upper confidence limit
NOTE: PMRs in **bold** are significantly different from 100 ($p<0.05$). PMRs in *italics* are based on fewer than five observed deaths. PMRs are based on underlying and contributing cause of death. Some values could not be calculated because the number of observed or expected deaths was zero; such values are indicated by ---. See appendices for source description, methods, ICD codes, and a list of selected states.
SOURCE: National Center for Health Statistics multiple-cause-of-death data

Tuberculosis Mortality within and by Agricultural Group

Table 2-42. Fishery workers: Proportionate mortality ratio (PMR) adjusted for age, sex, and race/ethnicity for tuberculosis, U.S. residents age 15 and over, selected states, 1988–1998

Disease Category (ICD Code)	Number of Deaths	PMR	95% Confidence Interval LCL	95% Confidence Interval UCL
Pulmonary tuberculosis (011)	10	156	75	287
Other respiratory tuberculosis (012)	0	*0*	---	---
Tuberculosis of meninges and central nervous system (013)	0	*0*	---	---
Tuberculosis of intestines, peritoneum and mesenteric glands (014)	0	*0*	---	---
Tuberculosis of bones and joints (015)	0	*0*	---	---
Tuberculosis of genitourinary system (016)	1	*1,771*	45	9,839
Tuberculosis of other organs (017)	1	*218*	6	1,211
Miliary tuberculosis (018)	2	*411*	50	1,484

ICD - International Classification of Diseases, 9th Revision   LCL - lower confidence limit   UCL - upper confidence limit
NOTE: PMRs in **bold** are significantly different from 100 (p<0.05). PMRs in *italics* are based on fewer than five observed deaths. PMRs are based on underlying and contributing cause of death. Some values could not be calculated because the number of observed or expected deaths was zero; such values are indicated by ---. See appendices for source description, methods, ICD codes, and a list of selected states.
SOURCE: National Center for Health Statistics multiple-cause-of-death data

Figure 2-31. Pulmonary tuberculosis: Proportionate mortality ratio (PMR) adjusted for age, sex, and race/ethnicity by agricultural group, U.S. residents age 15 and over, selected states, 1988–1998

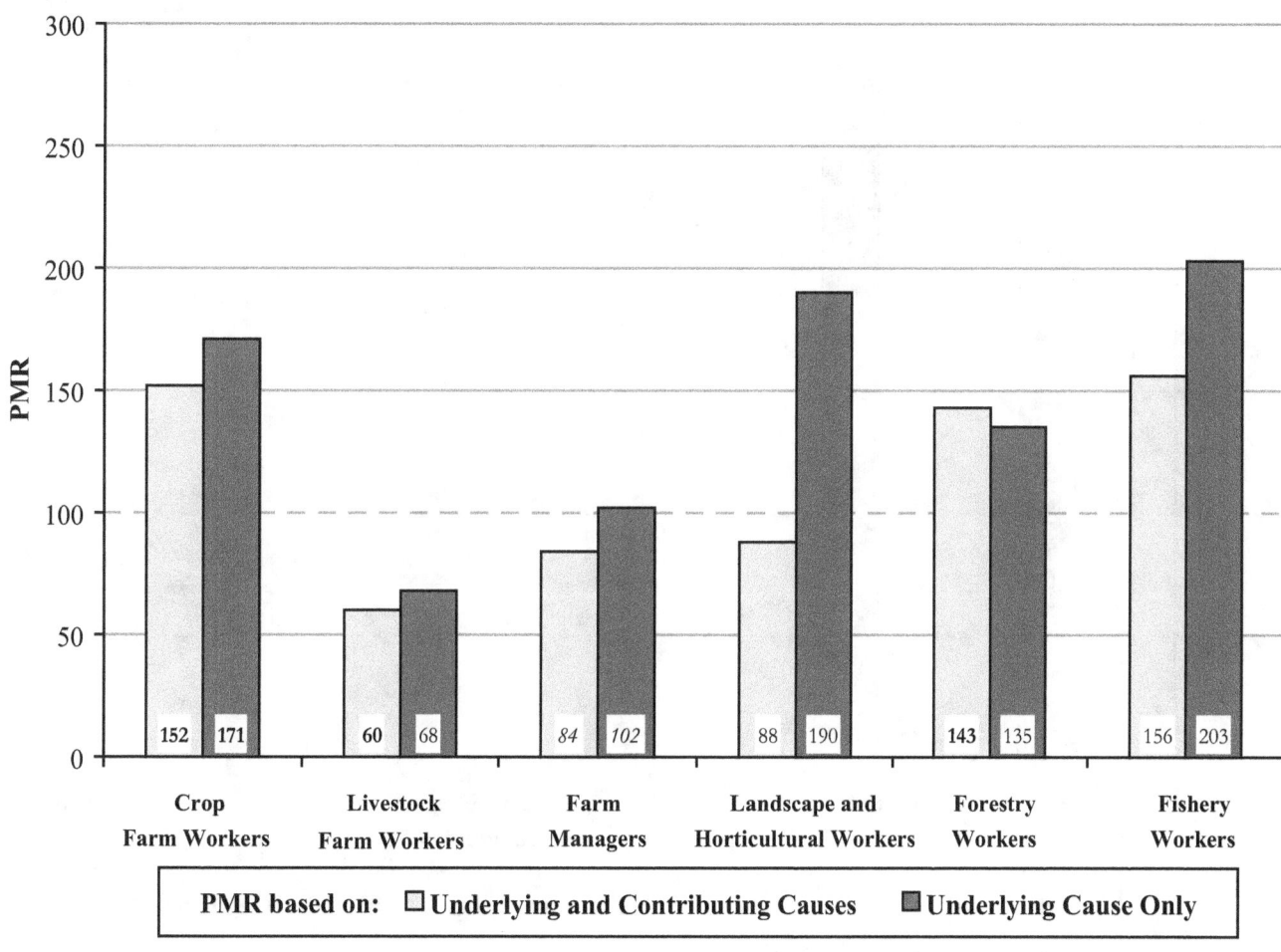

ICD - International Classification of Diseases, 9th Revision
NOTE: Pulmonary tuberculosis = ICD-9 code 011. PMRs in **bold** are significantly different from 100 ($p<0.05$). PMRs in *italics* are based on fewer than five observed deaths. PMRs are based on underlying and contributing cause of death. See appendices for source description, methods, ICD codes, and a list of selected states.
SOURCE: National Center for Health Statistics multiple-cause-of-death data

Tuberculosis Mortality within and by Agricultural Group

Figure 2-32. Other respiratory tuberculosis: Proportionate mortality ratio (PMR) adjusted for age, sex, and race/ethnicity by agricultural group, U.S. residents age 15 and over, selected states, 1988–1998

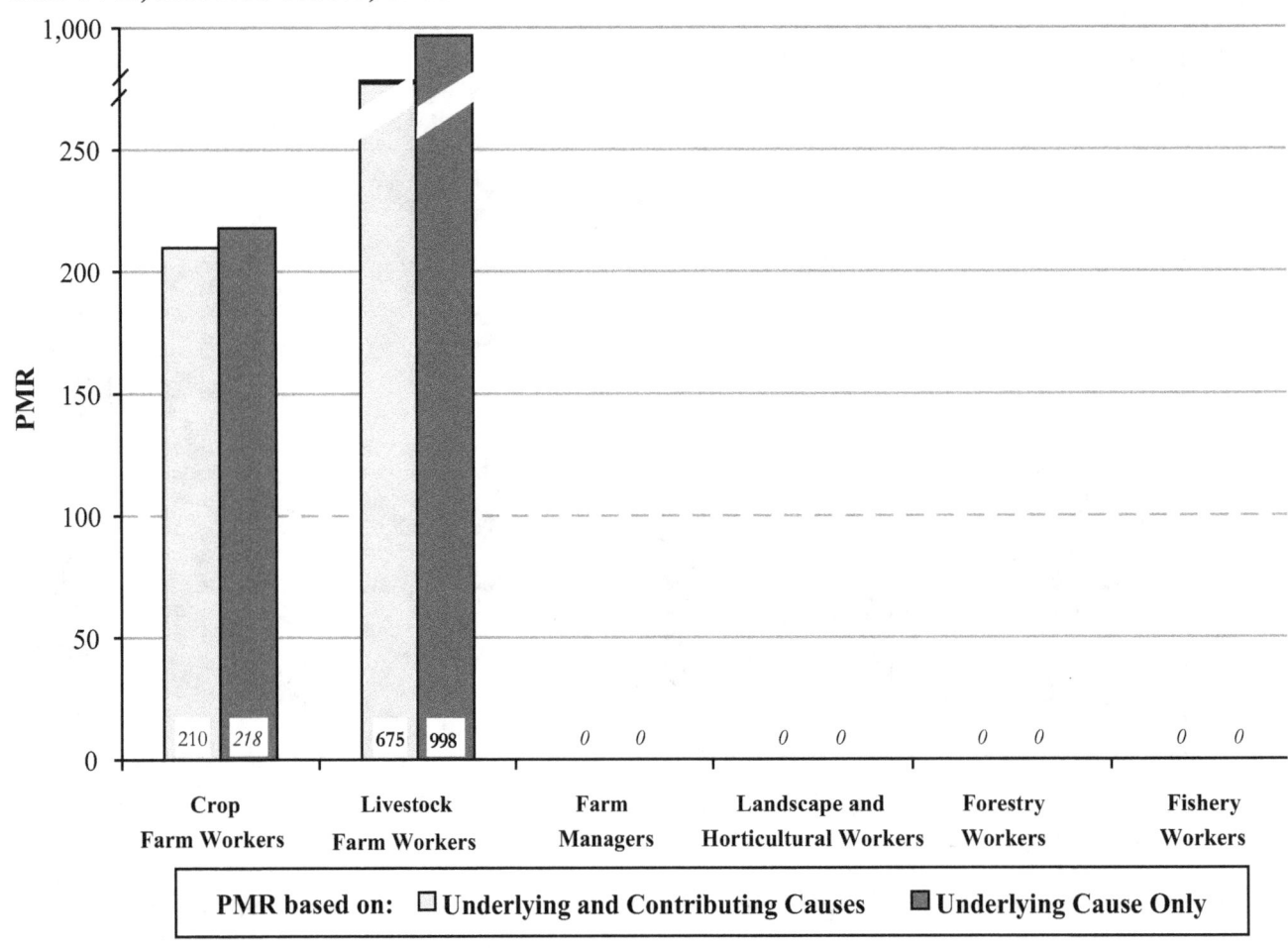

ICD - International Classification of Diseases, 9th Revision
NOTE: Other respiratory tuberculosis = ICD-9 code 012. PMRs in **bold** are significantly different from 100 (p<0.05). PMRs in *italics* are based on fewer than five observed deaths. PMRs are based on underlying and contributing cause of death. See appendices for source description, methods, ICD codes, and a list of selected states.
SOURCE: National Center for Health Statistics multiple-cause-of-death data

Figure 2-33. Tuberculosis of meninges and central nervous system: Proportionate mortality ratio (PMR) adjusted for age, sex, and race/ethnicity by agricultural group, U.S. residents age 15 and over, selected states, 1988–1998

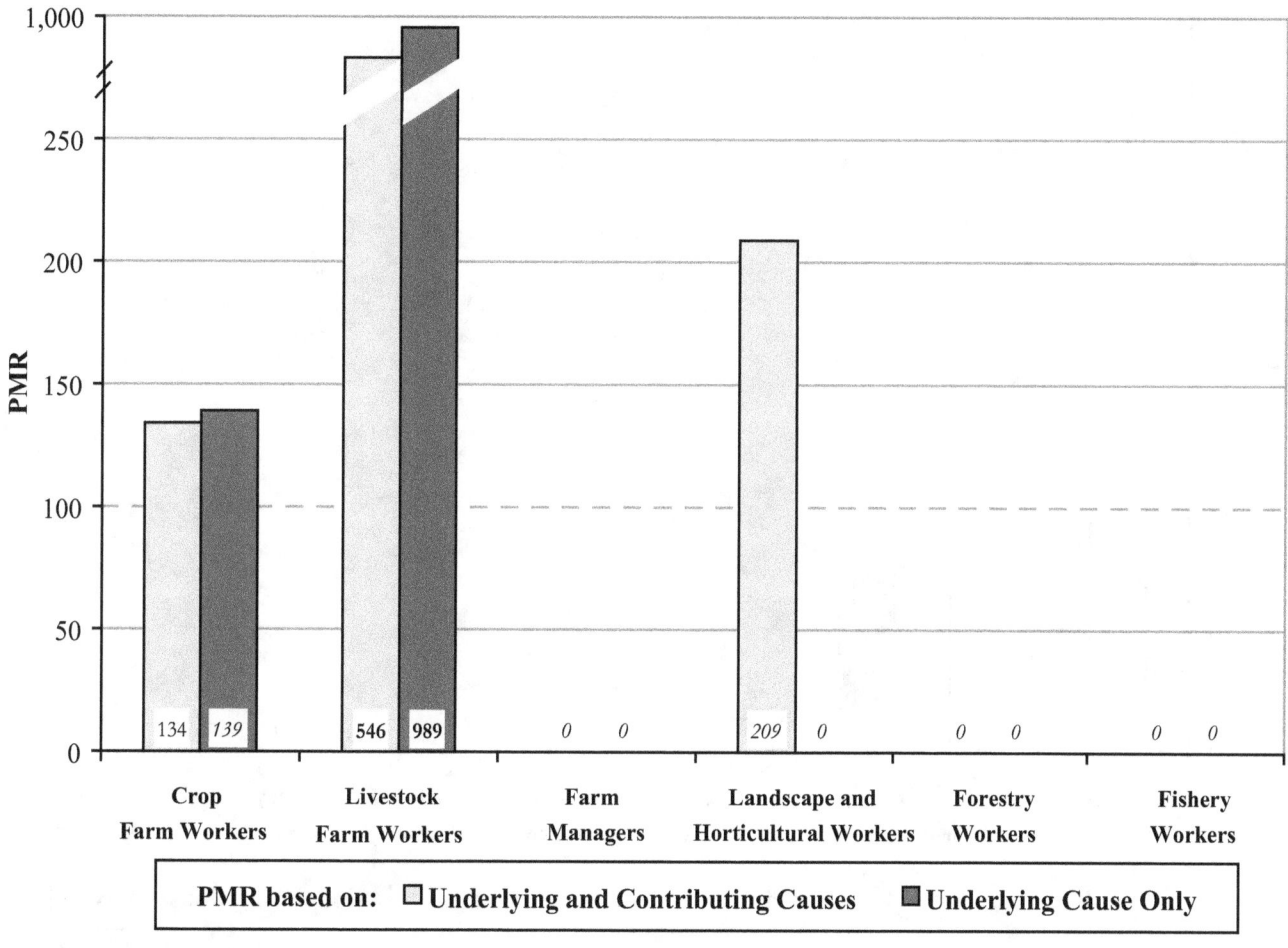

ICD - International Classification of Diseases, 9th Revision
NOTE: Tuberculosis of meninges and central nervous system = ICD-9 code 013. PMRs in **bold** are significantly different from 100 (p<0.05). PMRs in *italics* are based on fewer than five observed deaths. PMRs are based on underlying and contributing cause of death. See appendices for source description, methods, ICD codes, and a list of selected states.
SOURCE: National Center for Health Statistics multiple-cause-of-death data

Tuberculosis Mortality within and by Agricultural Group

Figure 2-34. Tuberculosis of bones and joints: Proportionate mortality ratio (PMR) adjusted for age, sex, and race/ethnicity by agricultural group, U.S. residents age 15 and over, selected states, 1988–1998

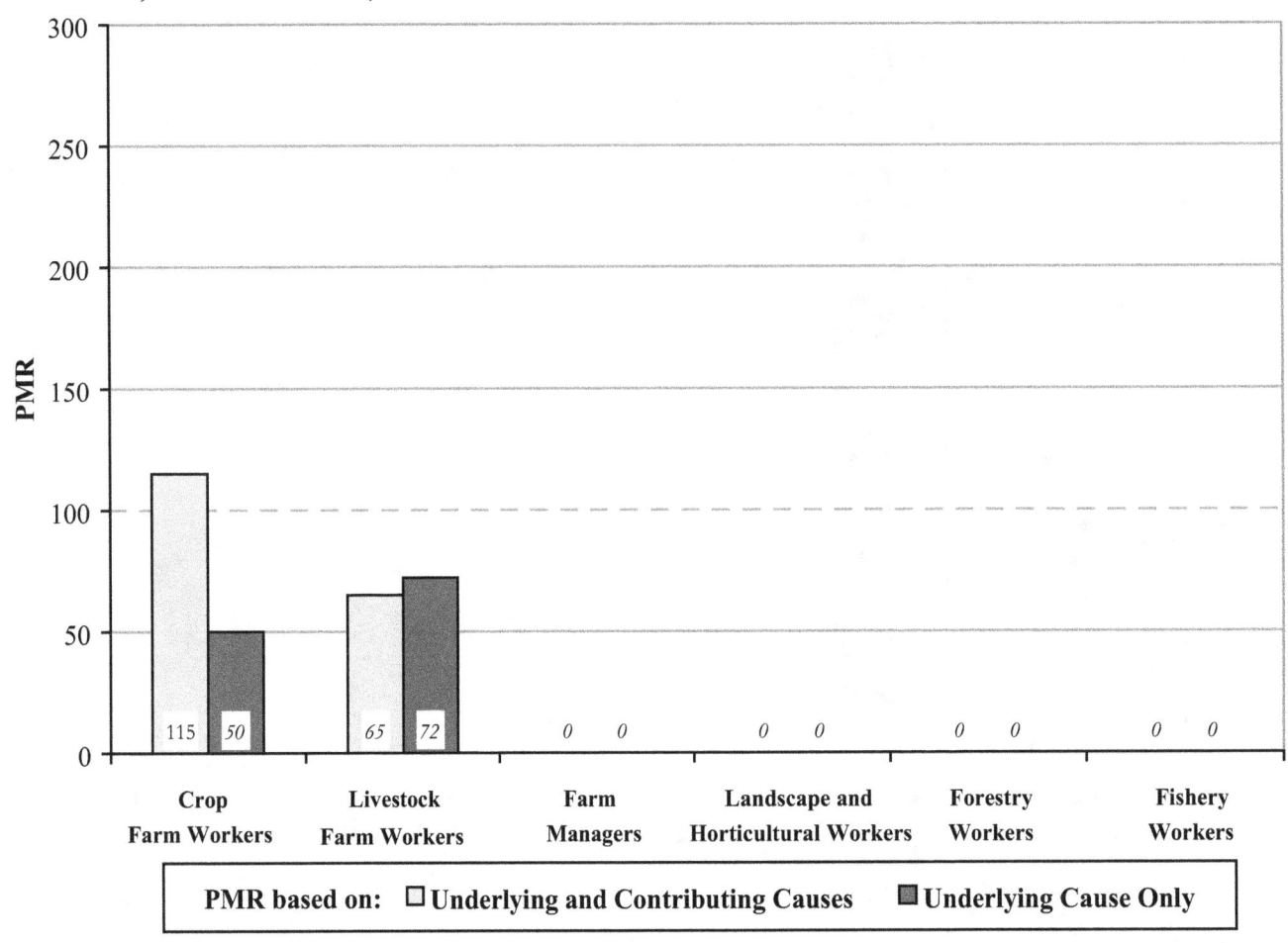

ICD - International Classification of Diseases, 9th Revision
NOTE: Tuberculosis of bones and joints = ICD-9 code 015. PMRs in **bold** are significantly different from 100 (p<0.05). PMRs in *italics* are based on fewer than five observed deaths. PMRs are based on underlying and contributing cause of death. See appendices for source description, methods, ICD codes, and a list of selected states.
SOURCE: National Center for Health Statistics multiple-cause-of-death data

Figure 2-35. Tuberculosis of other organs: Proportionate mortality ratio (PMR) adjusted for age, sex, and race/ethnicity by agricultural group, U.S. residents age 15 and over, selected states, 1988–1998

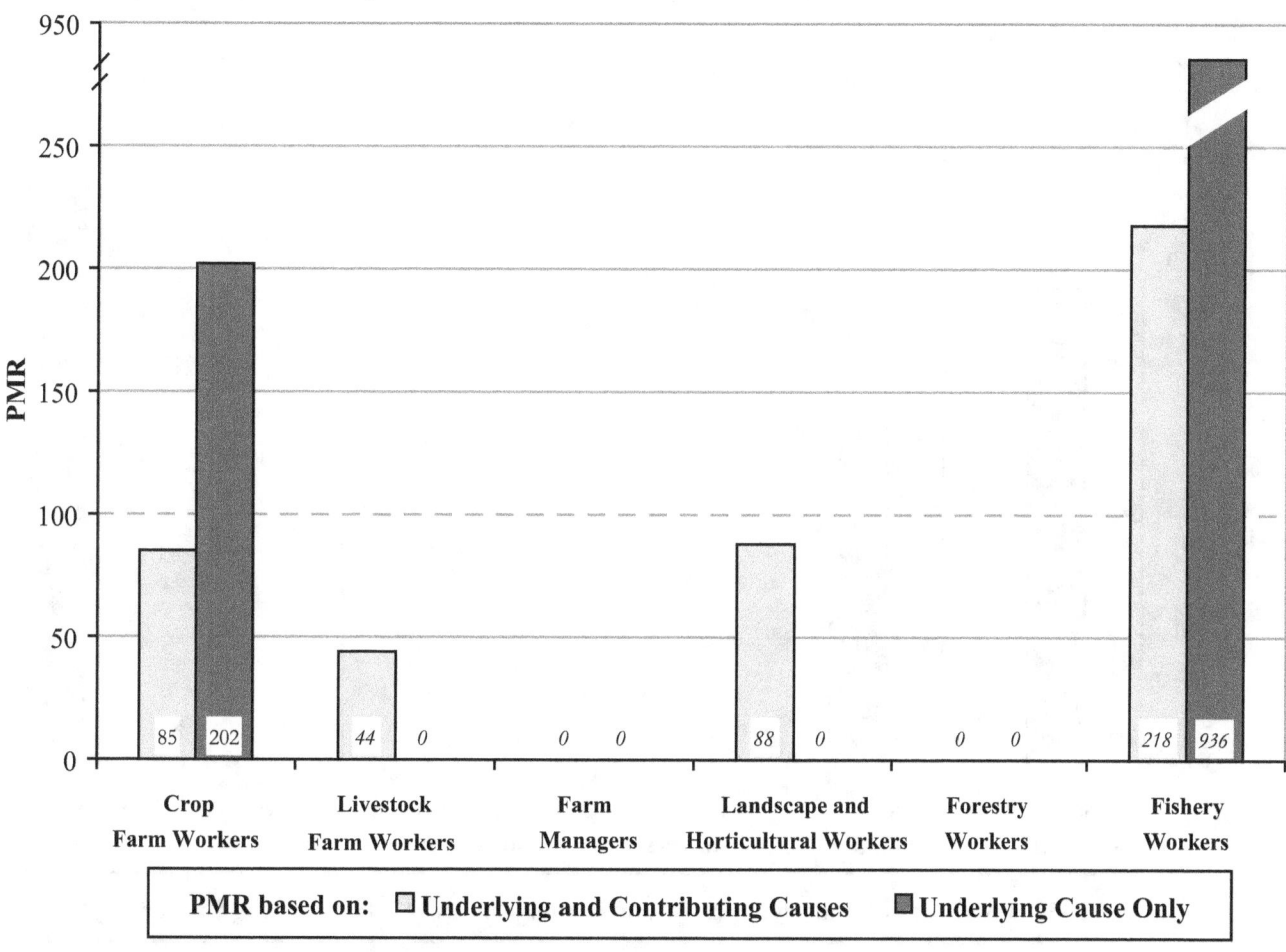

ICD - International Classification of Diseases, 9th Revision
NOTE: Tuberculosis of other organs = ICD-9 code 017. PMRs in **bold** are significantly different from 100 (p<0.05). PMRs in *italics* are based on fewer than five observed deaths. PMRs are based on underlying and contributing cause of death. See appendices for source description, methods, ICD codes, and a list of selected states.
SOURCE: National Center for Health Statistics multiple-cause-of-death data

Figure 2-36. Miliary tuberculosis: Proportionate mortality ratio (PMR) adjusted for age, sex, and race/ethnicity by agricultural group, U.S. residents age 15 and over, selected states, 1988–1998

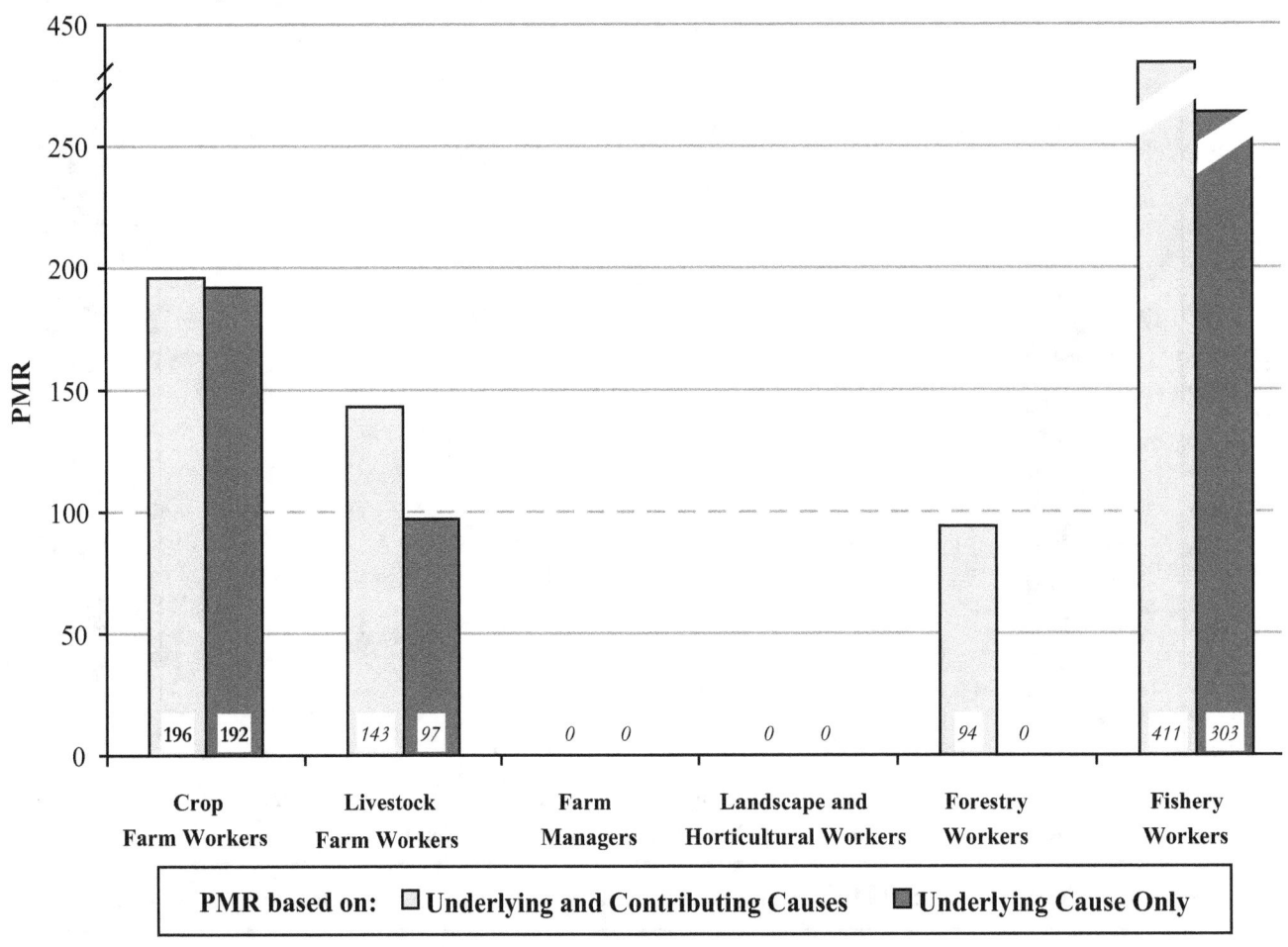

ICD - International Classification of Diseases, 9th Revision
NOTE: Miliary tuberculosis = ICD-9 code 018. PMRs in **bold** are significantly different from 100 (p<0.05). PMRs in *italics* are based on fewer than five observed deaths. PMRs are based on underlying and contributing cause of death. See appendices for source description, methods, ICD codes, and a list of selected states.
SOURCE: National Center for Health Statistics multiple-cause-of-death data

Mycoses Mortality within and by Agricultural Group

Table 2-43. Crop farm workers: Proportionate mortality ratio (PMR) adjusted for age, sex, and race/ethnicity for mycoses, U.S. residents age 15 and over, selected states, 1988–1998

Disease Category (ICD Code)	Number of Deaths	PMR	95% Confidence Interval	
			LCL	UCL
Dermatophytosis (110)	1	92	2	511
Dermatomycosis, other and unspecified (111)	2	465	56	1,679
Candidiasis (112)	134	98	82	116
Coccidioidomycosis (114)	2	62	8	224
Histoplasmosis (115)	27	**183**	120	266
Blastomycotic infection (116)	14	**245**	134	411
Other mycosis (117)	196	109	94	125

ICD - International Classification of Diseases, 9th Revision   LCL - lower confidence limit   UCL - upper confidence limit
NOTE: PMRs in **bold** are significantly different from 100 (p<0.05). PMRs in *italics* are based on fewer than five observed deaths. PMRs are based on underlying and contributing cause of death. Some values could not be calculated because the number of observed or expected deaths was zero; such values are indicated by ---. See appendices for source description, methods, ICD codes, and a list of selected states.
SOURCE: National Center for Health Statistics multiple-cause-of-death data

Mycoses Mortality within and by Agricultural Group

Table 2-44. Livestock farm workers: Proportionate mortality ratio (PMR) adjusted for age, sex, and race/ethnicity for mycoses, U.S. residents age 15 and over, selected states, 1988–1998

Disease Category (ICD Code)	Number of Deaths	PMR	95% Confidence Interval LCL	95% Confidence Interval UCL
Dermatophytosis (110)	0	0	---	---
Dermatomycosis, other and unspecified (111)	0	0	---	---
Candidiasis (112)	32	95	65	134
Coccidioidomycosis (114)	2	*240*	29	866
Histoplasmosis (115)	2	*49*	6	177
Blastomycotic infection (116)	2	*132*	16	477
Other mycosis (117)	41	94	69	128

ICD - International Classification of Diseases, 9th Revision   LCL - lower confidence limit   UCL - upper confidence limit
NOTE: PMRs in **bold** are significantly different from 100 (p<0.05). PMRs in *italics* are based on fewer than five observed deaths. PMRs are based on underlying and contributing cause of death. Some values could not be calculated because the number of observed or expected deaths was zero; such values are indicated by ---. See appendices for source description, methods, ICD codes, and a list of selected states.
SOURCE: National Center for Health Statistics multiple-cause-of-death data

Mycoses Mortality within and by Agricultural Group

Table 2-45. Farm managers: Proportionate mortality ratio (PMR) adjusted for age, sex, and race/ethnicity for mycoses, U.S. residents age 15 and over, selected states, 1988–1998

Disease Category (ICD Code)	Number of Deaths	PMR	95% Confidence Interval LCL	95% Confidence Interval UCL
Dermatophytosis (110)	0	*0*	---	---
Dermatomycosis, other and unspecified (111)	0	*0*	---	---
Candidiasis (112)	1	*44*	1	244
Coccidioidomycosis (114)	0	*0*	---	---
Histoplasmosis (115)	1	*362*	9	2,011
Blastomycotic infection (116)	0	*0*	---	---
Other mycosis (117)	3	*93*	19	272

ICD - International Classification of Diseases, 9th Revision　　LCL - lower confidence limit　　UCL - upper confidence limit
NOTE: PMRs in **bold** are significantly different from 100 (p<0.05). PMRs in *italics* are based on fewer than five observed deaths. PMRs are based on underlying and contributing cause of death. Some values could not be calculated because the number of observed or expected deaths was zero; such values are indicated by ---. See appendices for source description, methods, ICD codes, and a list of selected states.
SOURCE: National Center for Health Statistics multiple-cause-of-death data

Mycoses Mortality within and by Agricultural Group

Table 2-46. Landscape and horticultural workers: Proportionate mortality ratio (PMR) adjusted for age, sex, and race/ethnicity for mycoses, U.S. residents age 15 and over, selected states, 1988–1998

Disease Category (ICD Code)	Number of Deaths	PMR	95% Confidence Interval LCL	95% Confidence Interval UCL
Dermatophytosis (110)	1	*767*	19	4,261
Dermatomycosis, other and unspecified (111)	0	*0*	---	---
Candidiasis (112)	10	110	53	202
Coccidioidomycosis (114)	0	*0*	---	---
Histoplasmosis (115)	2	*130*	16	469
Blastomycotic infection (116)	0	*0*	---	---
Other mycosis (117)	14	100	55	168

ICD - International Classification of Diseases, 9th Revision   LCL - lower confidence limit   UCL - upper confidence limit
NOTE: PMRs in **bold** are significantly different from 100 (p<0.05). PMRs in *italics* are based on fewer than five observed deaths. PMRs are based on underlying and contributing cause of death. Some values could not be calculated because the number of observed or expected deaths was zero; such values are indicated by ---. See appendices for source description, methods, ICD codes, and a list of selected states.
SOURCE: National Center for Health Statistics multiple-cause-of-death data

Mycoses Mortality within and by Agricultural Group

Table 2-47. Forestry workers: Proportionate mortality ratio (PMR) adjusted for age, sex, and race/ethnicity for mycoses, U.S. residents age 15 and over, selected states, 1988–1998

Disease Category (ICD Code)	Number of Deaths	PMR	95% Confidence Interval LCL	95% Confidence Interval UCL
Dermatophytosis (110)	0	0	---	---
Dermatomycosis, other and unspecified (111)	0	0	---	---
Candidiasis (112)	7	50	20	103
Coccidioidomycosis (114)	1	254	6	1,411
Histoplasmosis (115)	1	*51*	1	283
Blastomycotic infection (116)	0	0	---	---
Other mycosis (117)	14	66	36	111

ICD - International Classification of Diseases, 9th Revision   LCL - lower confidence limit   UCL - upper confidence limit
NOTE: PMRs in **bold** are significantly different from 100 ($p<0.05$). PMRs in *italics* are based on fewer than five observed deaths. PMRs are based on underlying and contributing cause of death. Some values could not be calculated because the number of observed or expected deaths was zero; such values are indicated by ---. See appendices for source description, methods, ICD codes, and a list of selected states.
SOURCE: National Center for Health Statistics multiple-cause-of-death data

Mycoses Mortality within and by Agricultural Group

Table 2-48. Fishery workers: Proportionate mortality ratio (PMR) adjusted for age, sex, and race/ethnicity for mycoses, U.S. residents age 15 and over, selected states, 1988–1998

Disease Category (ICD Code)	Number of Deaths	PMR	95% Confidence Interval LCL	95% Confidence Interval UCL
Dermatophytosis (110)	0	0	---	---
Dermatomycosis, other and unspecified (111)	0	0	---	---
Candidiasis (112)	1	26	1	144
Coccidioidomycosis (114)	0	0	---	---
Histoplasmosis (115)	1	154	4	856
Blastomycotic infection (116)	0	0	---	---
Other mycosis (117)	5	82	27	192

ICD - International Classification of Diseases, 9th Revision   LCL - lower confidence limit   UCL - upper confidence limit
NOTE: PMRs in **bold** are significantly different from 100 (p<0.05). PMRs in *italics* are based on fewer than five observed deaths. PMRs are based on underlying and contributing cause of death. Some values could not be calculated because the number of observed or expected deaths was zero; such values are indicated by ---. See appendices for source description, methods, ICD codes, and a list of selected states.
SOURCE: National Center for Health Statistics multiple-cause-of-death data

Figure 2-37. Candidiasis: Proportionate mortality ratio (PMR) adjusted for age, sex, and race/ethnicity by agricultural group, U.S. residents age 15 and over, selected states, 1988–1998

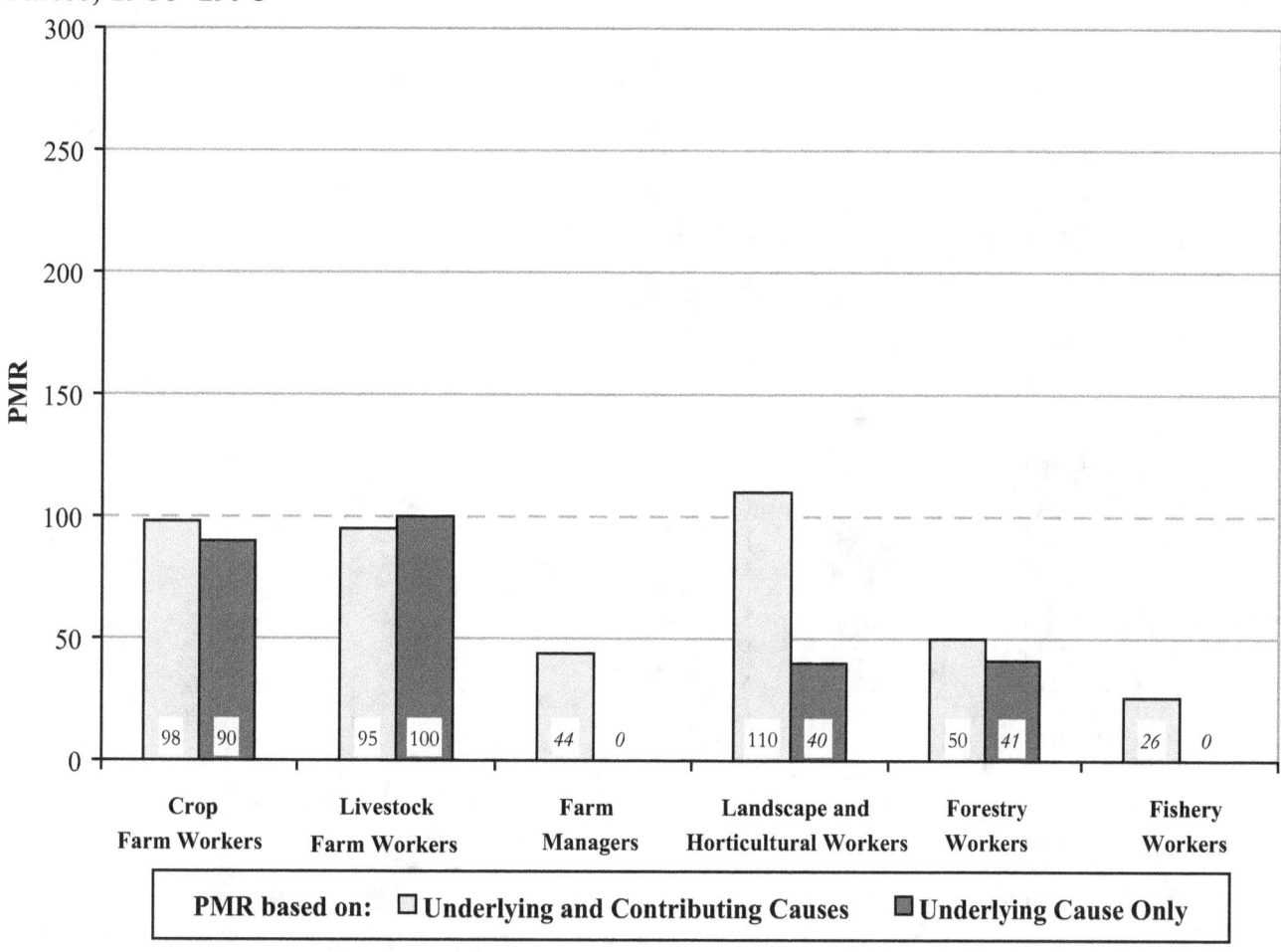

ICD - International Classification of Diseases, 9th Revision
NOTE: Candidiasis = ICD-9 code 112. PMRs in **bold** are significantly different from 100 (p<0.05). PMRs in *italics* are based on fewer than five observed deaths. PMRs are based on underlying and contributing cause of death. See appendices for source description, methods, ICD codes, and a list of selected states.
SOURCE: National Center for Health Statistics multiple-cause-of-death data

Mycoses Mortality within and by Agricultural Group

Figure 2-38. Histoplasmosis: Proportionate mortality ratio (PMR) adjusted for age, sex, and race/ethnicity by agricultural group, U.S. residents age 15 and over, selected states, 1988–1998

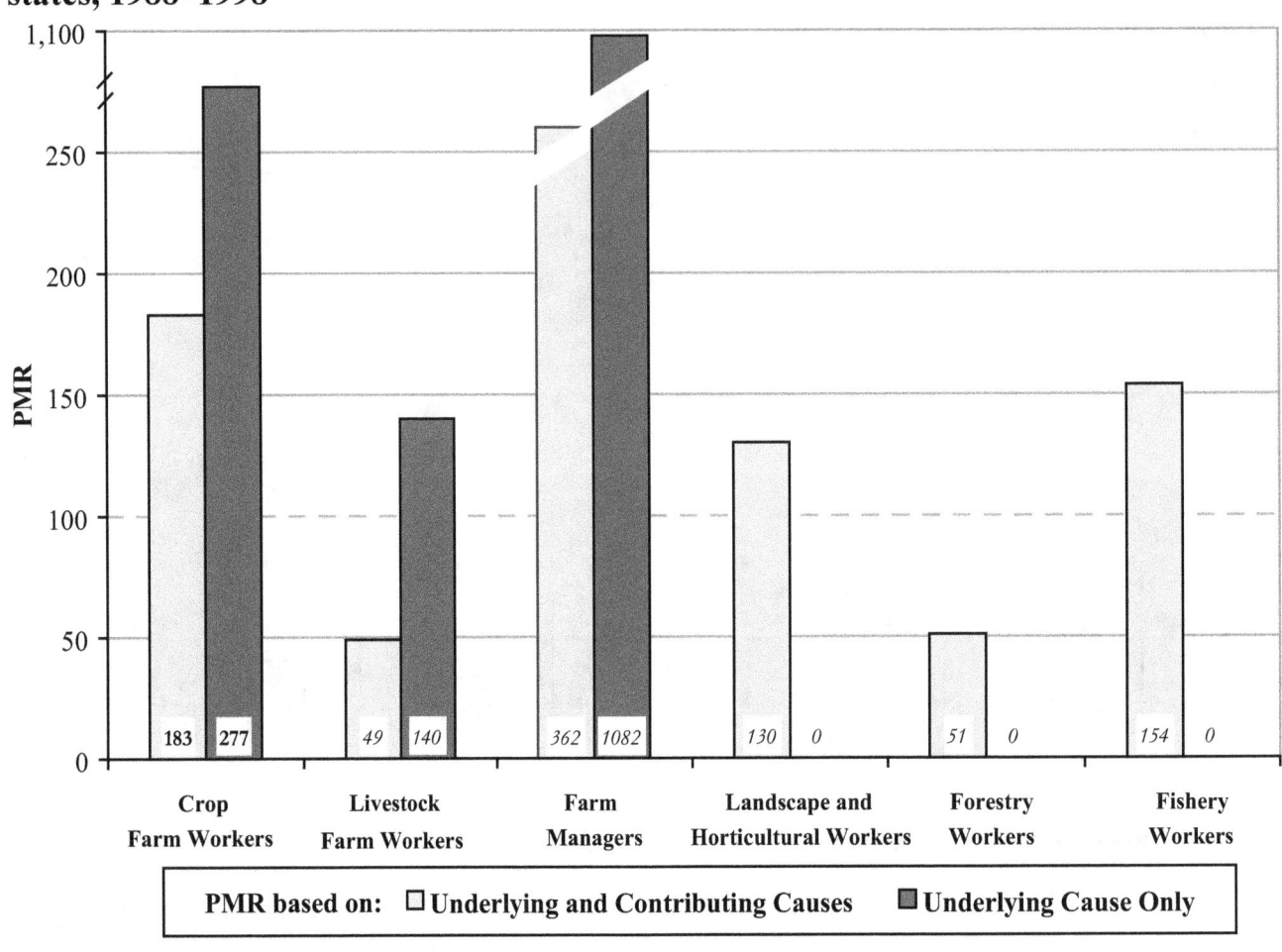

ICD - International Classification of Diseases, 9th Revision
NOTE: Histoplasmosis = ICD-9 code 115. PMRs in **bold** are significantly different from 100 (p<0.05). PMRs in *italics* are based on fewer than five observed deaths. PMRs are based on underlying and contributing cause of death. See appendices for source description, methods, ICD codes, and a list of selected states.
SOURCE: National Center for Health Statistics multiple-cause-of-death data

Figure 2-39. Blastomycotic infection: Proportionate mortality ratio (PMR) adjusted for age, sex, and race/ethnicity by agricultural group, U.S. residents age 15 and over, selected states, 1988–1998

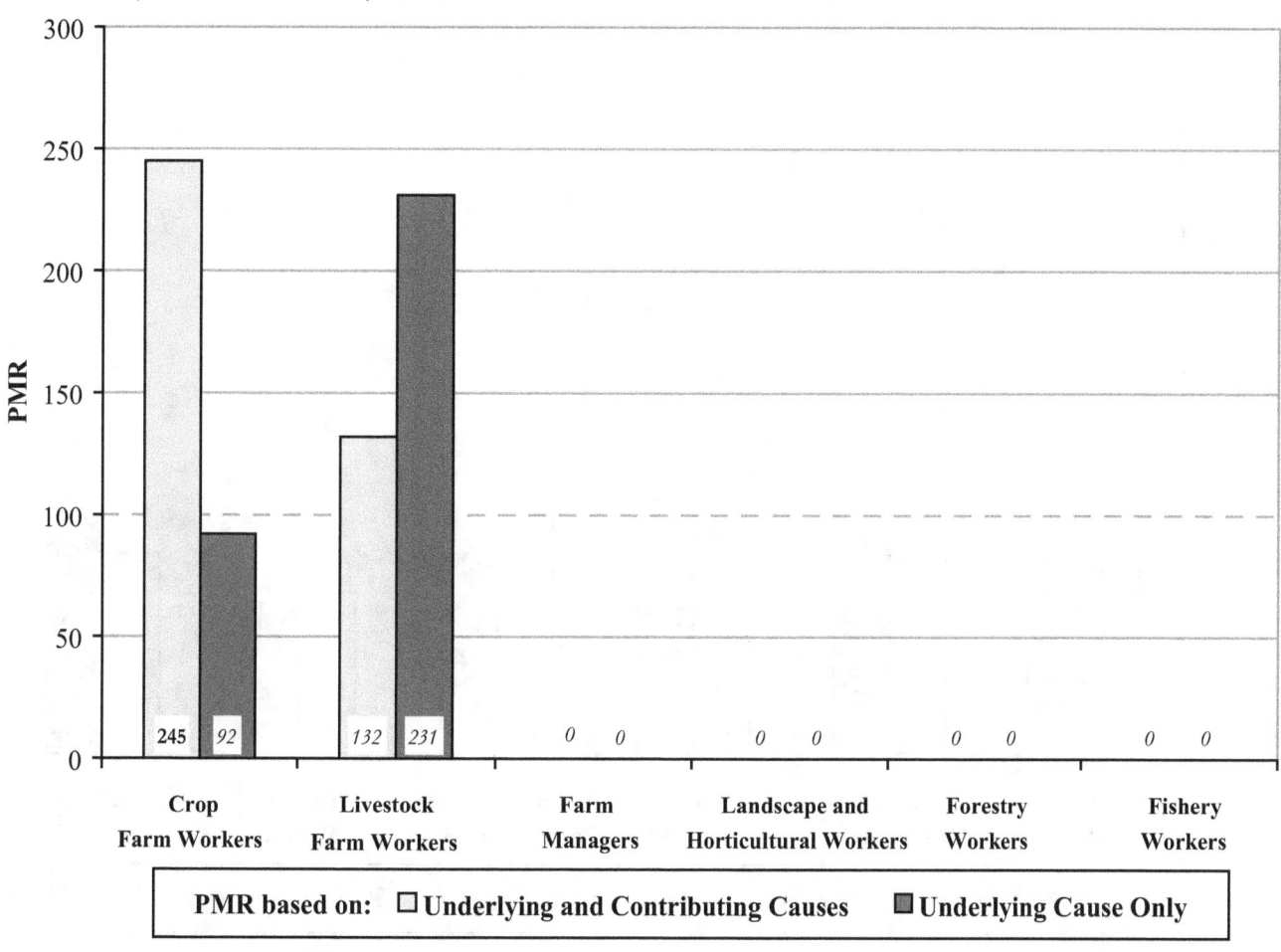

ICD - International Classification of Diseases, 9th Revision
NOTE: Blastomycotic infection = ICD-9 code 116. PMRs in **bold** are significantly different from 100 (p<0.05). PMRs in *italics* are based on fewer than five observed deaths. PMRs are based on underlying and contributing cause of death. See appendices for source description, methods, ICD codes, and a list of selected states.
SOURCE: National Center for Health Statistics multiple-cause-of-death data

Mycoses Mortality within and by Agricultural Group

Figure 2-40. Other mycoses: Proportionate mortality ratio (PMR) adjusted for age, sex, and race/ethnicity by agricultural group, U.S. residents age 15 and over, selected states, 1988–1998

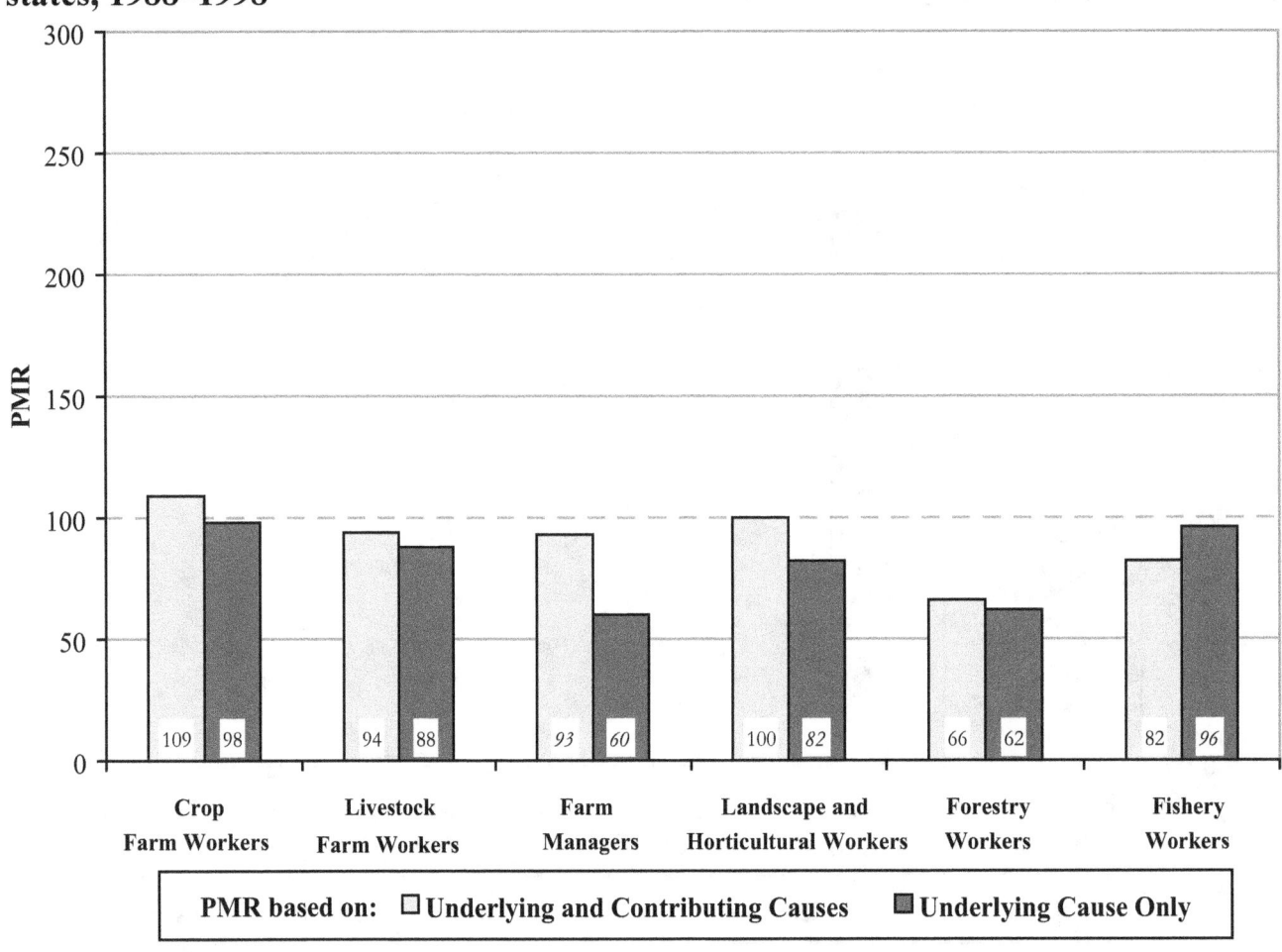

ICD - International Classification of Diseases, 9th Revision
NOTE: Other mycoses = ICD-9 code 117. PMRs in **bold** are significantly different from 100 (p<0.05). PMRs in *italics* are based on fewer than five observed deaths. PMRs are based on underlying and contributing cause of death. See appendices for source description, methods, ICD codes, and a list of selected states.
SOURCE: National Center for Health Statistics multiple-cause-of-death data

Malignant Neoplasm of Trachea/Bronchus/Lung/Pleura Mortality within and by Agricultural Group

Table 2-49. Crop farm workers: Proportionate mortality ratio (PMR) adjusted for age, sex, and race/ethnicity for malignant neoplasms of trachea/bronchus/lung/pleura, U.S. residents age 15 and over, selected states, 1988–1998

Disease Category (ICD Code)	Number of Deaths	PMR	95% Confidence Interval	
			LCL	UCL
Malignant neoplasm of trachea, bronchus, and lung (162)	13,080	**80**	78	82
Malignant neoplasm of pleura (163)	*19*	*30*	*18*	*47*

ICD - International Classification of Diseases, 9th Revision   LCL - lower confidence limit   UCL - upper confidence limit
NOTE: PMRs in **bold** are significantly different from 100 (p<0.05). PMRs in *italics* are based on fewer than five observed deaths. PMRs are based on underlying and contributing cause of death. Some values could not be calculated because the number of observed or expected deaths was zero; such values are indicated by ---. See appendices for source description, methods, ICD codes, and a list of selected states.
SOURCE: National Center for Health Statistics multiple-cause-of-death data

Malignant Neoplasm of Trachea/Bronchus/Lung/Pleura Mortality within and by Agricultural Group

Table 2-50. Livestock farm workers: Proportionate mortality ratio (PMR) adjusted for age, sex, and race/ethnicity for malignant neoplasms of trachea/bronchus/lung/pleura, U.S. residents age 15 and over, selected states, 1988–1998

Disease Category (ICD Code)	Number of Deaths	PMR	95% Confidence Interval LCL	UCL
Malignant neoplasm of trachea, bronchus, and lung (162)	2,949	**68**	66	70
Malignant neoplasm of pleura (163)	11	62	31	111

ICD - International Classification of Diseases, 9th Revision   LCL - lower confidence limit   UCL - upper confidence limit
NOTE: PMRs in **bold** are significantly different from 100 (p<0.05). PMRs in *italics* are based on fewer than five observed deaths. PMRs are based on underlying and contributing cause of death. Some values could not be calculated because the number of observed or expected deaths was zero; such values are indicated by ---. See appendices for source description, methods, ICD codes, and a list of selected states.
SOURCE: National Center for Health Statistics multiple-cause-of-death data

Malignant Neoplasm of Trachea/Bronchus/Lung/Pleura Mortality within and by Agricultural Group

Table 2-51. Farm managers: Proportionate mortality ratio (PMR) adjusted for age, sex, and race/ethnicity for malignant neoplasms of trachea/bronchus/lung/pleura, U.S. residents age 15 and over, selected states, 1988–1998

Disease Category (ICD Code)	Number of Deaths	PMR	95% Confidence Interval LCL	UCL
Malignant neoplasm of trachea, bronchus, and lung (162)	250	94	83	106
Malignant neoplasm of pleura (163)	1	*98*	*2*	*544*

ICD - International Classification of Diseases, 9th Revision   LCL - lower confidence limit   UCL - upper confidence limit
NOTE: PMRs in **bold** are significantly different from 100 (p<0.05). PMRs in *italics* are based on fewer than five observed deaths. PMRs are based on underlying and contributing cause of death. Some values could not be calculated because the number of observed or expected deaths was zero; such values are indicated by ---. See appendices for source description, methods, ICD codes, and a list of selected states.
SOURCE: National Center for Health Statistics multiple-cause-of-death data

Malignant Neoplasm of Trachea/Bronchus/Lung/Pleura Mortality within and by Agricultural Group

Table 2-52. Landscape and horticultural workers: Proportionate mortality ratio (PMR) adjusted for age, sex, and race/ethnicity for malignant neoplasms of trachea/bronchus/lung/pleura, U.S. residents age 15 and over, selected states, 1988–1998

Disease Category (ICD Code)	Number of Deaths	PMR	95% Confidence Interval LCL	UCL
Malignant neoplasm of trachea, bronchus, and lung (162)	642	97	90	105
Malignant neoplasm of pleura (163)	5	*235*	*76*	*235*

ICD - International Classification of Diseases, 9th Revision   LCL - lower confidence limit   UCL - upper confidence limit
NOTE: PMRs in **bold** are significantly different from 100 (p<0.05). PMRs in *italics* are based on fewer than five observed deaths. PMRs are based on underlying and contributing cause of death. Some values could not be calculated because the number of observed or expected deaths was zero; such values are indicated by ---. See appendices for source description, methods, ICD codes, and a list of selected states.
SOURCE: National Center for Health Statistics multiple-cause-of-death data

Malignant Neoplasm of Trachea/Bronchus/Lung/Pleura Mortality within and by Agricultural Group

Table 2-53. Forestry workers: Proportionate mortality ratio (PMR) adjusted for age, sex, and race/ethnicity for malignant neoplasms of trachea/bronchus/lung/pleura, U.S. residents age 15 and over, selected states, 1988–1998

Disease Category (ICD Code)	Number of Deaths	PMR	95% Confidence Interval LCL	UCL
Malignant neoplasm of trachea, bronchus, and lung (162)	1,552	102	97	107
Malignant neoplasm of pleura (163)	1	*19*	0	106

ICD - International Classification of Diseases, 9th Revision   LCL - lower confidence limit   UCL - upper confidence limit
NOTE: PMRs in **bold** are significantly different from 100 (p<0.05). PMRs in *italics* are based on fewer than five observed deaths. PMRs are based on underlying and contributing cause of death. Some values could not be calculated because the number of observed or expected deaths was zero; such values are indicated by ---. See appendices for source description, methods, ICD codes, and a list of selected states.
SOURCE: National Center for Health Statistics multiple-cause-of-death data

Malignant Neoplasm of Trachea/Bronchus/Lung/Pleura Mortality within and by Agricultural Group

Table 2-54. Fishery workers: Proportionate mortality ratio (PMR) adjusted for age, sex, and race/ethnicity for malignant neoplasms of trachea/bronchus/lung/pleura, U.S. residents age 15 and over, selected states, 1988–1998

Disease Category (ICD Code)	Number of Deaths	PMR	95% Confidence Interval LCL	UCL
Malignant neoplasm of trachea, bronchus, and lung (162)	426	108	98	119
Malignant neoplasm of pleura (163)	0	*0*	---	---

ICD - International Classification of Diseases, 9th Revision   LCL - lower confidence limit   UCL - upper confidence limit
NOTE: PMRs in **bold** are significantly different from 100 (p<0.05). PMRs in *italics* are based on fewer than five observed deaths. PMRs are based on underlying and contributing cause of death. Some values could not be calculated because the number of observed or expected deaths was zero; such values are indicated by ---. See appendices for source description, methods, ICD codes, and a list of selected states.
SOURCE: National Center for Health Statistics multiple-cause-of-death data

Malignant Neoplasm of Trachea/Bronchus/Lung/Pleura Mortality within and by Agricultural Group

Figure 2-41. Malignant neoplasm of trachea, bronchus, and lung: Proportionate mortality ratio (PMR) adjusted for age, sex, and race/ethnicity by agricultural group, U.S. residents age 15 and over, selected states, 1988–1998

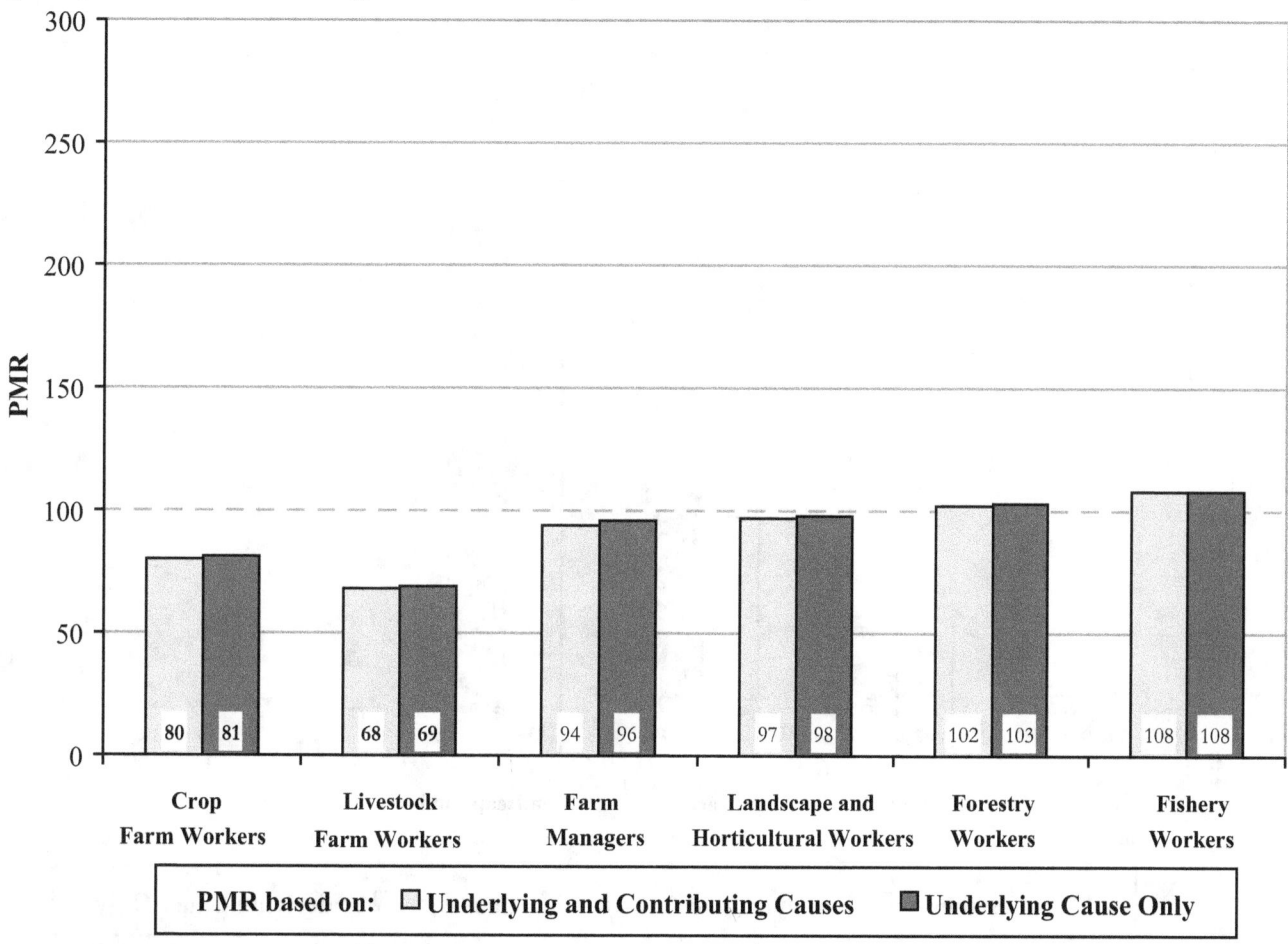

ICD - International Classification of Diseases, 9th Revision
NOTE: Malignant neoplasm of trachea, bronchus, and lung = ICD-9 code 162. PMRs in **bold** are significantly different from 100 (p<0.05). PMRs in *italics* are based on fewer than five observed deaths. PMRs are based on underlying and contributing cause of death. See appendices for source description, methods, ICD codes, and a list of selected states.
SOURCE: National Center for Health Statistics multiple-cause-of-death data

Malignant Neoplasm of Trachea/Bronchus/Lung/Pleura Mortality within and by Agricultural Group

Figure 2-42. Malignant neoplasm of pleura: Proportionate mortality ratio (PMR) adjusted for age, sex, and race/ethnicity by agricultural group, U.S. residents age 15 and over, selected states, 1988–1998

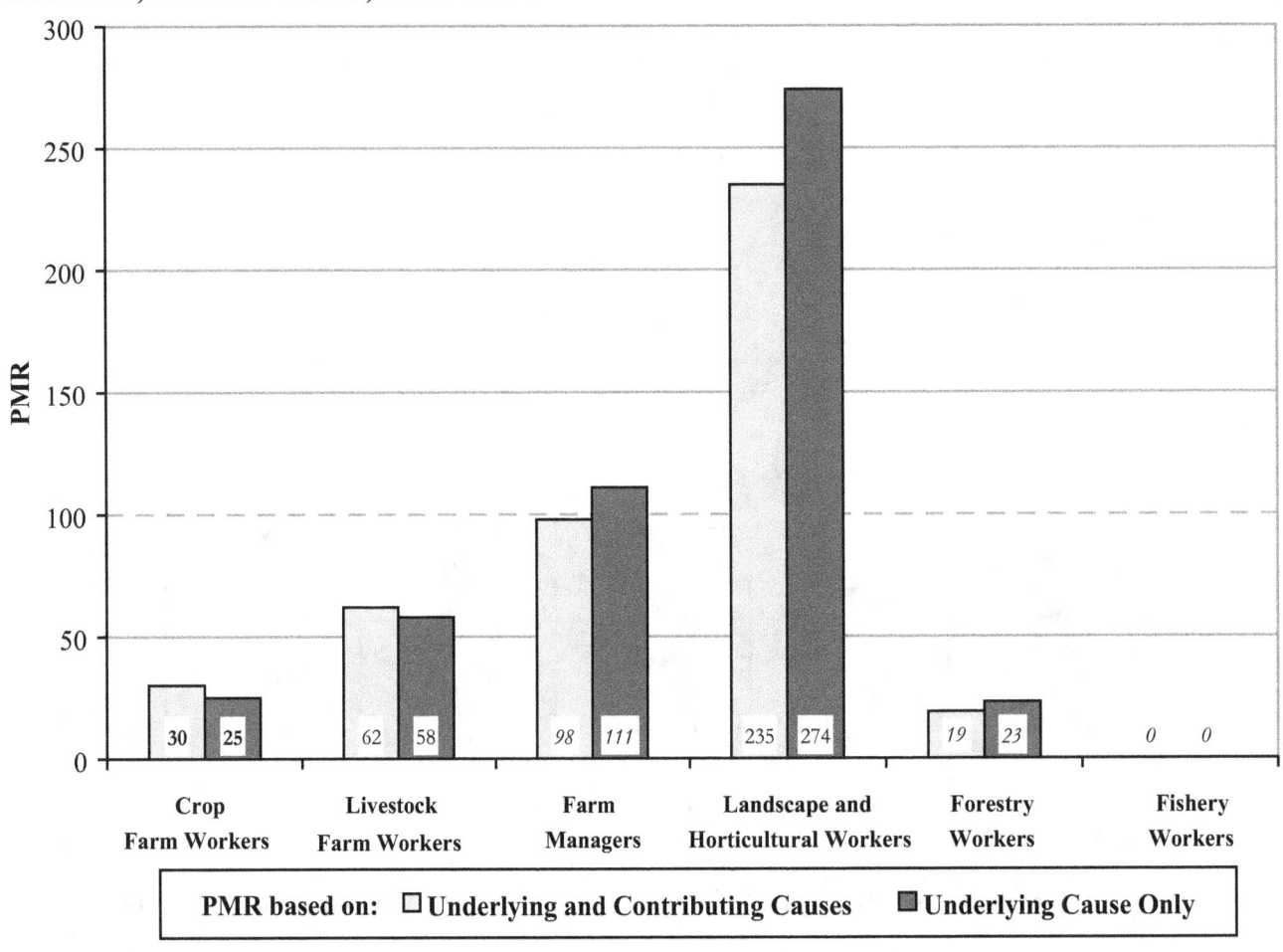

ICD - International Classification of Diseases, 9th Revision
NOTE: Malignant neoplasm of pleura = ICD-9 code 163. PMRs in **bold** are significantly different from 100 (p<0.05). PMRs in *italics* are based on fewer than five observed deaths. PMRs are based on underlying and contributing cause of death. See appendices for source description, methods, ICD codes, and a list of selected states.
SOURCE: National Center for Health Statistics multiple-cause-of-death data

Table 2-55. Crop farm workers: Proportionate mortality ratio (PMR) adjusted for age, sex, and race/ethnicity for acute respiratory infections, U.S. residents age 15 and over, selected states, 1988–1998

Disease Category (ICD Code)	Number of Deaths	PMR	95% Confidence Interval LCL	95% Confidence Interval UCL
Acute nasopharyngitis (460)	3	*143*	29	418
Acute pharyngitis (462)	5	99	32	231
Acute tonsillitis (463)	1	*205*	5	1,139
Acute laryngitis and tracheitis (464)	7	92	37	190
Acute upper respiratory infections of multiple or unspecified sites (465)	87	**160**	129	197
Acute bronchitis and bronchiolitis (466)	226	**117**	103	133

ICD - International Classification of Diseases, 9th Revision LCL - lower confidence limit UCL - upper confidence limit
NOTE: PMRs in **bold** are significantly different from 100 (p<0.05). PMRs in *italics* are based on fewer than five observed deaths. PMRs are based on underlying and contributing cause of death. Some values could not be calculated because the number of observed or expected deaths was zero; such values are indicated by ---. See appendices for source description, methods, ICD codes, and a list of selected states.
SOURCE: National Center for Health Statistics multiple-cause-of-death data

Acute Respiratory Infection Mortality within and by Agricultural Group

Table 2-56. Livestock farm workers: Proportionate mortality ratio (PMR) adjusted for age, sex, and race/ethnicity for acute respiratory infections, U.S. residents age 15 and over, selected states, 1988–1998

Disease Category (ICD Code)	Number of Deaths	PMR	95% Confidence Interval LCL	95% Confidence Interval UCL
Acute nasopharyngitis (460)	2	*348*	42	1,256
Acute pharyngitis (462)	1	*76*	2	422
Acute tonsillitis (463)	0	*0*	---	---
Acute laryngitis and tracheitis (464)	1	*52*	1	289
Acute upper respiratory infections of multiple or unspecified sites (465)	20	129	79	199
Acute bronchitis and bronchiolitis (466)	35	**65**	45	90

ICD - International Classification of Diseases, 9th Revision   LCL - lower confidence limit   UCL - upper confidence limit
NOTE: PMRs in **bold** are significantly different from 100 (p<0.05). PMRs in *italics* are based on fewer than five observed deaths. PMRs are based on underlying and contributing cause of death. Some values could not be calculated because the number of observed or expected deaths was zero; such values are indicated by ---. See appendices for source description, methods, ICD codes, and a list of selected states.
SOURCE: National Center for Health Statistics multiple-cause-of-death data

Table 2-57. Farm managers: Proportionate mortality ratio (PMR) adjusted for age, sex, and race/ethnicity for acute respiratory infections, U.S. residents age 15 and over, selected states, 1988–1998

Disease Category (ICD Code)	Number of Deaths	PMR	95% Confidence Interval LCL	95% Confidence Interval UCL
Acute nasopharyngitis (460)	0	0	---	---
Acute pharyngitis (462)	0	0	---	---
Acute tonsillitis (463)	0	0	---	---
Acute laryngitis and tracheitis (464)	0	0	---	---
Acute upper respiratory infections of multiple or unspecified sites (465)	0	0	---	---
Acute bronchitis and bronchiolitis (466)	2	*66*	*8*	*238*

ICD - International Classification of Diseases, 9th Revision   LCL - lower confidence limit   UCL - upper confidence limit
NOTE: PMRs in **bold** are significantly different from 100 (p<0.05). PMRs in *italics* are based on fewer than five observed deaths. PMRs are based on underlying and contributing cause of death. Some values could not be calculated because the number of observed or expected deaths was zero; such values are indicated by ---. See appendices for source description, methods, ICD codes, and a list of selected states.
SOURCE: National Center for Health Statistics multiple-cause-of-death data

Acute Respiratory Infection Mortality within and by Agricultural Group

Table 2-58. Landscape and horticultural workers: Proportionate mortality ratio (PMR) adjusted for age, sex, and race/ethnicity for acute respiratory infections, U.S. residents age 15 and over, selected states, 1988–1998

Disease Category (ICD Code)	Number of Deaths	PMR	95% Confidence Interval LCL	95% Confidence Interval UCL
Acute nasopharyngitis (460)	0	*0*	---	---
Acute pharyngitis (462)	1	*425*	11	2,361
Acute tonsillitis (463)	0	*0*	---	---
Acute laryngitis and tracheitis (464)	1	*179*	5	994
Acute upper respiratory infections of multiple or unspecified sites (465)	3	*186*	38	544
Acute bronchitis and bronchiolitis (466)	4	*70*	19	179

ICD - International Classification of Diseases, 9th Revision LCL - lower confidence limit UCL - upper confidence limit
NOTE: PMRs in **bold** are significantly different from 100 (p<0.05). PMRs in *italics* are based on fewer than five observed deaths. PMRs are based on underlying and contributing cause of death. Some values could not be calculated because the number of observed or expected deaths was zero; such values are indicated by ---. See appendices for source description, methods, ICD codes, and a list of selected states.
SOURCE: National Center for Health Statistics multiple-cause-of-death data

Acute Respiratory Infection Mortality within and by Agricultural Group

Table 2-59. Forestry workers: Proportionate mortality ratio (PMR) adjusted for age, sex, and race/ethnicity for acute respiratory infections, U.S. residents age 15 and over, selected states, 1988–1998

Disease Category (ICD Code)	Number of Deaths	PMR	95% Confidence Interval LCL	95% Confidence Interval UCL
Acute nasopharyngitis (460)	0	0	---	---
Acute pharyngitis (462)	0	0	---	---
Acute tonsillitis (463)	0	0	---	---
Acute laryngitis and tracheitis (464)	0	0	---	---
Acute upper respiratory infections of multiple or unspecified sites (465)	3	89	18	260
Acute bronchitis and bronchiolitis (466)	12	91	47	159

ICD - International Classification of Diseases, 9th Revision   LCL - lower confidence limit   UCL - upper confidence limit
NOTE: PMRs in **bold** are significantly different from 100 (p<0.05). PMRs in *italics* are based on fewer than five observed deaths. PMRs are based on underlying and contributing cause of death. Some values could not be calculated because the number of observed or expected deaths was zero; such values are indicated by ---. See appendices for source description, methods, ICD codes, and a list of selected states.
SOURCE: National Center for Health Statistics multiple-cause-of-death data

Table 2-60. Fishery workers: Proportionate mortality ratio (PMR) adjusted for age, sex, and race/ethnicity for acute respiratory infections, U.S. residents age 15 and over, selected states, 1988–1998

Disease Category (ICD Code)	Number of Deaths	PMR	95% Confidence Interval LCL	95% Confidence Interval UCL
Acute nasopharyngitis (460)	0	*0*	---	---
Acute pharyngitis (462)	0	*0*	---	---
Acute tonsillitis (463)	0	*0*	---	---
Acute laryngitis and tracheitis (464)	0	*0*	---	---
Acute upper respiratory infections of multiple or unspecified sites (465)	1	*102*	*3*	*567*
Acute bronchitis and bronchiolitis (466)	3	*81*	*17*	*237*

ICD - International Classification of Diseases, 9th Revision LCL - lower confidence limit UCL - upper confidence limit
NOTE: PMRs in **bold** are significantly different from 100 (p<0.05). PMRs in *italics* are based on fewer than five observed deaths. PMRs are based on underlying and contributing cause of death. Some values could not be calculated because the number of observed or expected deaths was zero; such values are indicated by ---. See appendices for source description, methods, ICD codes, and a list of selected states.
SOURCE: National Center for Health Statistics multiple-cause-of-death data

Figure 2-43. Acute laryngitis and tracheitis: Proportionate mortality ratio (PMR) adjusted for age, sex, and race/ethnicity by agricultural group, U.S. residents age 15 and over, selected states, 1988–1998

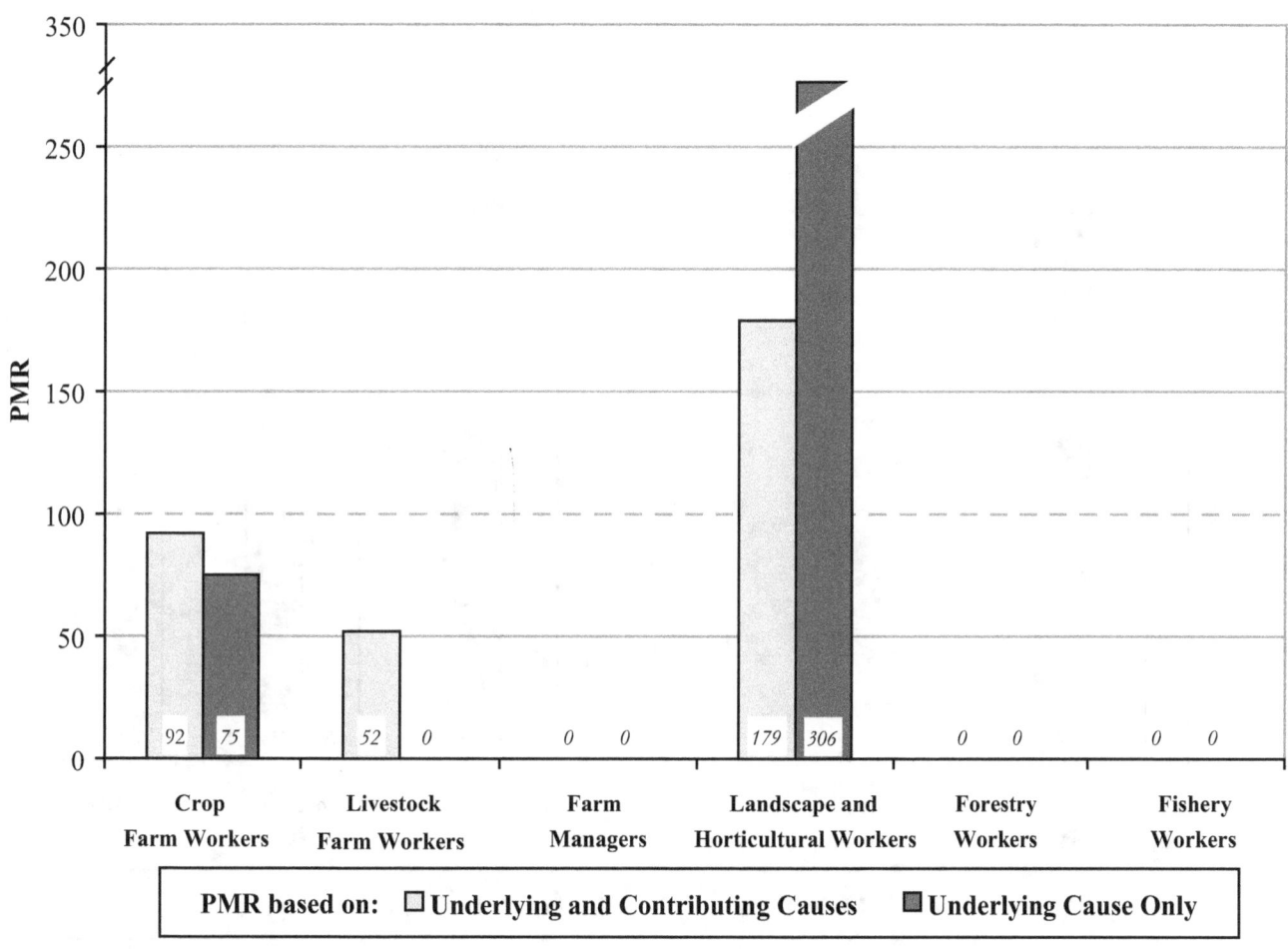

ICD - International Classification of Diseases, 9th Revision
NOTE: Acute laryngitis and tracheitis = ICD-9 code 464. PMRs in **bold** are significantly different from 100 (p<0.05). PMRs in *italics* are based on fewer than five observed deaths. PMRs are based on underlying and contributing cause of death. See appendices for source description, methods, ICD codes, and a list of selected states.
SOURCE: National Center for Health Statistics multiple-cause-of-death data

Acute Respiratory Infection Mortality within and by Agricultural Group

Figure 2-44. Acute upper respiratory infections of multiple or unspecified sites: Proportionate mortality ratio (PMR) adjusted for age, sex, and race/ethnicity by agricultural group, U.S. residents age 15 and over, selected states, 1988–1998

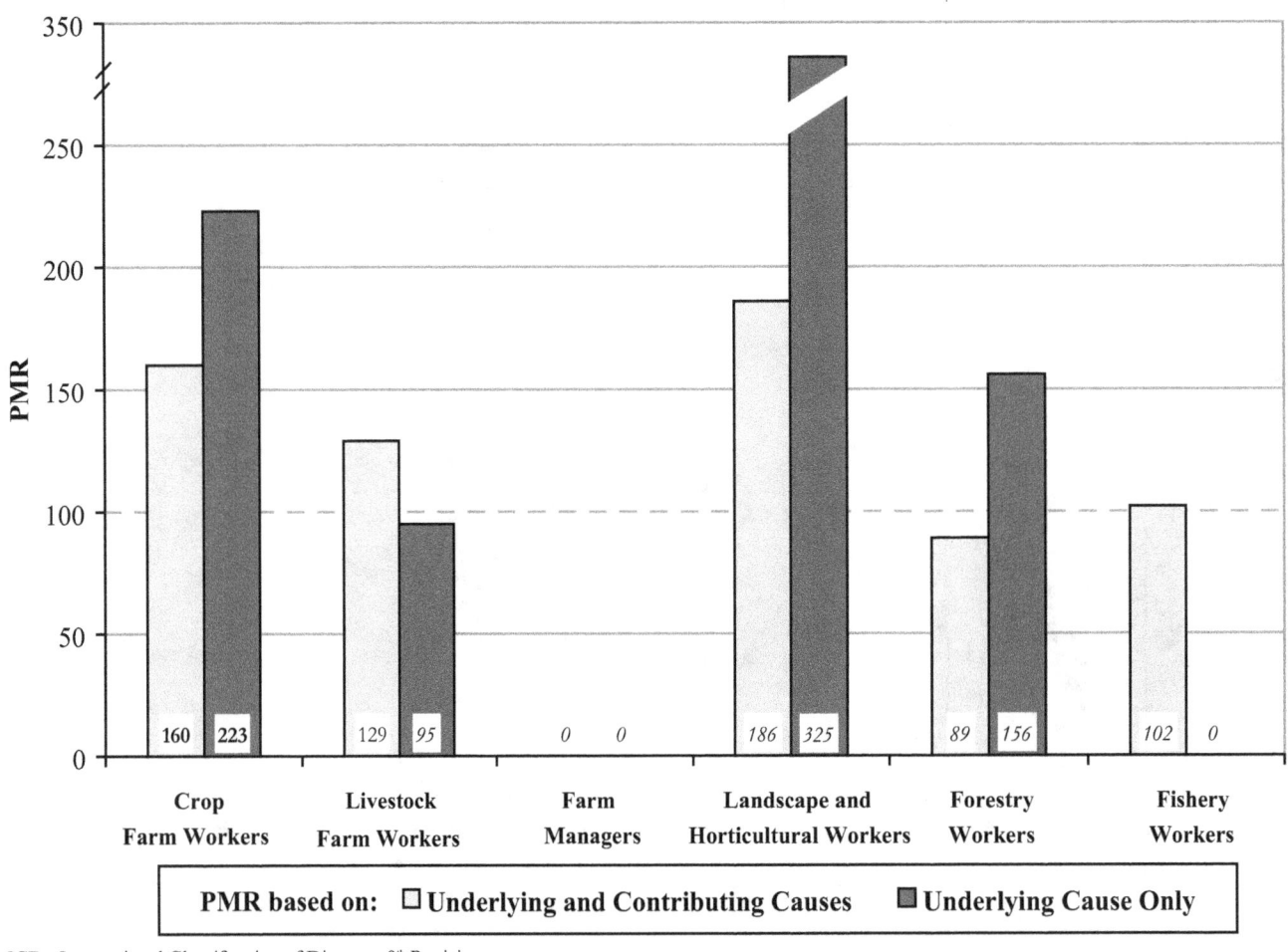

ICD - International Classification of Diseases, 9th Revision
NOTE: Acute upper respiratory infections of multiple or unspecified sites = ICD-9 code 465. PMRs in **bold** are significantly different from 100 (p<0.05). PMRs in *italics* are based on fewer than five observed deaths. PMRs are based on underlying and contributing cause of death. See appendices for source description, methods, ICD codes, and a list of selected states.
SOURCE: National Center for Health Statistics multiple-cause-of-death data

Figure 2-45. Acute bronchitis and bronchiolitis: Proportionate mortality ratio (PMR) adjusted for age, sex, and race/ethnicity by agricultural group, U.S. residents age 15 and over, selected states, 1988–1998

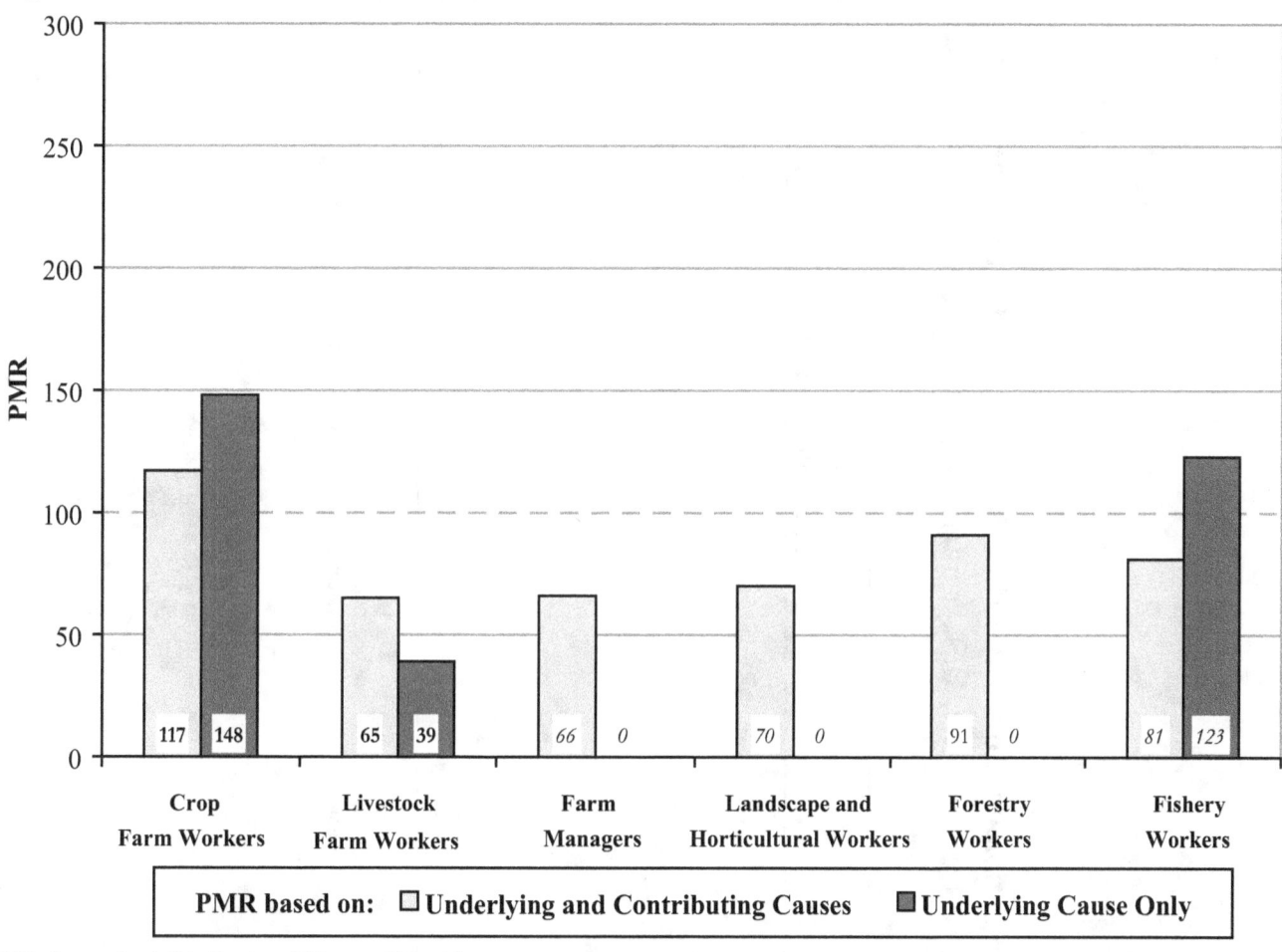

ICD - International Classification of Diseases, 9th Revision
NOTE: Acute bronchitis and bronchiolitis = ICD-9 code 466. PMRs in **bold** are significantly different from 100 ($p<0.05$). PMRs in *italics* are based on fewer than five observed deaths. PMRs are based on underlying and contributing cause of death. See appendices for source description, methods, ICD codes, and a list of selected states.
SOURCE: National Center for Health Statistics multiple-cause-of-death data

Other Diseases of Upper Respiratory Tract Mortality within and by Agricultural Group

Table 2-61. Crop farm workers: Proportionate mortality ratio (PMR) adjusted for age, sex, and race/ethnicity for other diseases of upper respiratory tract, U.S. residents age 15 and over, selected states, 1988–1998

Disease Category (ICD Code)	Number of Deaths	PMR	95% Confidence Interval LCL	95% Confidence Interval UCL
Nasal polyps (471)	1	*91*	2	506
Chronic pharyngitis and nasopharyngitis (472)	1	*68*	2	378
Chronic sinusitis (473)	19	72	43	113
Chronic disease of tonsils and adenoids (474)	3	*161*	33	471
Peritonsillar abscess (475)	2	*150*	18	542
Allergic rhinitis (477)	6	214	78	466
Other diseases of upper respiratory tract (478)	65	90	70	115

ICD - International Classification of Diseases, 9th Revision   LCL - lower confidence limit   UCL - upper confidence limit
NOTE: PMRs in **bold** are significantly different from 100 (p<0.05). PMRs in *italics* are based on fewer than five observed deaths. PMRs are based on underlying and contributing cause of death. Some values could not be calculated because the number of observed or expected deaths was zero; such values are indicated by ---. See appendices for source description, methods, ICD codes, and a list of selected states.
SOURCE: National Center for Health Statistics multiple-cause-of-death data

Table 2-62. Livestock farm workers: Proportionate mortality ratio (PMR) adjusted for age, sex, and race/ethnicity for other diseases of upper respiratory tract, U.S. residents age 15 and over, selected states, 1988–1998

Disease Category (ICD Code)	Number of Deaths	PMR	95% Confidence Interval LCL	95% Confidence Interval UCL
Nasal polyps (471)	0	*0*	---	---
Chronic pharyngitis and nasopharyngitis (472)	0	*0*	---	---
Chronic sinusitis (473)	8	117	50	230
Chronic disease of tonsils and adenoids (474)	1	*207*	5	1,150
Peritonsillar abscess (475)	1	*283*	7	1,572
Allergic rhinitis (477)	1	*117*	3	650
Other diseases of upper respiratory tract (478)	9	**49**	22	93

ICD - International Classification of Diseases, 9th Revision   LCL - lower confidence limit   UCL - upper confidence limit
NOTE: PMRs in **bold** are significantly different from 100 (p<0.05). PMRs in *italics* are based on fewer than five observed deaths. PMRs are based on underlying and contributing cause of death. Some values could not be calculated because the number of observed or expected deaths was zero; such values are indicated by ---. See appendices for source description, methods, ICD codes, and a list of selected states.
SOURCE: National Center for Health Statistics multiple-cause-of-death data

Other Diseases of Upper Respiratory Tract Mortality within and by Agricultural Group

Table 2-63. Farm managers: Proportionate mortality ratio (PMR) adjusted for age, sex, and race/ethnicity for other diseases of upper respiratory tract, U.S. residents age 15 and over, selected states, 1988–1998

Disease Category (ICD Code)	Number of Deaths	PMR	95% Confidence Interval LCL	95% Confidence Interval UCL
Nasal polyps (471)	0	*0*	---	---
Chronic pharyngitis and nasopharyngitis (472)	0	*0*	---	---
Chronic sinusitis (473)	1	*205*	5	1,139
Chronic disease of tonsils and adenoids (474)	0	*0*	---	---
Peritonsillar abscess (475)	0	*0*	---	---
Allergic rhinitis (477)	0	*0*	---	---
Other diseases of upper respiratory tract (478)	0	*0*	---	---

ICD - International Classification of Diseases, 9th Revision   LCL - lower confidence limit   UCL - upper confidence limit
NOTE: PMRs in **bold** are significantly different from 100 (p<0.05). PMRs in *italics* are based on fewer than five observed deaths. PMRs are based on underlying and contributing cause of death. Some values could not be calculated because the number of observed or expected deaths was zero; such values are indicated by ---. See appendices for source description, methods, ICD codes, and a list of selected states.
SOURCE: National Center for Health Statistics multiple-cause-of-death data

Table 2-64. Landscape and horticultural workers: Proportionate mortality ratio (PMR) adjusted for age, sex, and race/ethnicity for other diseases of upper respiratory tract, U.S. residents age 15 and over, selected states, 1988–1998

Disease Category (ICD Code)	Number of Deaths	PMR	95% Confidence Interval LCL	UCL
Nasal polyps (471)	0	0	---	---
Chronic pharyngitis and nasopharyngitis (472)	0	0	---	---
Chronic sinusitis (473)	1	*57*	1	317
Chronic disease of tonsils and adenoids (474)	0	0	---	---
Peritonsillar abscess (475)	0	0	---	---
Allergic rhinitis (477)	0	0	---	---
Other diseases of upper respiratory tract (478)	3	*99*	20	289

ICD - International Classification of Diseases, 9th Revision   LCL - lower confidence limit   UCL - upper confidence limit
NOTE: PMRs in **bold** are significantly different from 100 (p<0.05). PMRs in *italics* are based on fewer than five observed deaths. PMRs are based on underlying and contributing cause of death. Some values could not be calculated because the number of observed or expected deaths was zero; such values are indicated by ---. See appendices for source description, methods, ICD codes, and a list of selected states.
SOURCE: National Center for Health Statistics multiple-cause-of-death data

Other Diseases of Upper Respiratory Tract Mortality within and by Agricultural Group

Table 2-65. Forestry workers: Proportionate mortality ratio (PMR) adjusted for age, sex, and race/ethnicity for other diseases of upper respiratory tract, U.S. residents age 15 and over, selected states, 1988–1998

Disease Category (ICD Code)	Number of Deaths	PMR	95% Confidence Interval	
			LCL	UCL
Nasal polyps (471)	1	*1,095*	28	6,083
Chronic pharyngitis and nasopharyngitis (472)	0	0	---	---
Chronic sinusitis (473)	1	*38*	1	211
Chronic disease of tonsils and adenoids (474)	0	0	---	---
Peritonsillar abscess (475)	0	0	---	---
Allergic rhinitis (477)	0	0	---	---
Other diseases of upper respiratory tract (478)	6	102	37	222

ICD - International Classification of Diseases, 9th Revision   LCL - lower confidence limit   UCL - upper confidence limit
NOTE: PMRs in **bold** are significantly different from 100 (p<0.05). PMRs in *italics* are based on fewer than five observed deaths. PMRs are based on underlying and contributing cause of death. Some values could not be calculated because the number of observed or expected deaths was zero; such values are indicated by ---. See appendices for source description, methods, ICD codes, and a list of selected states.
SOURCE: National Center for Health Statistics multiple-cause-of-death data

Table 2-66. Fishery workers: Proportionate mortality ratio (PMR) adjusted for age, sex, and race/ethnicity for other diseases of upper respiratory tract, U.S. residents age 15 and over, selected states, 1988–1998

Disease Category (ICD Code)	Number of Deaths	PMR	95% Confidence Interval LCL	95% Confidence Interval UCL
Nasal polyps (471)	0	*0*	---	---
Chronic pharyngitis and nasopharyngitis (472)	0	*0*	---	---
Chronic sinusitis (473)	0	*0*	---	---
Chronic disease of tonsils and adenoids (474)	1	*1,376*	35	7,644
Peritonsillar abscess (475)	0	*0*	---	---
Allergic rhinitis (477)	1	*1,619*	41	8,994
Other diseases of upper respiratory tract (478)	3	*191*	39	558

ICD - International Classification of Diseases, 9th Revision   LCL - lower confidence limit   UCL - upper confidence limit
NOTE: PMRs in **bold** are significantly different from 100 (p<0.05). PMRs in *italics* are based on fewer than five observed deaths. PMRs are based on underlying and contributing cause of death. Some values could not be calculated because the number of observed or expected deaths was zero; such values are indicated by ---. See appendices for source description, methods, ICD codes, and a list of selected states.
SOURCE: National Center for Health Statistics multiple-cause-of-death data

Other Diseases of Upper Respiratory Tract Mortality within and by Agricultural Group

Figure 2-46. Chronic sinusitis: Proportionate mortality ratio (PMR) adjusted for age, sex, and race/ethnicity by agricultural group, U.S. residents age 15 and over, selected states, 1988–1998

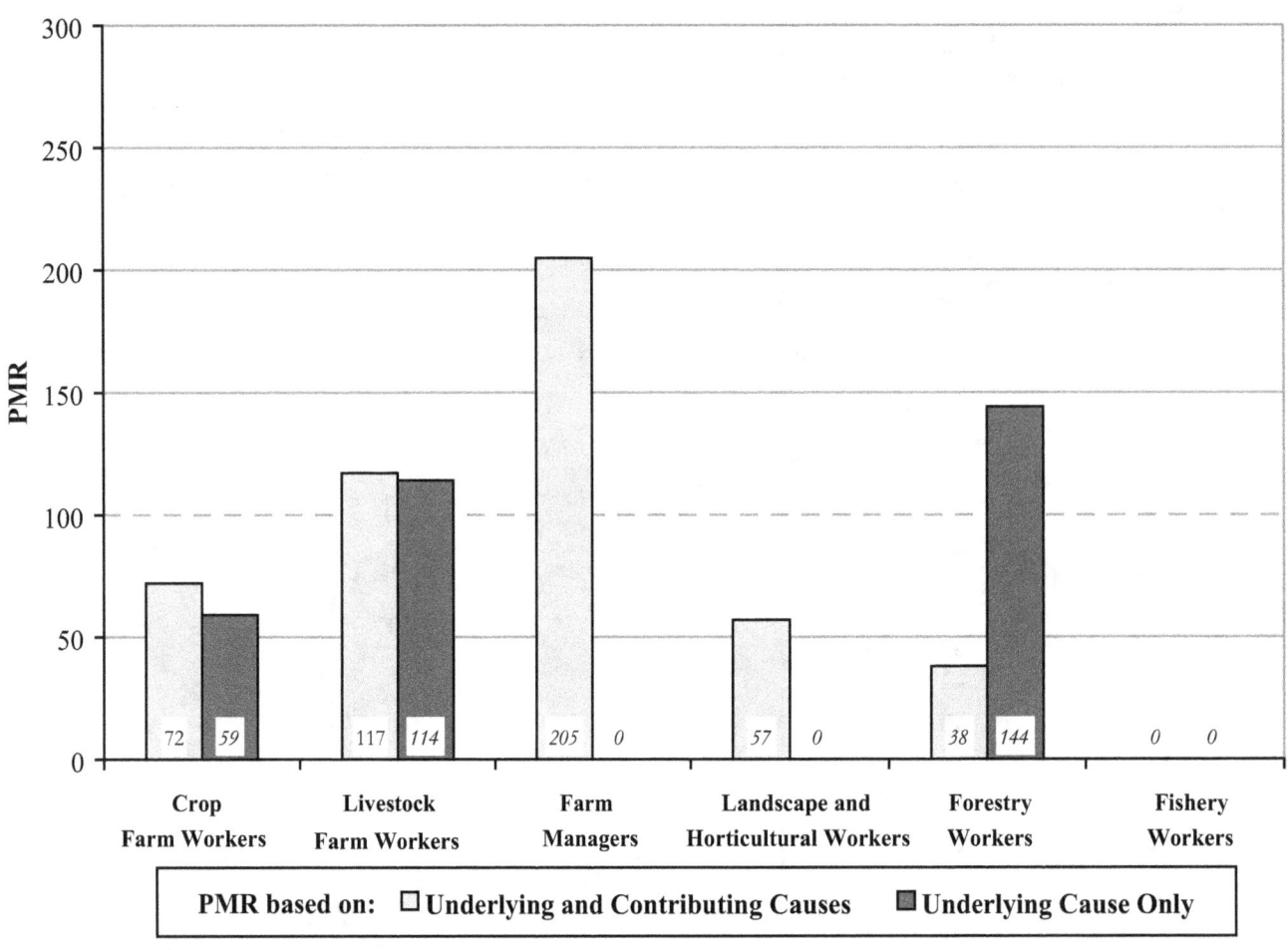

ICD - International Classification of Diseases, 9th Revision
NOTE: Chronic sinusitis = ICD-9 code 473. PMRs in **bold** are significantly different from 100 (p<0.05). PMRs in *italics* are based on fewer than five observed deaths. PMRs are based on underlying and contributing cause of death. See appendices for source description, methods, ICD codes, and a list of selected states.
SOURCE: National Center for Health Statistics multiple-cause-of-death data

Other Diseases of Upper Respiratory Tract Mortality within and by Agricultural Group

Figure 2-47. Other diseases of upper respiratory tract: Proportionate mortality ratio (PMR) adjusted for age, sex, and race/ethnicity by agricultural group, U.S. residents age 15 and over, selected states, 1988–1998

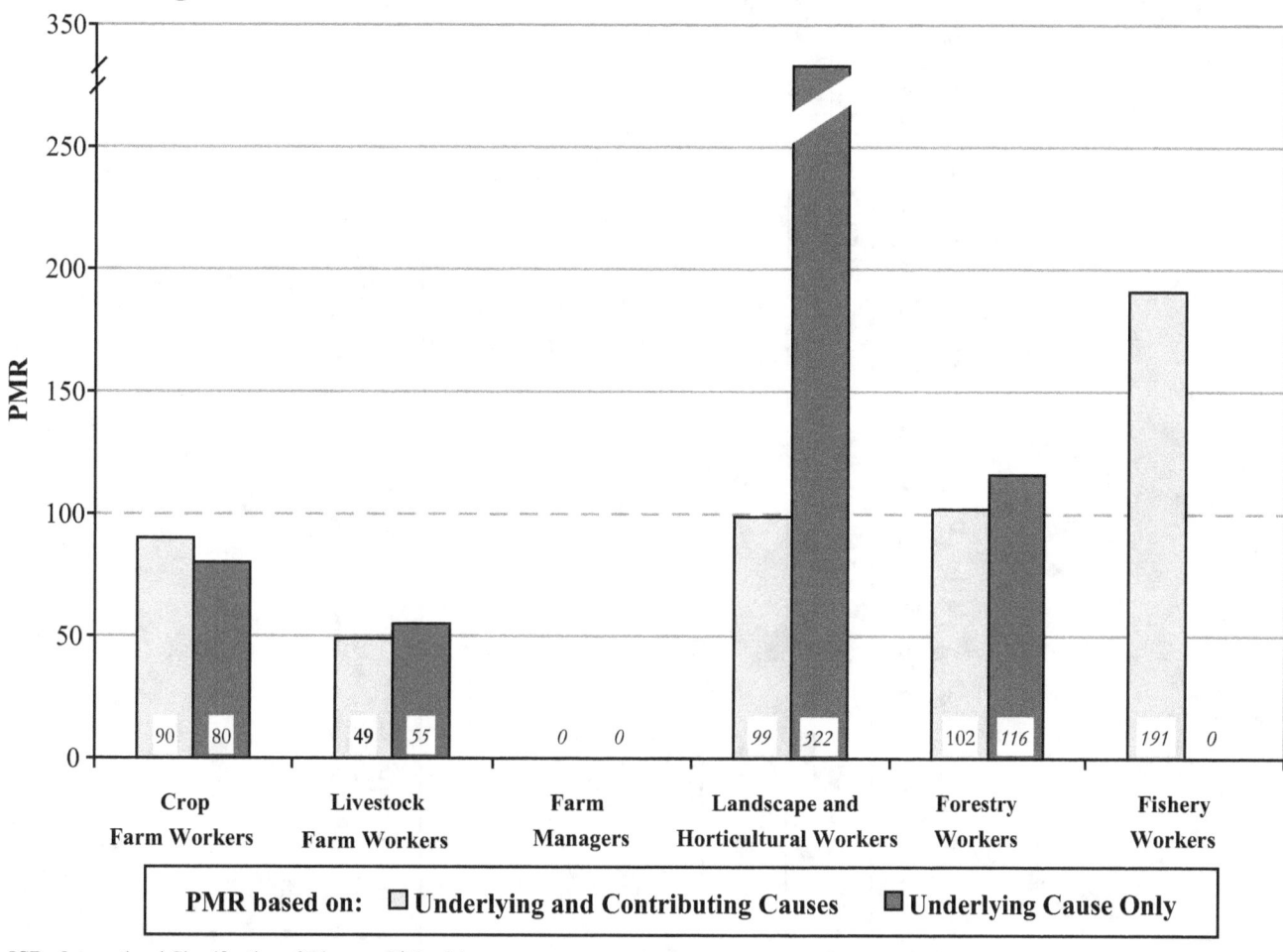

ICD - International Classification of Diseases, 9th Revision
NOTE: Other diseases of upper respiratory tract = ICD-9 code 478. PMRs in **bold** are significantly different from 100 (p<0.05). PMRs in *italics* are based on fewer than five observed deaths. PMRs are based on underlying and contributing cause of death. See appendices for source description, methods, ICD codes, and a list of selected states.
SOURCE: National Center for Health Statistics multiple-cause-of-death data

Table 2-67. Crop farm workers: Proportionate mortality ratio (PMR) adjusted for age, sex, and race/ethnicity for pneumonia and influenza, U.S. residents age 15 and over, selected states, 1988–1998

Disease Category (ICD Code)	Number of Deaths	PMR	95% Confidence Interval LCL	95% Confidence Interval UCL
Viral pneumonia (480)	60	110	84	142
Pneumococcal pneumonia [Streptococcus pneumoniae pneumonia] (481)	347	98	88	109
Other bacterial pneumonia (482)	955	**120**	113	128
Pneumonia due to other specified organism (483)	20	113	69	175
Bronchopneumonia, organism unspecified (485)	1,365	106	100	112
Pneumonia, organism unspecified (486)	23,135	**109**	107	111
Influenza (487)	232	**142**	125	162

ICD - International Classification of Diseases, 9th Revision   LCL - lower confidence limit   UCL - upper confidence limit
NOTE: PMRs in **bold** are significantly different from 100 (p<0.05). PMRs in *italics* are based on fewer than five observed deaths. PMRs are based on underlying and contributing cause of death. Some values could not be calculated because the number of observed or expected deaths was zero; such values are indicated by ---. See appendices for source description, methods, ICD codes, and a list of selected states.
SOURCE: National Center for Health Statistics multiple-cause-of-death data

Pneumonia and Influenza Mortality within and by Agricultural Group

Table 2-68. Livestock farm workers: Proportionate mortality ratio (PMR) adjusted for age, sex, and race/ethnicity for pneumonia and influenza, U.S. residents age 15 and over, selected states, 1988–1998

Disease Category (ICD Code)	Number of Deaths	PMR	95% Confidence Interval LCL	95% Confidence Interval UCL
Viral pneumonia (480)	21	132	81	202
Pneumococcal pneumonia [Streptococcus pneumoniae pneumonia] (481)	83	90	73	112
Other bacterial pneumonia (482)	170	79	68	92
Pneumonia due to other specified organism (483)	4	*86*	*23*	*220*
Bronchopneumonia, organism unspecified (485)	317	**89**	80	99
Pneumonia, organism unspecified (486)	5,723	100	97	103
Influenza (487)	73	**150**	119	189

ICD - International Classification of Diseases, 9th Revision   LCL - lower confidence limit   UCL - upper confidence limit
NOTE: PMRs in **bold** are significantly different from 100 (p<0.05). PMRs in *italics* are based on fewer than five observed deaths. PMRs are based on underlying and contributing cause of death. Some values could not be calculated because the number of observed or expected deaths was zero; such values are indicated by ---.   See appendices for source description, methods, ICD codes, and a list of selected states.
SOURCE: National Center for Health Statistics multiple-cause-of-death data

Table 2-69. Farm managers: Proportionate mortality ratio (PMR) adjusted for age, sex, and race/ethnicity for pneumonia and influenza, U.S. residents age 15 and over, selected states, 1988–1998

Disease Category (ICD Code)	Number of Deaths	PMR	95% Confidence Interval LCL	95% Confidence Interval UCL
Viral pneumonia (480)	2	*208*	13	390
Pneumococcal pneumonia [Streptococcus pneumoniae pneumonia] (481)	6	109	40	237
Other bacterial pneumonia (482)	10	83	40	153
Pneumonia due to other specified organism (483)	0	*0*	---	---
Bronchopneumonia, organism unspecified (485)	14	72	39	121
Pneumonia, organism unspecified (486)	337	106	95	118
Influenza (487)	4	*155*	42	396

ICD - International Classification of Diseases, 9th Revision   LCL - lower confidence limit   UCL - upper confidence limit
NOTE: PMRs in **bold** are significantly different from 100 (p<0.05). PMRs in *italics* are based on fewer than five observed deaths. PMRs are based on underlying and contributing cause of death. Some values could not be calculated because the number of observed or expected deaths was zero; such values are indicated by ---. See appendices for source description, methods, ICD codes, and a list of selected states.
SOURCE: National Center for Health Statistics multiple-cause-of-death data

Table 2-70. Landscape and horticultural workers: Proportionate mortality ratio (PMR) adjusted for age, sex, and race/ethnicity for pneumonia and influenza, U.S. residents age 15 and over, selected states, 1988–1998

Disease Category (ICD Code)	Number of Deaths	PMR	95% Confidence Interval	
			LCL	UCL
Viral pneumonia (480)	1	*50*	1	278
Pneumococcal pneumonia [Streptococcus pneumoniae pneumonia] (481)	16	108	62	175
Other bacterial pneumonia (482)	29	106	71	152
Pneumonia due to other specified organism (483)	0	0	---	---
Bronchopneumonia, organism unspecified (485)	40	96	69	131
Pneumonia, organism unspecified (486)	518	93	86	101
Influenza (487)	3	73	15	213

ICD - International Classification of Diseases, 9th Revision   LCL - lower confidence limit   UCL - upper confidence limit
NOTE: PMRs in **bold** are significantly different from 100 (p<0.05). PMRs in *italics* are based on fewer than five observed deaths. PMRs are based on underlying and contributing cause of death. Some values could not be calculated because the number of observed or expected deaths was zero; such values are indicated by ---. See appendices for source description, methods, ICD codes, and a list of selected states.
SOURCE: National Center for Health Statistics multiple-cause-of-death data

Table 2-71. Forestry workers: Proportionate mortality ratio (PMR) adjusted for age, sex, and race/ethnicity for pneumonia and influenza, U.S. residents age 15 and over, selected states, 1988–1998

Disease Category (ICD Code)	Number of Deaths	PMR	95% Confidence Interval LCL	95% Confidence Interval UCL
Viral pneumonia (480)	3	*78*	16	228
Pneumococcal pneumonia [Streptococcus pneumoniae pneumonia] (481)	28	102	68	147
Other bacterial pneumonia (482)	65	112	88	143
Pneumonia due to other specified organism (483)	2	*133*	16	480
Bronchopneumonia, organism unspecified (485)	93	104	85	127
Pneumonia, organism unspecified (486)	1,564	**117**	111	123
Influenza (487)	16	169	97	274

ICD - International Classification of Diseases, 9th Revision   LCL - lower confidence limit   UCL - upper confidence limit
NOTE: PMRs in **bold** are significantly different from 100 (p<0.05). PMRs in *italics* are based on fewer than five observed deaths. PMRs are based on underlying and contributing cause of death. Some values could not be calculated because the number of observed or expected deaths was zero; such values are indicated by ---. See appendices for source description, methods, ICD codes, and a list of selected states.
SOURCE: National Center for Health Statistics multiple-cause-of-death data

Pneumonia and Influenza Mortality within and by Agricultural Group

Table 2-72. Fishery workers: Proportionate mortality ratio (PMR) adjusted for age, sex, and race/ethnicity for pneumonia and influenza, U.S. residents age 15 and over, selected states, 1988–1998

Disease Category (ICD Code)	Number of Deaths	PMR	95% Confidence Interval LCL	95% Confidence Interval UCL
Viral pneumonia (480)	1	*86*	2	478
Pneumococcal pneumonia [Streptococcus pneumoniae pneumonia] (481)	5	*67*	22	157
Other bacterial pneumonia (482)	12	77	40	134
Pneumonia due to other specified organism (483)	0	0	---	---
Bronchopneumonia, organism unspecified (485)	32	133	91	188
Pneumonia, organism unspecified (486)	370	104	94	115
Influenza (487)	2	*73*	9	264

ICD - International Classification of Diseases, 9th Revision   LCL - lower confidence limit   UCL - upper confidence limit
NOTE: PMRs in **bold** are significantly different from 100 (p<0.05). PMRs in *italics* are based on fewer than five observed deaths. PMRs are based on underlying and contributing cause of death. Some values could not be calculated because the number of observed or expected deaths was zero; such values are indicated by ---. See appendices for source description, methods, ICD codes, and a list of selected states.
SOURCE: National Center for Health Statistics multiple-cause-of-death data

Pneumonia and Influenza Mortality within and by Agricultural Group

Figure 2-48. Viral pneumonia: Proportionate mortality ratio (PMR) adjusted for age, sex, and race/ethnicity by agricultural group, U.S. residents age 15 and over, selected states, 1988–1998

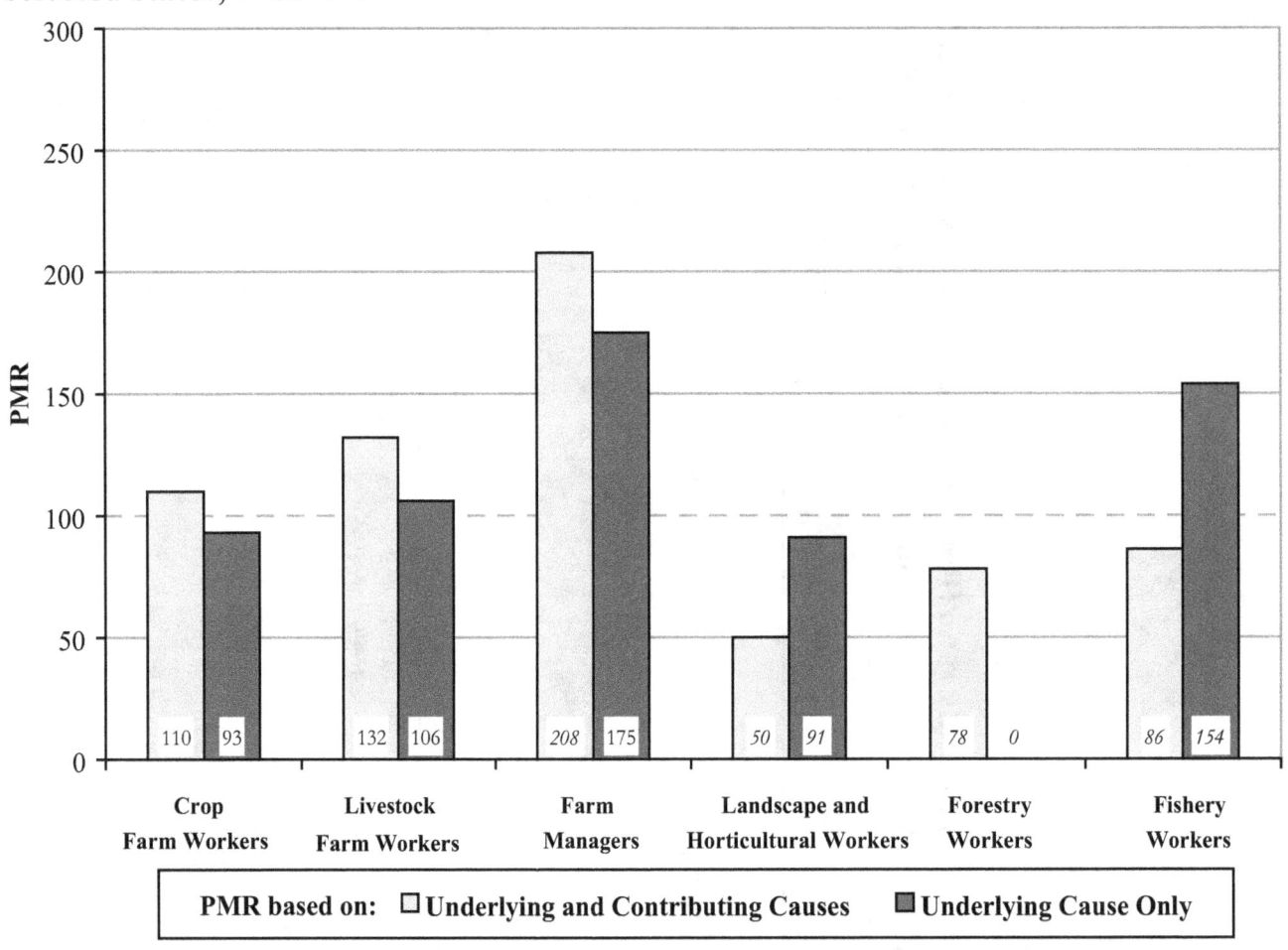

ICD - International Classification of Diseases, 9th Revision
NOTE: Viral pneumonia = ICD-9 code 480. PMRs in **bold** are significantly different from 100 (p<0.05). PMRs in *italics* are based on fewer than five observed deaths. PMRs are based on underlying and contributing cause of death. See appendices for source description, methods, ICD codes, and a list of selected states.
SOURCE: National Center for Health Statistics multiple-cause-of-death data

Figure 2-49. Pneumococcal pneumonia: Proportionate mortality ratio (PMR) adjusted for age, sex, and race/ethnicity by agricultural group, U.S. residents age 15 and over, selected states, 1988–1998

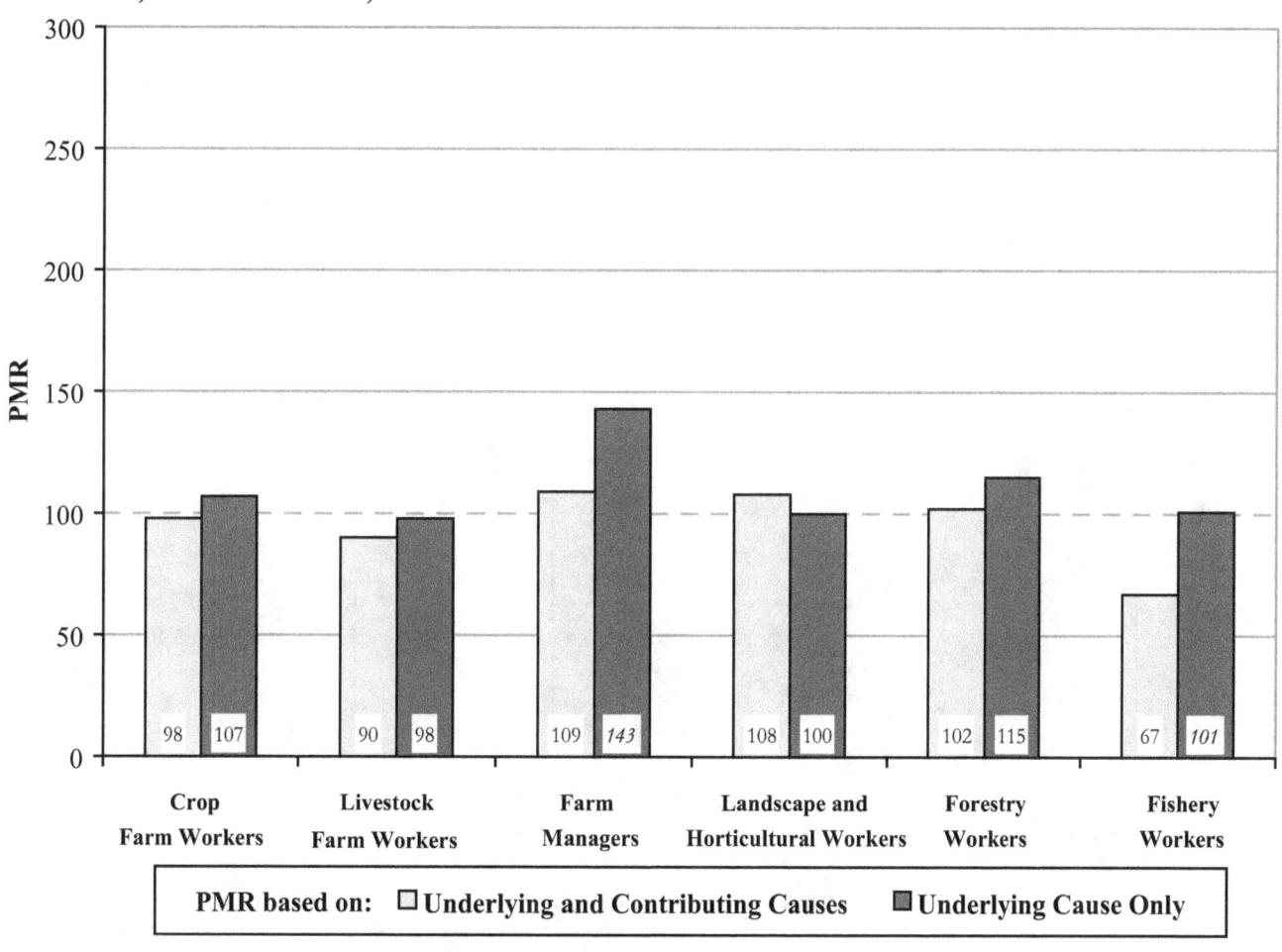

ICD - International Classification of Diseases, 9th Revision
NOTE: Pneumococcal pneumonia = ICD-9 code 481. PMRs in **bold** are significantly different from 100 (p<0.05). PMRs in *italics* are based on fewer than five observed deaths. PMRs are based on underlying and contributing cause of death. See appendices for source description, methods, ICD codes, and a list of selected states.
SOURCE: National Center for Health Statistics multiple-cause-of-death data

Pneumonia and Influenza Mortality within and by Agricultural Group

Figure 2-50. Other bacterial pneumonia: Proportionate mortality ratio (PMR) adjusted for age, sex, and race/ethnicity by agricultural group, U.S. residents age 15 and over, selected states, 1988–1998

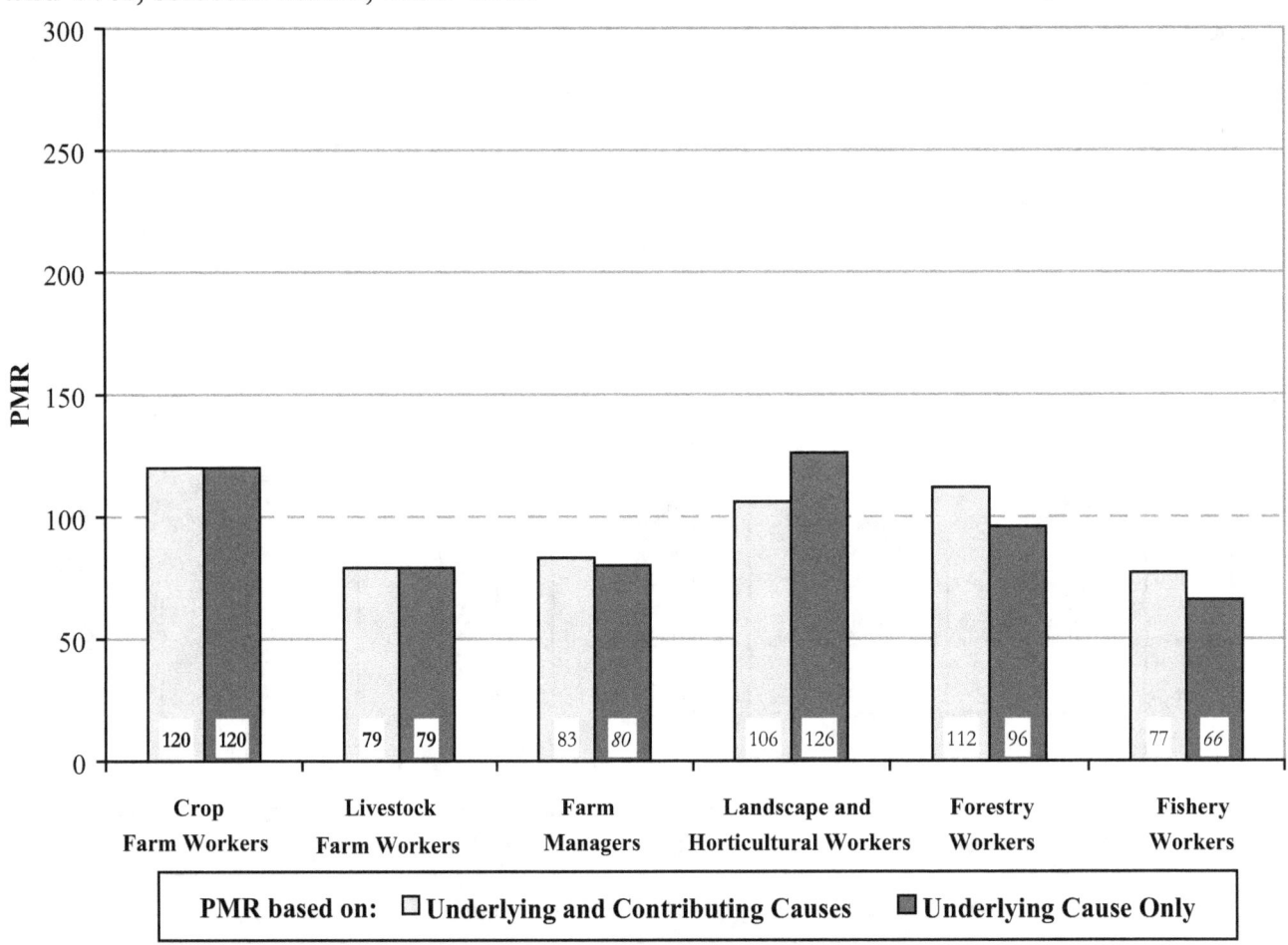

ICD - International Classification of Diseases, 9th Revision
NOTE: Other bacterial pneumonia = ICD-9 code 482. PMRs in **bold** are significantly different from 100 (p<0.05). PMRs in *italics* are based on fewer than five observed deaths. PMRs are based on underlying and contributing cause of death. See appendices for source description, methods, ICD codes, and a list of selected states.
SOURCE: National Center for Health Statistics multiple-cause-of-death data

Figure 2-51. Pneumonia due to other specified organism: Proportionate mortality ratio (PMR) adjusted for age, sex, and race/ethnicity by agricultural group, U.S. residents age 15 and over, selected states, 1988–1998

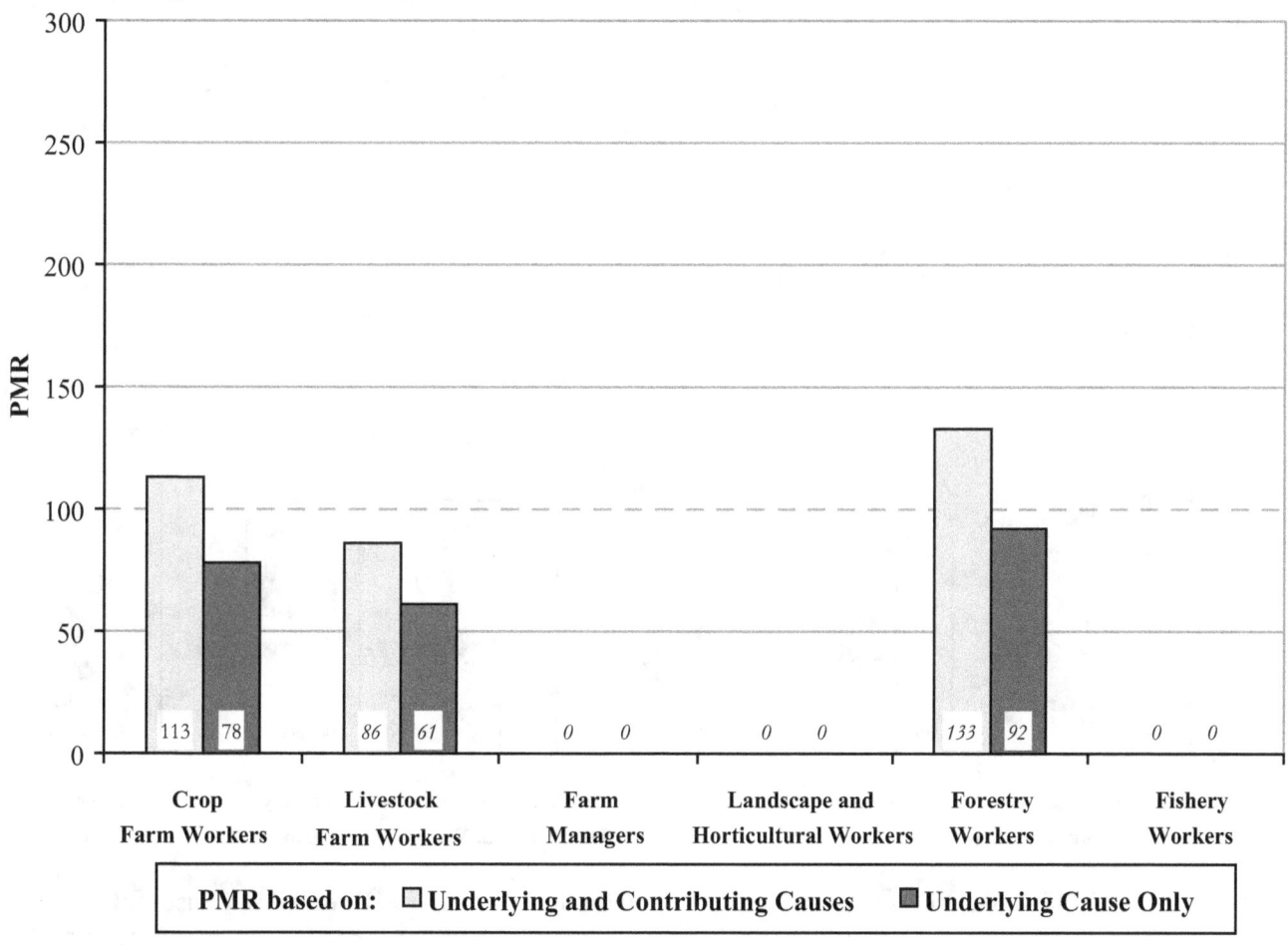

ICD - International Classification of Diseases, 9th Revision
NOTE: Other bacterial pneumonia = ICD-9 code 482. PMRs in **bold** are significantly different from 100 (p<0.05). PMRs in *italics* are based on fewer than five observed deaths. PMRs are based on underlying and contributing cause of death. See appendices for source description, methods, ICD codes, and a list of selected states.
SOURCE: National Center for Health Statistics multiple-cause-of-death data

Pneumonia and Influenza Mortality within and by Agricultural Group

Figure 2-52. Bronchopneumonia, organism unspecified: Proportionate mortality ratio (PMR) adjusted for age, sex, and race/ethnicity by agricultural group, U.S. residents age 15 and over, selected states, 1988–1998

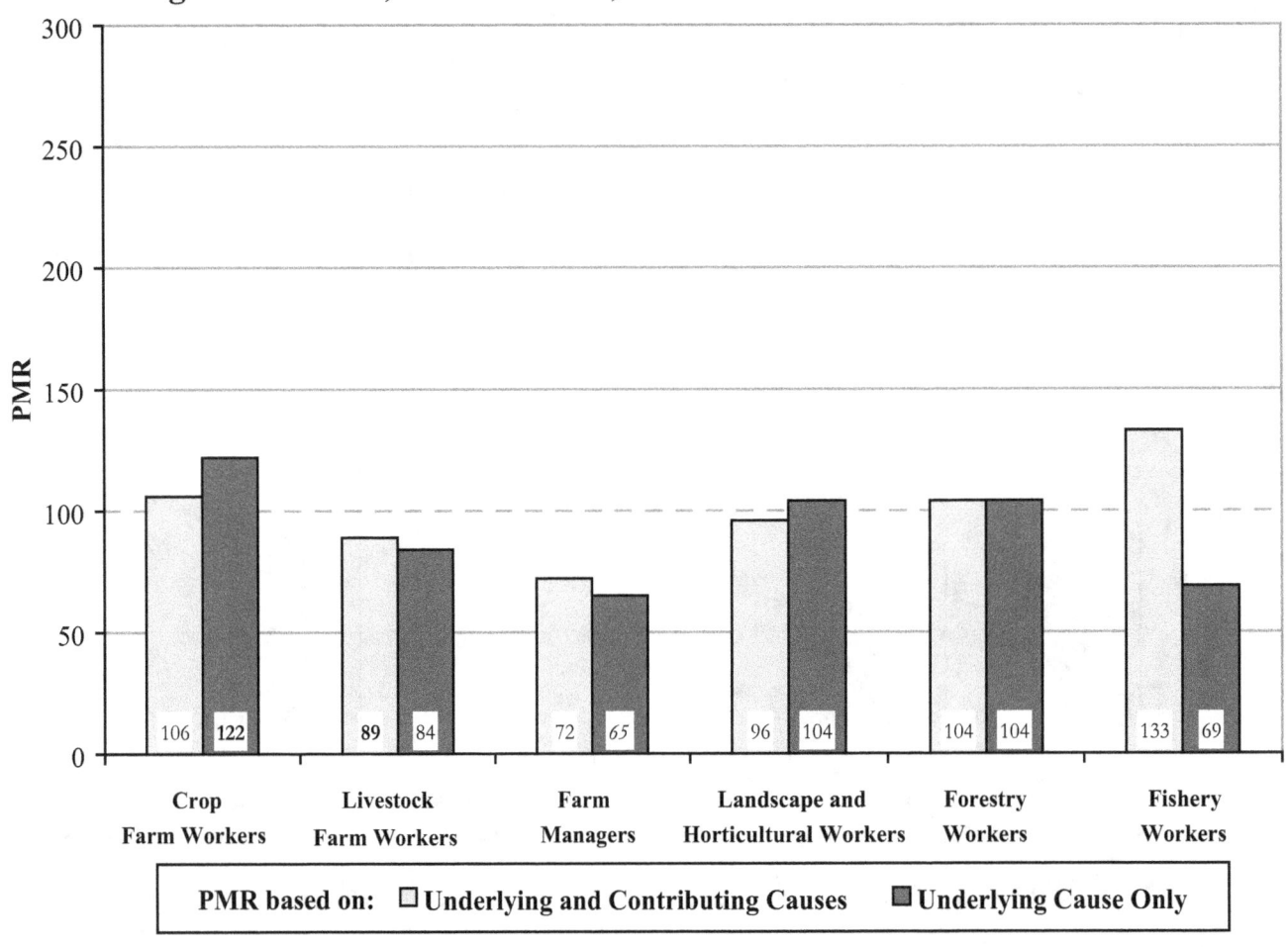

ICD - International Classification of Diseases, 9th Revision
NOTE: Bronchopneumonia, organism unspecified = ICD-9 code 485. PMRs in **bold** are significantly different from 100 (p<0.05). PMRs in *italics* are based on fewer than five observed deaths. PMRs are based on underlying and contributing cause of death. See appendices for source description, methods, ICD codes, and a list of selected states.
SOURCE: National Center for Health Statistics multiple-cause-of-death data

Figure 2-53. Pneumonia, organism unspecified: Proportionate mortality ratio (PMR) adjusted for age, sex, and race/ethnicity by agricultural group, U.S. residents age 15 and over, selected states, 1988–1998

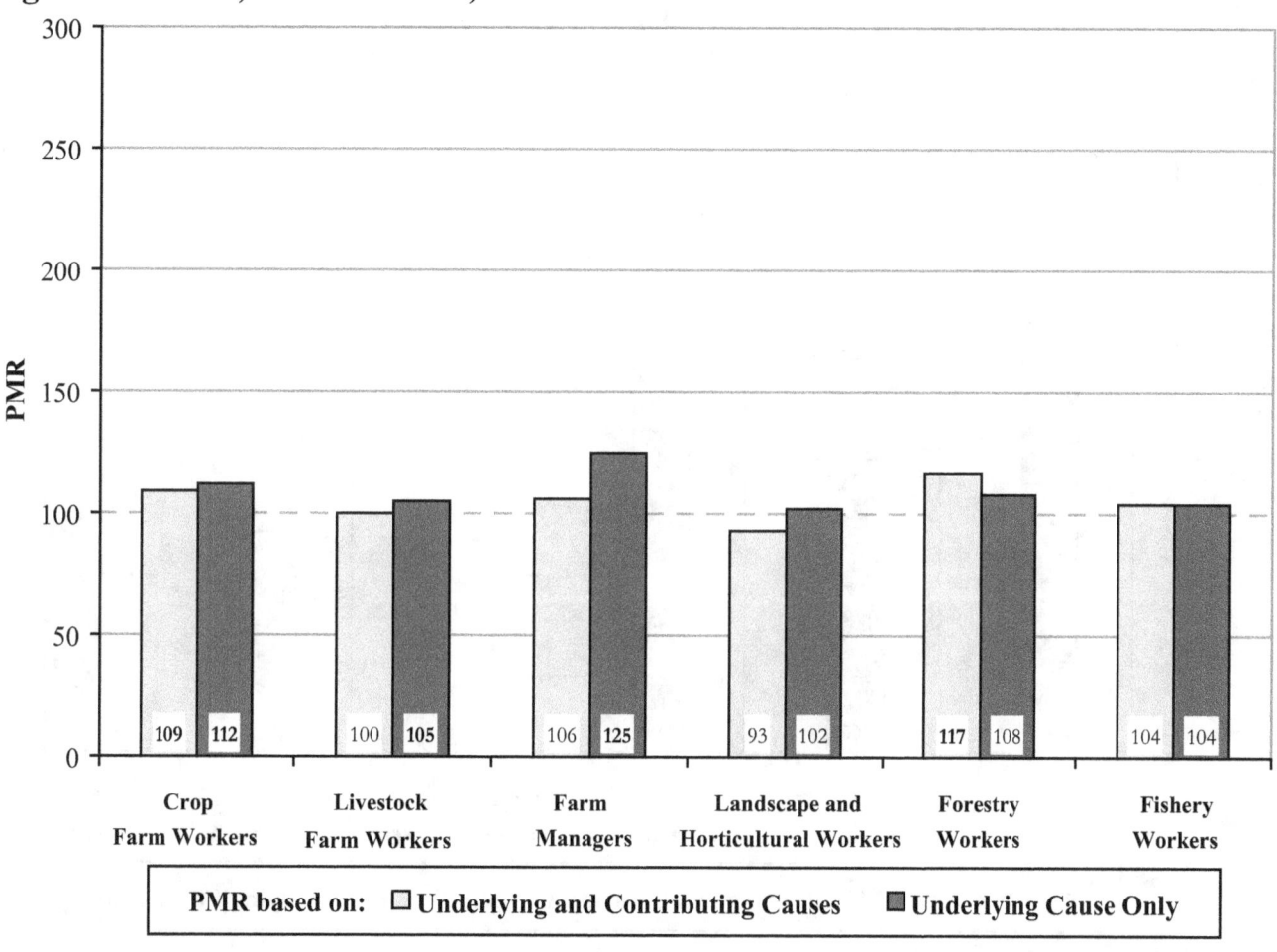

ICD - International Classification of Diseases, 9th Revision
NOTE: Pneumonia, organism unspecified = ICD-9 code 486. PMRs in **bold** are significantly different from 100 (p<0.05). PMRs in *italics* are based on fewer than five observed deaths. PMRs are based on underlying and contributing cause of death. See appendices for source description, methods, ICD codes, and a list of selected states.
SOURCE: National Center for Health Statistics multiple-cause-of-death data

Pneumonia and Influenza Mortality within and by Agricultural Group

Figure 2-54. Influenza: Proportionate mortality ratio (PMR) adjusted for age, sex, and race/ethnicity by agricultural group, U.S. residents age 15 and over, selected states, 1988–1998

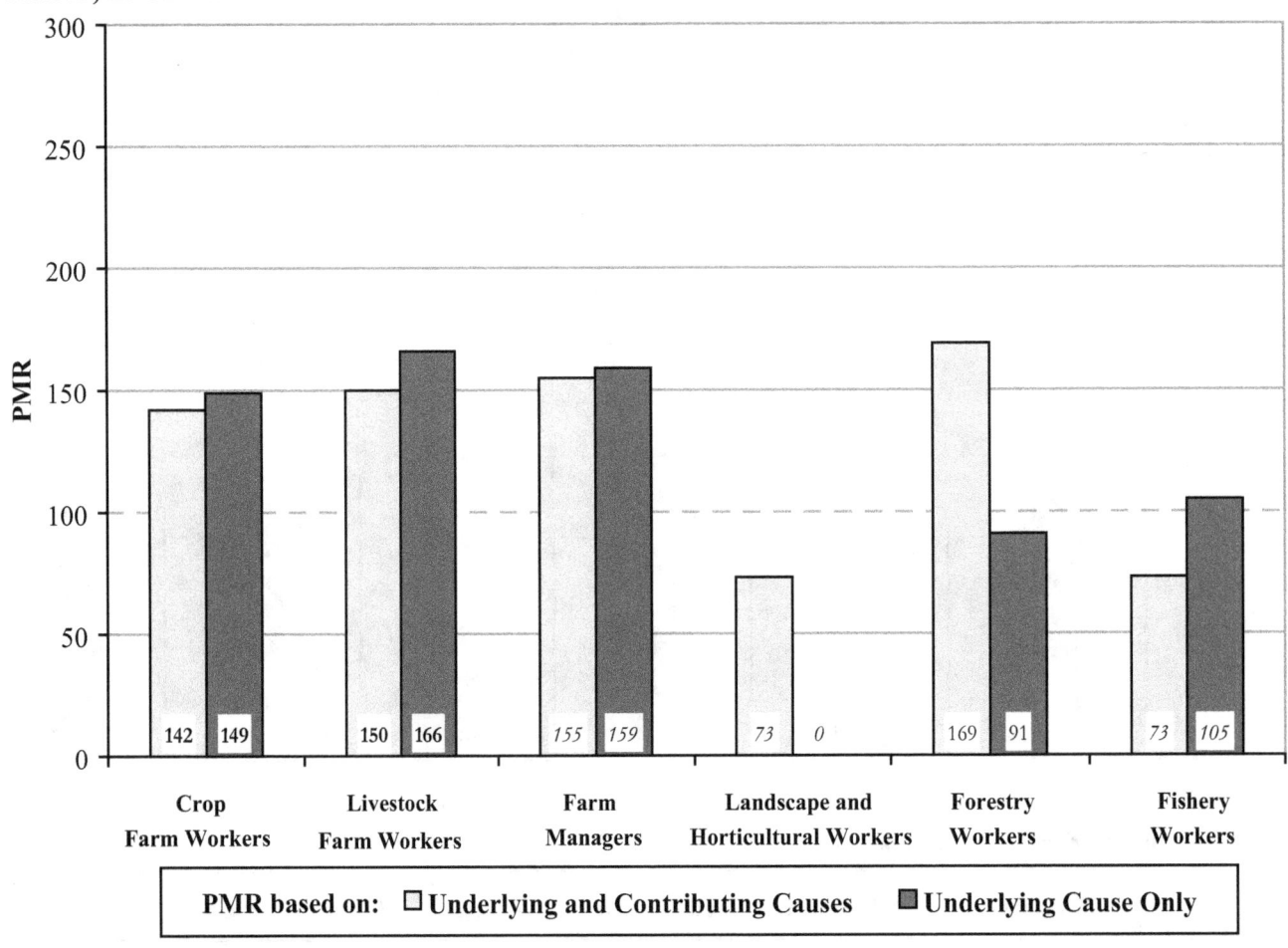

ICD - International Classification of Diseases, 9th Revision
NOTE: Influenza = ICD-9 code 487. PMRs in **bold** are significantly different from 100 (p<0.05). PMRs in *italics* are based on fewer than five observed deaths. PMRs are based on underlying and contributing cause of death. See appendices for source description, methods, ICD codes, and a list of selected states.
SOURCE: National Center for Health Statistics multiple-cause-of-death data

Table 2-73. Crop farm workers: Proportionate mortality ratio (PMR) adjusted for age, sex, and race/ethnicity for chronic obstructive pulmonary disease and allied conditions, U.S. residents age 15 and over, selected states, 1988–1998

Disease Category (ICD Code)	Number of Deaths	PMR	95% Confidence Interval	
			LCL	UCL
Bronchitis, not specified as acute or chronic (490)	269	**134**	119	151
Chronic bronchitis (491)	679	103	96	111
Emphysema (492)	3,265	**86**	83	89
Asthma (493)	813	111	104	119
Bronchiectasis (494)	139	90	76	106
Hypersensitivity pneumonitis (495)	23	**1,228**	777	1,844
Chronic airway obstruction, not elsewhere classified (496)	20,998	97	95	99

ICD - International Classification of Diseases, 9th Revision   LCL - lower confidence limit   UCL - upper confidence limit
NOTE: PMRs in **bold** are significantly different from 100 (p<0.05). PMRs in *italics* are based on fewer than five observed deaths. PMRs are based on underlying and contributing cause of death. Some values could not be calculated because the number of observed or expected deaths was zero; such values are indicated by ---. See appendices for source description, methods, ICD codes, and a list of selected states.
SOURCE: National Center for Health Statistics multiple-cause-of-death data

COPD Mortality within and by Agricultural Group

Table 2-74. Livestock farm workers: Proportionate mortality ratio (PMR) adjusted for age, sex, and race/ethnicity for chronic obstructive pulmonary disease and allied conditions, U.S. residents age 15 and over, selected states, 1988–1998

Disease Category (ICD Code)	Number of Deaths	PMR	95% Confidence Interval LCL	95% Confidence Interval UCL
Bronchitis, not specified as acute or chronic (490)	58	103	79	133
Chronic bronchitis (491)	161	88	76	103
Emphysema (492)	956	91	85	97
Asthma (493)	276	**150**	133	169
Bronchiectasis (494)	35	79	55	110
Hypersensitivity pneumonitis (495)	31	**5,563**	3,779	7,904
Chronic airway obstruction, not elsewhere classified (496)	5,439	**91**	89	93

ICD - International Classification of Diseases, 9th Revision   LCL - lower confidence limit   UCL - upper confidence limit
NOTE: PMRs in **bold** are significantly different from 100 (p<0.05). PMRs in *italics* are based on fewer than five observed deaths. PMRs are based on underlying and contributing cause of death. Some values could not be calculated because the number of observed or expected deaths was zero; such values are indicated by ---. See appendices for source description, methods, ICD codes, and a list of selected states.
SOURCE: National Center for Health Statistics multiple-cause-of-death data

COPD Mortality within and by Agricultural Group

Table 2-75. Farm managers: Proportionate mortality ratio (PMR) adjusted for age, sex, and race/ethnicity for chronic obstructive pulmonary disease and allied conditions, U.S. residents age 15 and over, selected states, 1988–1998

Disease Category (ICD Code)	Number of Deaths	PMR	95% Confidence Interval LCL	95% Confidence Interval UCL
Bronchitis, not specified as acute or chronic (490)	0	0	---	---
Chronic bronchitis (491)	11	105	53	188
Emphysema (492)	67	114	89	145
Asthma (493)	14	107	58	180
Bronchiectasis (494)	1	34	1	189
Hypersensitivity pneumonitis (495)	0	0	---	---
Chronic airway obstruction, not elsewhere classified (496)	314	97	87	108

ICD - International Classification of Diseases, 9th Revision   LCL - lower confidence limit   UCL - upper confidence limit
NOTE: PMRs in **bold** are significantly different from 100 (p<0.05). PMRs in *italics* are based on fewer than five observed deaths. PMRs are based on underlying and contributing cause of death. Some values could not be calculated because the number of observed or expected deaths was zero; such values are indicated by ---. See appendices for source description, methods, ICD codes, and a list of selected states.
SOURCE: National Center for Health Statistics multiple-cause-of-death data

Table 2-76. Landscape and horticultural workers: Proportionate mortality ratio (PMR) adjusted for age, sex, and race/ethnicity for chronic obstructive pulmonary disease and allied conditions, U.S. residents age 15 and over, selected states, 1988–1998

Disease Category (ICD Code)	Number of Deaths	PMR	95% Confidence Interval LCL	95% Confidence Interval UCL
Bronchitis, not specified as acute or chronic (490)	5	85	28	199
Chronic bronchitis (491)	23	124	78	186
Emphysema (492)	114	105	87	126
Asthma (493)	33	92	63	129
Bronchiectasis (494)	0	*0*	---	---
Hypersensitivity pneumonitis (495)	0	*0*	---	---
Chronic airway obstruction, not elsewhere classified (496)	624	**111**	103	120

ICD - International Classification of Diseases, 9th Revision   LCL - lower confidence limit   UCL - upper confidence limit
NOTE: PMRs in **bold** are significantly different from 100 (p<0.05). PMRs in *italics* are based on fewer than five observed deaths. PMRs are based on underlying and contributing cause of death. Some values could not be calculated because the number of observed or expected deaths was zero; such values are indicated by ---. See appendices for source description, methods, ICD codes, and a list of selected states.
SOURCE: National Center for Health Statistics multiple-cause-of-death data

Table 2-77. Forestry workers: Proportionate mortality ratio (PMR) adjusted for age, sex, and race/ethnicity for chronic obstructive pulmonary disease and allied conditions, U.S. residents age 15 and over, selected states, 1988–1998

Disease Category (ICD Code)	Number of Deaths	PMR	95% Confidence Interval LCL	95% Confidence Interval UCL
Bronchitis, not specified as acute or chronic (490)	8	62	27	122
Chronic bronchitis (491)	45	98	72	131
Emphysema (492)	293	106	94	119
Asthma (493)	70	112	88	142
Bronchiectasis (494)	12	115	59	201
Hypersensitivity pneumonitis (495)	0	0	---	---
Chronic airway obstruction, not elsewhere classified (496)	1,890	**127**	122	133

ICD - International Classification of Diseases, 9th Revision   LCL - lower confidence limit   UCL - upper confidence limit
NOTE: PMRs in **bold** are significantly different from 100 (p<0.05). PMRs in *italics* are based on fewer than five observed deaths. PMRs are based on underlying and contributing cause of death. Some values could not be calculated because the number of observed or expected deaths was zero; such values are indicated by ---. See appendices for source description, methods, ICD codes, and a list of selected states.
SOURCE: National Center for Health Statistics multiple-cause-of-death data

Table 2-78. Fishery workers: Proportionate mortality ratio (PMR) adjusted for age, sex, and race/ethnicity for chronic obstructive pulmonary disease and allied conditions, U.S. residents age 15 and over, selected states, 1988–1998

Disease Category (ICD Code)	Number of Deaths	PMR	95% Confidence Interval LCL	95% Confidence Interval UCL
Bronchitis, not specified as acute or chronic (490)	4	*111*	30	284
Chronic bronchitis (491)	13	103	55	176
Emphysema (492)	87	117	94	144
Asthma (493)	9	53	24	101
Bronchiectasis (494)	0	*0*	---	---
Hypersensitivity pneumonitis (495)	0	*0*	---	---
Chronic airway obstruction, not elsewhere classified (496)	455	**116**	106	127

ICD - International Classification of Diseases, 9th Revision   LCL - lower confidence limit   UCL - upper confidence limit
NOTE: PMRs in **bold** are significantly different from 100 (p<0.05). PMRs in *italics* are based on fewer than five observed deaths. PMRs are based on underlying and contributing cause of death. Some values could not be calculated because the number of observed or expected deaths was zero; such values are indicated by ---. See appendices for source description, methods, ICD codes, and a list of selected states.
SOURCE: National Center for Health Statistics multiple-cause-of-death data

Figure 2-55. Bronchitis, not specified as acute or chronic: Proportionate mortality ratio (PMR) adjusted for age, sex, and race/ethnicity by agricultural group, U.S. residents age 15 and over, selected states, 1988–1998

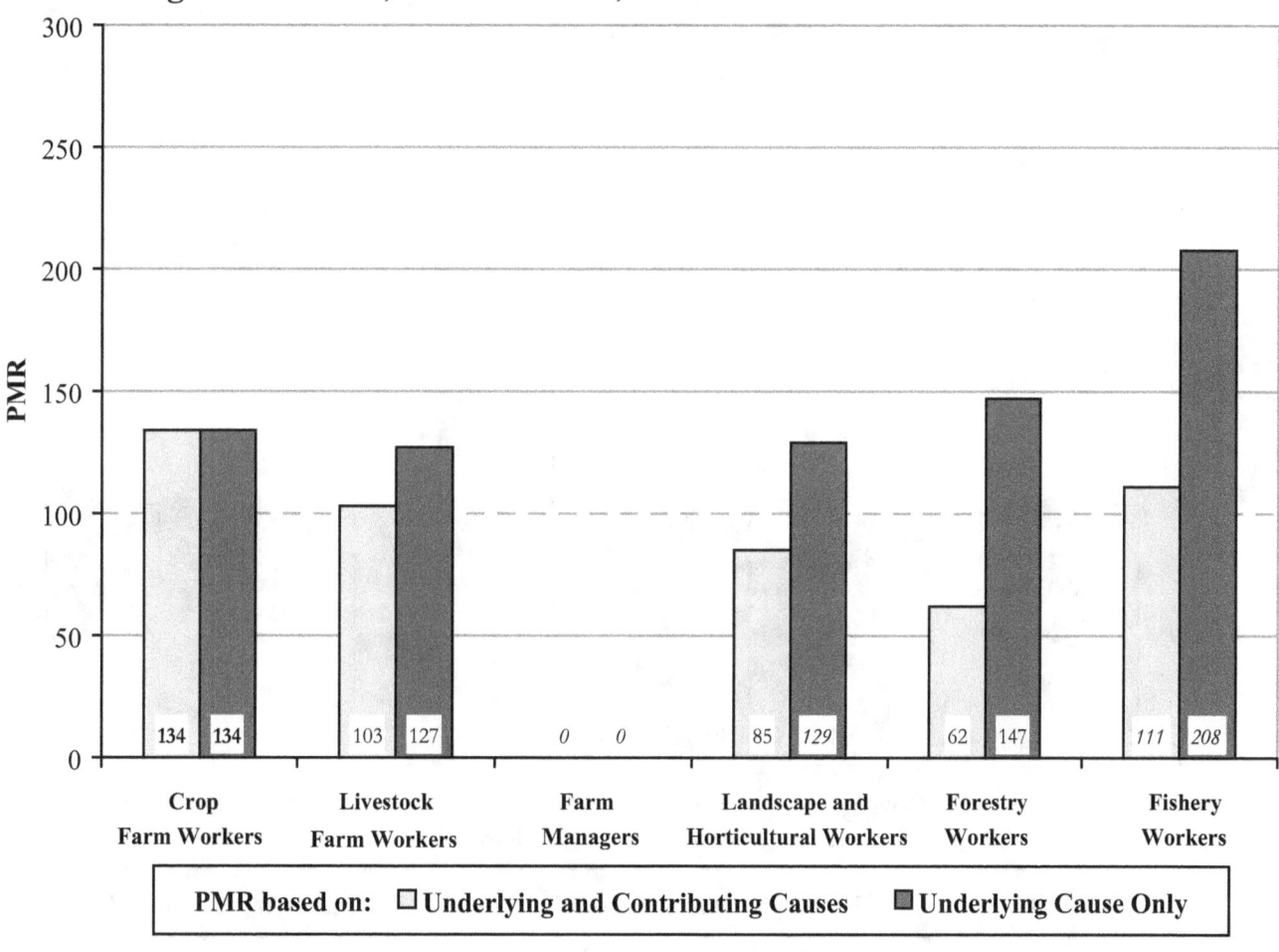

ICD - International Classification of Diseases, 9th Revision
NOTE: Bronchitis, not specified as acute or chronic = ICD-9 code 490. PMRs in **bold** are significantly different from 100 (p<0.05). PMRs in *italics* are based on fewer than five observed deaths. PMRs are based on underlying and contributing cause of death. See appendices for source description, methods, ICD codes, and a list of selected states.
SOURCE: National Center for Health Statistics multiple-cause-of-death data

COPD Mortality within and by Agricultural Group

Figure 2-56. Chronic bronchitis: Proportionate mortality ratio (PMR) adjusted for age, sex, and race/ethnicity by agricultural group, U.S. residents age 15 and over, selected states, 1988–1998

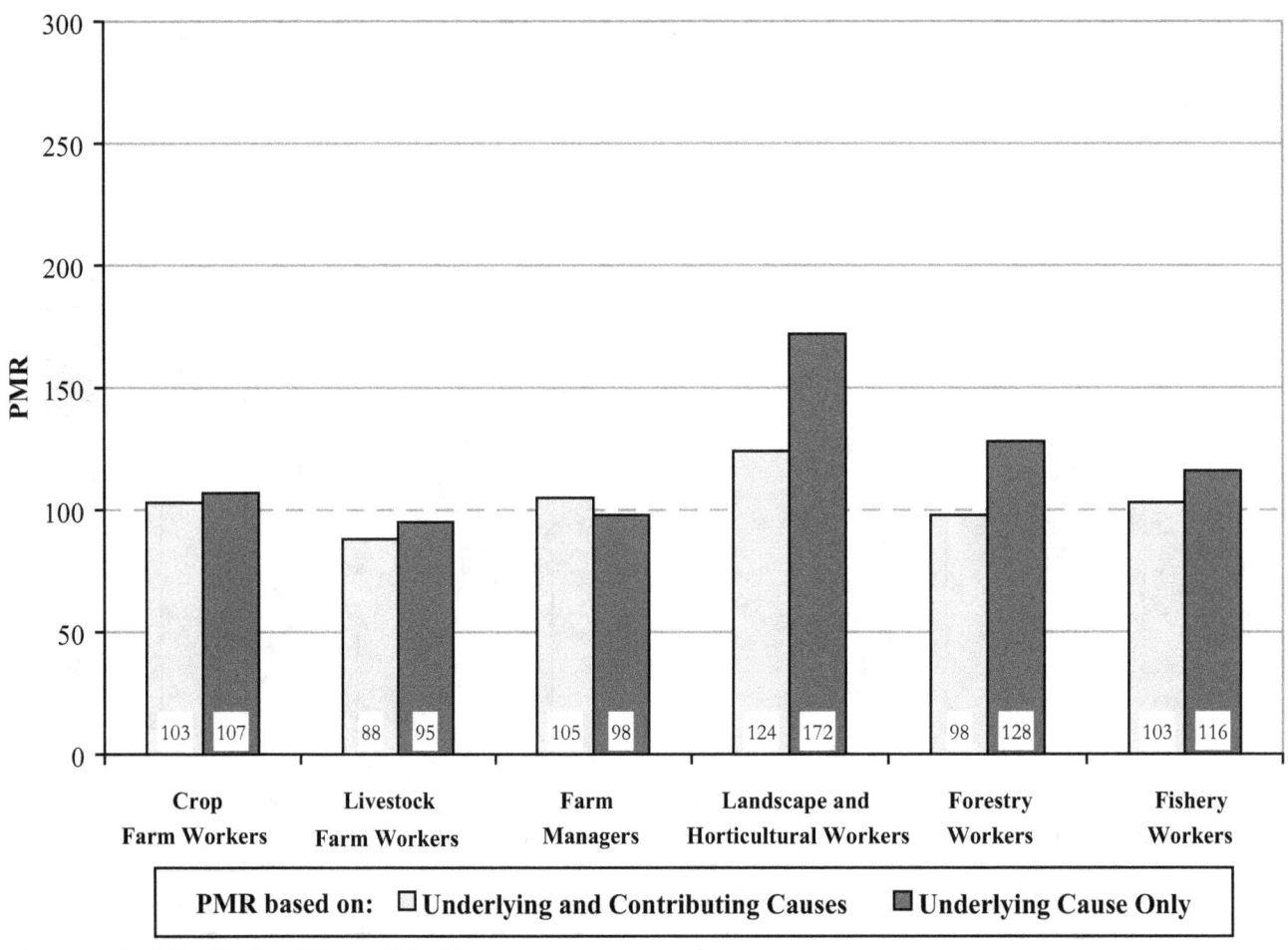

ICD - International Classification of Diseases, 9th Revision
NOTE: Chronic bronchitis = ICD-9 code 491. PMRs in **bold** are significantly different from 100 (p<0.05). PMRs in *italics* are based on fewer than five observed deaths. PMRs are based on underlying and contributing cause of death. See appendices for source description, methods, ICD codes, and a list of selected states.
SOURCE: National Center for Health Statistics multiple-cause-of-death data

Figure 2-57. Emphysema: Proportionate mortality ratio (PMR) adjusted for age, sex, and race/ethnicity by agricultural group, U.S. residents age 15 and over, selected states, 1988–1998

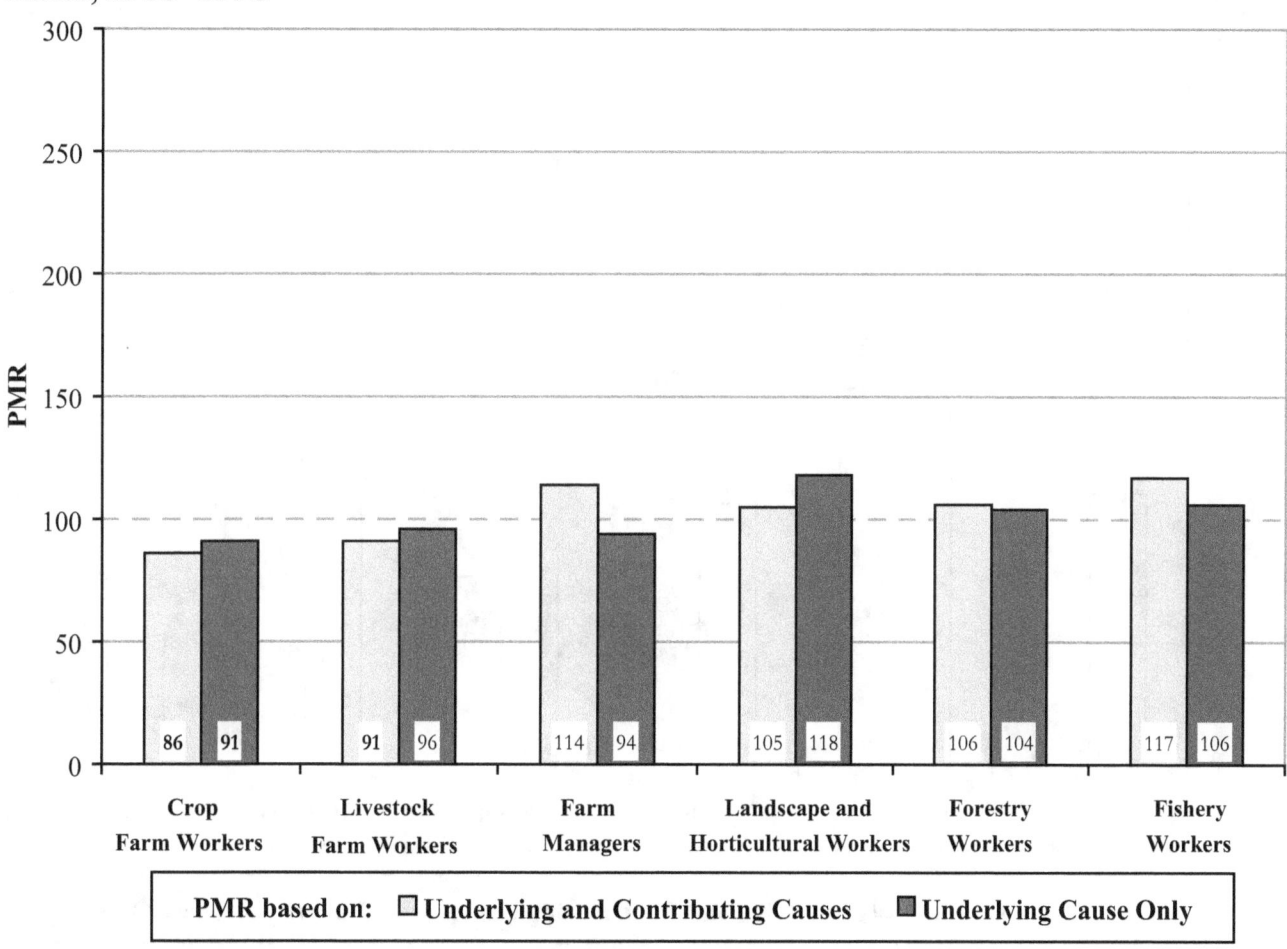

ICD - International Classification of Diseases, 9th Revision
NOTE: Emphysema = ICD-9 code 492. PMRs in **bold** are significantly different from 100 (p<0.05). PMRs in *italics* are based on fewer than five observed deaths. PMRs are based on underlying and contributing cause of death. See appendices for source description, methods, ICD codes, and a list of selected states.
SOURCE: National Center for Health Statistics multiple-cause-of-death data

COPD Mortality within and by Agricultural Group

Figure 2-58. Asthma: Proportionate mortality ratio (PMR) adjusted for age, sex, and race/ethnicity by agricultural group, U.S. residents age 15 and over, selected states, 1988–1998

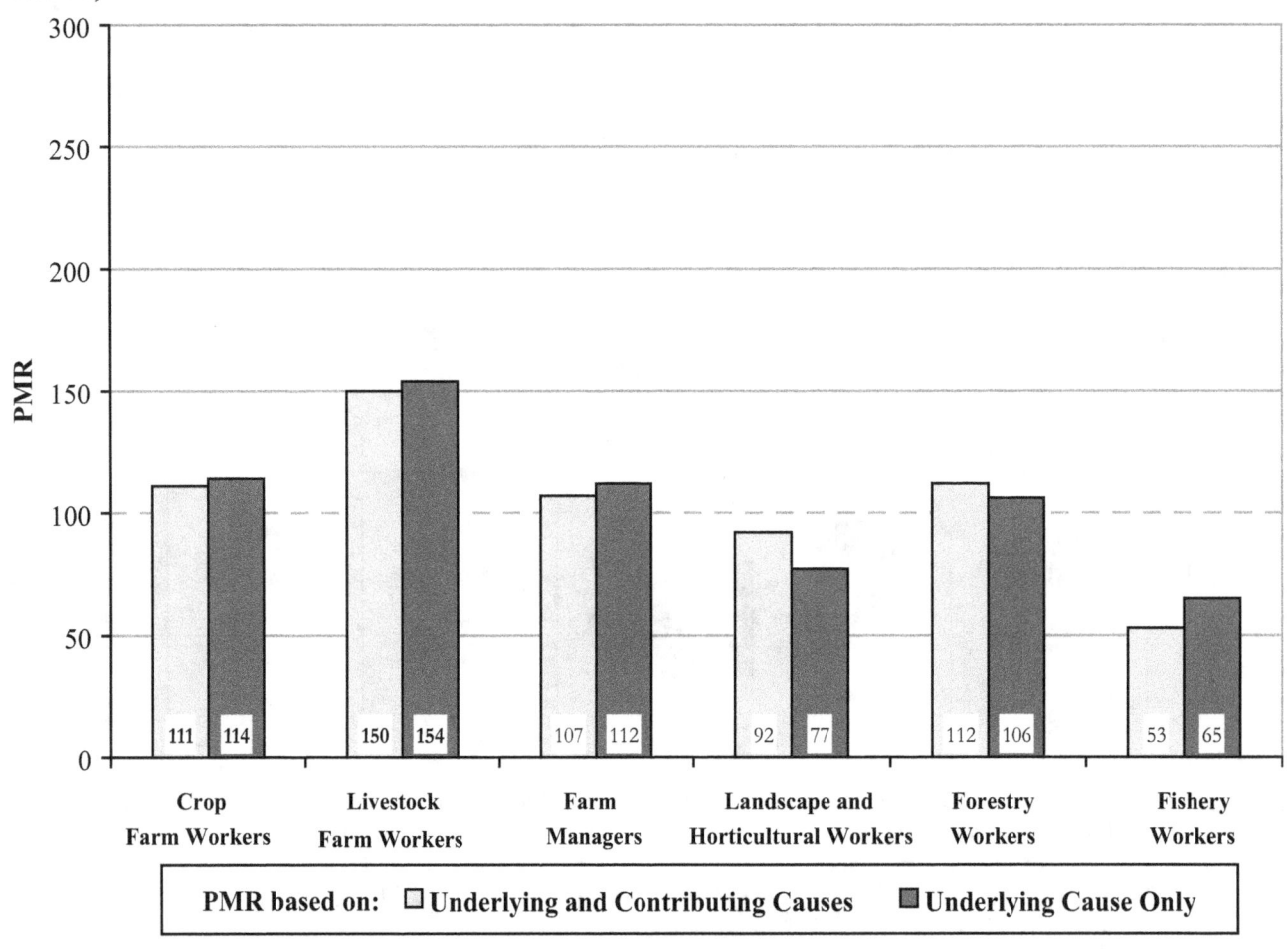

ICD - International Classification of Diseases, 9th Revision
NOTE: Asthma = ICD-9 code 493. PMRs in **bold** are significantly different from 100 (p<0.05). PMRs in *italics* are based on fewer than five observed deaths. PMRs are based on underlying and contributing cause of death. See appendices for source description, methods, ICD codes, and a list of selected states.
SOURCE: National Center for Health Statistics multiple-cause-of-death data

Figure 2-59. Bronchiectasis: Proportionate mortality ratio (PMR) adjusted for age, sex, and race/ethnicity by agricultural group, U.S. residents age 15 and over, selected states, 1988–1998

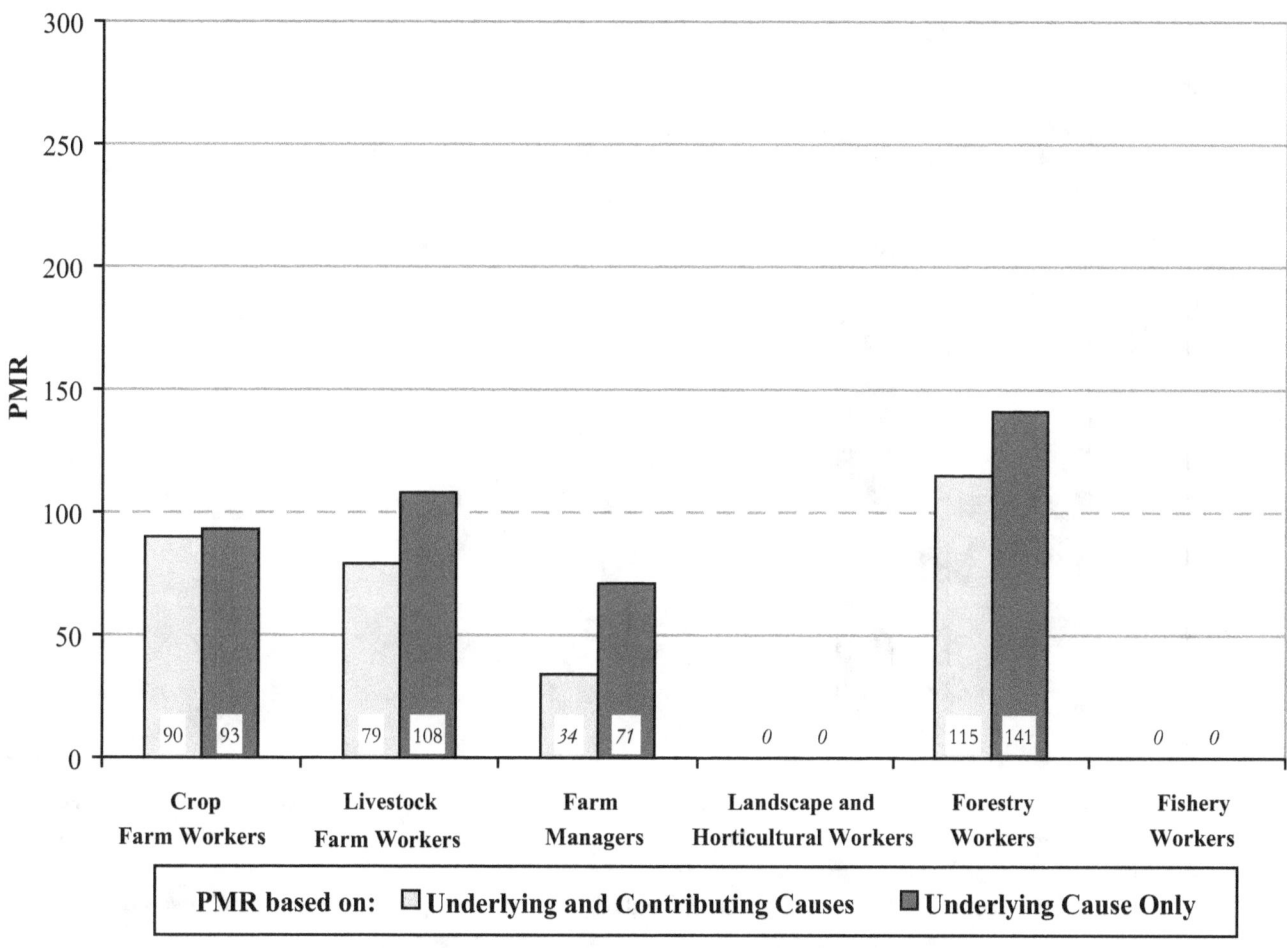

ICD - International Classification of Diseases, 9th Revision
NOTE: Bronchiectasis = ICD-9 code 494. PMRs in **bold** are significantly different from 100 (p<0.05). PMRs in *italics* are based on fewer than five observed deaths. PMRs are based on underlying and contributing cause of death. See appendices for source description, methods, ICD codes, and a list of selected states.
SOURCE: National Center for Health Statistics multiple-cause-of-death data

COPD Mortality within and by Agricultural Group

Figure 2-60. Hypersensitivity pneumonitis: Proportionate mortality ratio (PMR) adjusted for age, sex, and race/ethnicity by agricultural group, U.S. residents age 15 and over, selected states, 1988–1998

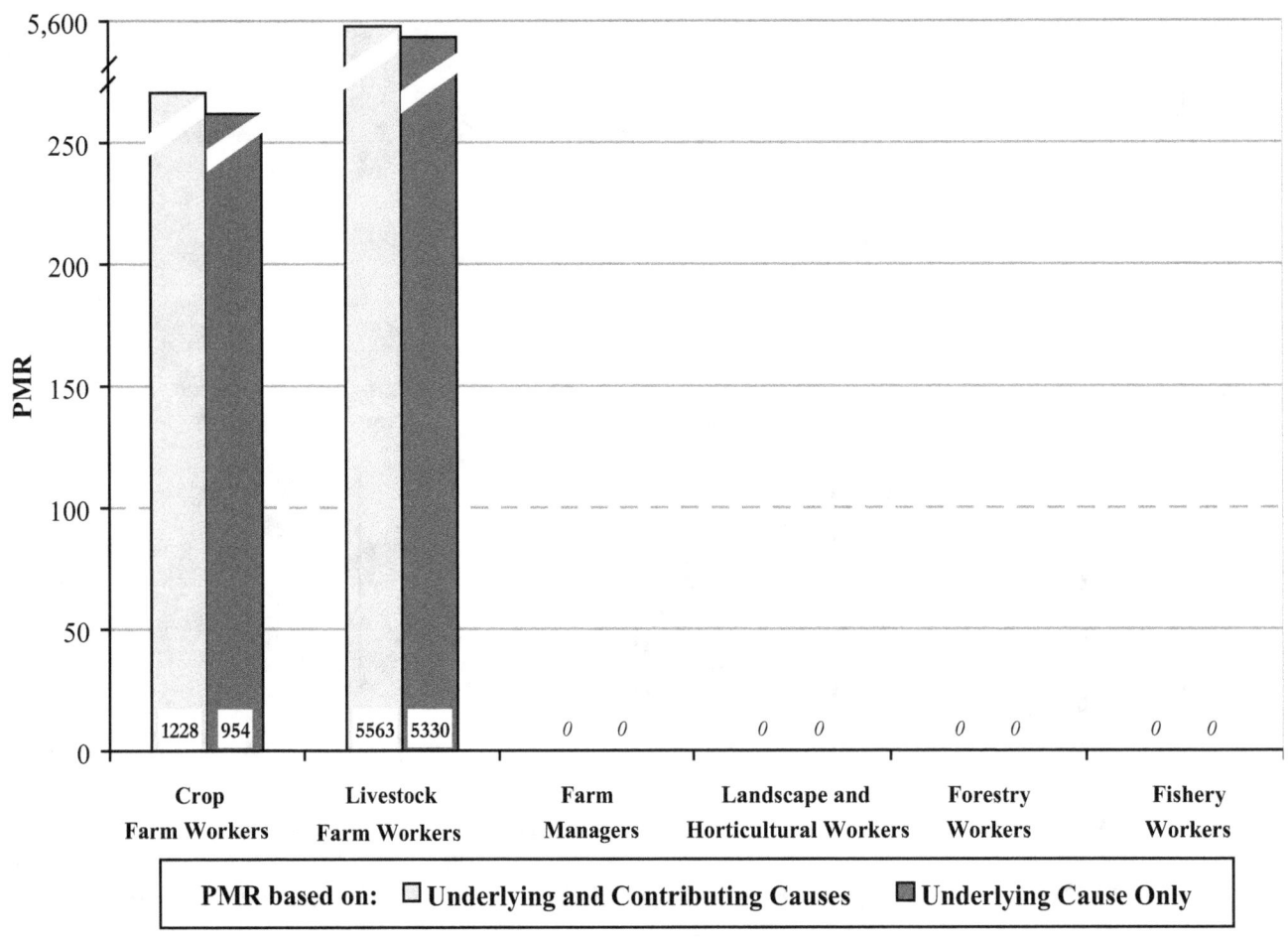

ICD - International Classification of Diseases, 9th Revision
NOTE: Hypersensitivity pneumonitis = ICD-9 code 495. PMRs in **bold** are significantly different from 100 (p<0.05). PMRs in *italics* are based on fewer than five observed deaths. PMRs are based on underlying and contributing cause of death. See appendices for source description, methods, ICD codes, and a list of selected states.
SOURCE: National Center for Health Statistics multiple-cause-of-death data

Figure 2-61. Chronic airway obstruction, not elsewhere classified: Proportionate mortality ratio (PMR) adjusted for age, sex, and race/ethnicity by agricultural group, U.S. residents age 15 and over, selected states, 1988–1998

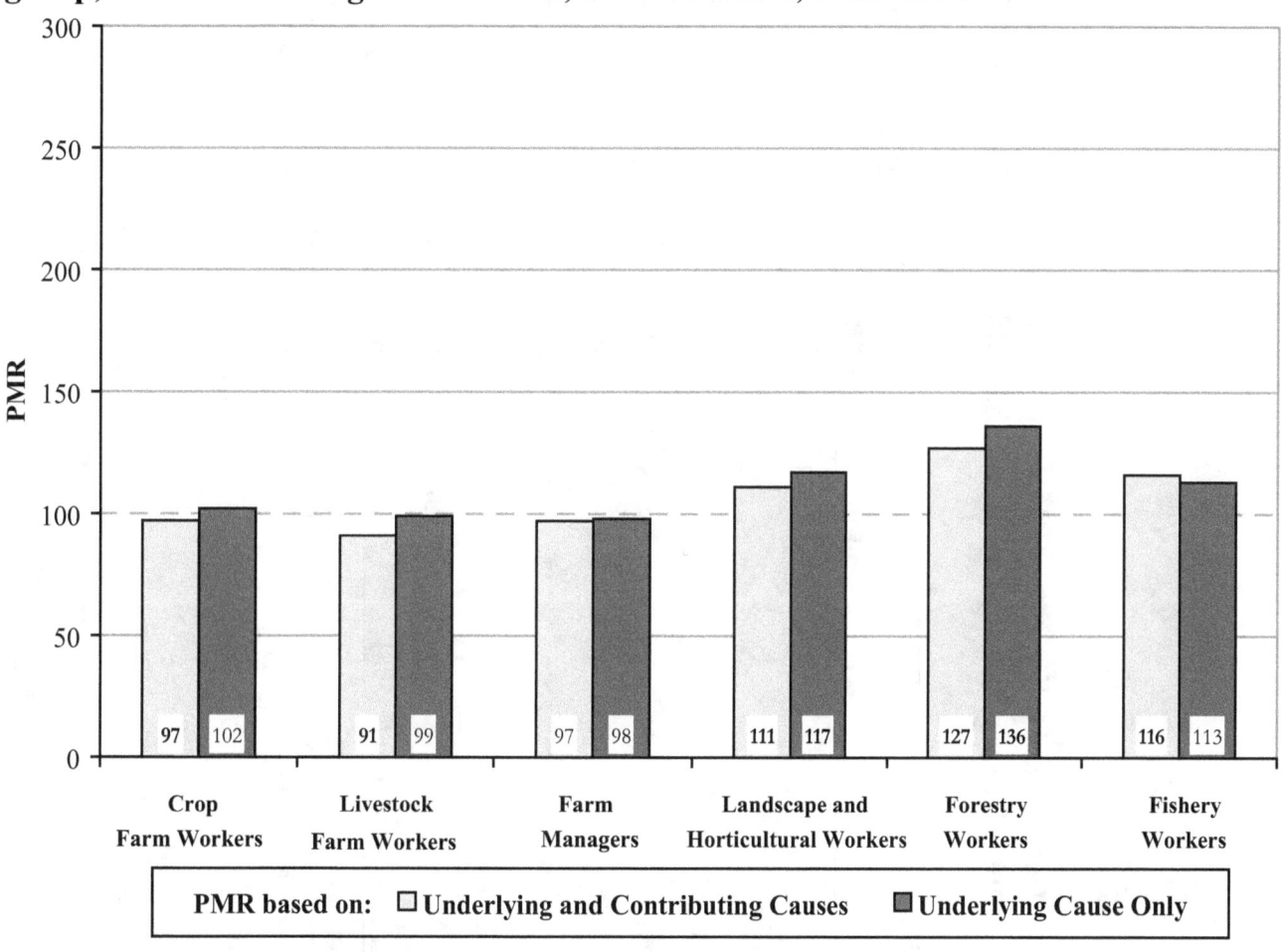

ICD - International Classification of Diseases, 9th Revision
NOTE: Chronic airway obstruction, not elsewhere classified = ICD-9 code 496. PMRs in **bold** are significantly different from 100 (p<0.05). PMRs in *italics* are based on fewer than five observed deaths. PMRs are based on underlying and contributing cause of death. See appendices for source description, methods, ICD codes, and a list of selected states.
SOURCE: National Center for Health Statistics multiple-cause-of-death data

Table 2-79. Crop farm workers: Proportionate mortality ratio (PMR) adjusted for age, sex, and race/ethnicity for pneumoconiosis and other lung diseases–external agents, U.S. residents age 15 and over, selected states, 1988–1998

Disease Category (ICD Code)	Number of Deaths	PMR	95% Confidence Interval LCL	95% Confidence Interval UCL
Coal workers' pneumoconiosis (500)	41	9	7	12
Asbestosis (501)	22	11	7	17
Pneumoconiosis due to other silica or silicates (502)	17	**22**	13	35
Pneumoconiosis due to other inorganic dust (503)	1	*63*	2	350
Pneumoconiosis due to inhalation of other dust (504)	1	*28*	1	156
Pneumoconiosis, unspecified (505)	19	**17**	10	27
Respiratory conditions due to chemical fumes and vapors (506)	0	*0*	---	---
Pneumonitis due to solids and liquids (507)	5,094	**95**	93	98
Respiratory conditions due to other and unspecified external agents (508)	29	66	44	95

ICD - International Classification of Diseases, 9th Revision   LCL - lower confidence limit   UCL - upper confidence limit
NOTE: PMRs in **bold** are significantly different from 100 (p<0.05). PMRs in *italics* are based on fewer than five observed deaths. PMRs are based on underlying and contributing cause of death. Some values could not be calculated because the number of observed or expected deaths was zero; such values are indicated by ---.   See appendices for source description, methods, ICD codes, and a list of selected states.
SOURCE: National Center for Health Statistics multiple-cause-of-death data

Pneumoconiosis and Other Lung Disease Mortality within and by Agricultural Group

Table 2-80. Livestock farm workers: Proportionate mortality ratio (PMR) adjusted for age, sex, and race/ethnicity for pneumoconiosis and other lung diseases–external agents, U.S. residents age 15 and over, selected states, 1988–1998

Disease Category (ICD Code)	Number of Deaths	PMR	95% Confidence Interval LCL	95% Confidence Interval UCL
Coal workers' pneumoconiosis (500)	7	6	2	12
Asbestosis (501)	9	**16**	7	30
Pneumoconiosis due to other silica or silicates (502)	6	**30**	11	65
Pneumoconiosis due to other inorganic dust (503)	0	0	---	---
Pneumoconiosis due to inhalation of other dust (504)	0	0	---	---
Pneumoconiosis, unspecified (505)	6	**20**	7	44
Respiratory conditions due to chemical fumes and vapors (506)	3	*364*	75	1,064
Pneumonitis due to solids and liquids (507)	1,346	95	90	100
Respiratory conditions due to other and unspecified external agents (508)	4	*33*	9	84

ICD - International Classification of Diseases, 9th Revision   LCL - lower confidence limit   UCL - upper confidence limit
NOTE: PMRs in **bold** are significantly different from 100 ($p<0.05$). PMRs in *italics* are based on fewer than five observed deaths. PMRs are based on underlying and contributing cause of death. Some values could not be calculated because the number of observed or expected deaths was zero; such values are indicated by ---. See appendices for source description, methods, ICD codes, and a list of selected states.
SOURCE: National Center for Health Statistics multiple-cause-of-death data

Pneumoconiosis and Other Lung Disease Mortality within and by Agricultural Group

Table 2-81. Farm managers: Proportionate mortality ratio (PMR) adjusted for age, sex, and race/ethnicity for pneumoconiosis and other lung diseases–external agents, U.S. residents age 15 and over, selected states, 1988–1998

Disease Category (ICD Code)	Number of Deaths	PMR	95% Confidence Interval	
			LCL	UCL
Coal workers' pneumoconiosis (500)	0	*0*	---	---
Asbestosis (501)	1	*33*	1	183
Pneumoconiosis due to other silica or silicates (502)	1	*84*	2	467
Pneumoconiosis due to other inorganic dust (503)	0	*0*	---	---
Pneumoconiosis due to inhalation of other dust (504)	1	*1,969*	50	10,939
Pneumoconiosis, unspecified (505)	0	*0*	---	---
Respiratory conditions due to chemical fumes and vapors (506)	0	*0*	---	---
Pneumonitis due to solids and liquids (507)	61	**76**	59	98
Respiratory conditions due to other and unspecified external agents (508)	1	*134*	3	744

ICD - International Classification of Diseases, 9th Revision   LCL - lower confidence limit   UCL - upper confidence limit
NOTE: PMRs in **bold** are significantly different from 100 (p<0.05). PMRs in *italics* are based on fewer than five observed deaths. PMRs are based on underlying and contributing cause of death. Some values could not be calculated because the number of observed or expected deaths was zero; such values are indicated by ---. See appendices for source description, methods, ICD codes, and a list of selected states.
SOURCE: National Center for Health Statistics multiple-cause-of-death data

Table 2-82. Landscape and horticultural workers: Proportionate mortality ratio (PMR) adjusted for age, sex, and race/ethnicity for pneumoconiosis and other lung diseases–external agents, U.S. residents age 15 and over, selected states, 1988–1998

Disease Category (ICD Code)	Number of Deaths	PMR	95% Confidence Interval LCL	95% Confidence Interval UCL
Coal workers' pneumoconiosis (500)	2	*20*	2	72
Asbestosis (501)	4	*79*	22	202
Pneumoconiosis due to other silica or silicates (502)	0	*0*	---	---
Pneumoconiosis due to other inorganic dust (503)	0	*0*	---	---
Pneumoconiosis due to inhalation of other dust (504)	0	*0*	---	---
Pneumoconiosis, unspecified (505)	1	*38*	1	211
Respiratory conditions due to chemical fumes and vapors (506)	0	*0*	---	---
Pneumonitis due to solids and liquids (507)	146	108	92	127
Respiratory conditions due to other and unspecified external agents (508)	1	*58*	1	322

ICD - International Classification of Diseases, 9th Revision   LCL - lower confidence limit   UCL - upper confidence limit
NOTE: PMRs in **bold** are significantly different from 100 ($p<0.05$). PMRs in *italics* are based on fewer than five observed deaths. PMRs are based on underlying and contributing cause of death. Some values could not be calculated because the number of observed or expected deaths was zero; such values are indicated by ---. See appendices for source description, methods, ICD codes, and a list of selected states.
SOURCE: National Center for Health Statistics multiple-cause-of-death data

Pneumoconiosis and Other Lung Disease Mortality within and by Agricultural Group

Table 2-83. Forestry workers: Proportionate mortality ratio (PMR) adjusted for age, sex, and race/ethnicity for pneumoconiosis and other lung diseases–external agents, U.S. residents age 15 and over, selected states, 1988–1998

Disease Category (ICD Code)	Number of Deaths	PMR	95% Confidence Interval LCL	95% Confidence Interval UCL
Coal workers' pneumoconiosis (500)	10	**34**	16	63
Asbestosis (501)	6	**42**	15	92
Pneumoconiosis due to other silica or silicates (502)	1	*17*	0	94
Pneumoconiosis due to other inorganic dust (503)	0	0	---	---
Pneumoconiosis due to inhalation of other dust (504)	0	0	---	---
Pneumoconiosis, unspecified (505)	6	81	30	176
Respiratory conditions due to chemical fumes and vapors (506)	0	0	---	---
Pneumonitis due to solids and liquids (507)	330	99	89	110
Respiratory conditions due to other and unspecified external agents (508)	1	25	1	139

ICD - International Classification of Diseases, 9th Revision   LCL - lower confidence limit   UCL - upper confidence limit
NOTE: PMRs in **bold** are significantly different from 100 (p<0.05). PMRs in *italics* are based on fewer than five observed deaths. PMRs are based on underlying and contributing cause of death. Some values could not be calculated because the number of observed or expected deaths was zero; such values are indicated by ---. See appendices for source description, methods, ICD codes, and a list of selected states.
SOURCE: National Center for Health Statistics multiple-cause-of-death data

Table 2-84. Fishery workers: Proportionate mortality ratio (PMR) adjusted for age, sex, and race/ethnicity for pneumoconiosis and other lung diseases–external agents, U.S. residents age 15 and over, selected states, 1988–1998

Disease Category (ICD Code)	Number of Deaths	PMR	95% Confidence Interval LCL	95% Confidence Interval UCL
Coal workers' pneumoconiosis (500)	0	0	---	---
Asbestosis (501)	4	*104*	28	266
Pneumoconiosis due to other silica or silicates (502)	0	0	---	---
Pneumoconiosis due to other inorganic dust (503)	0	0	---	---
Pneumoconiosis due to inhalation of other dust (504)	0	0	---	---
Pneumoconiosis, unspecified (505)	0	0	---	---
Respiratory conditions due to chemical fumes and vapors (506)	0	0	---	---
Pneumonitis due to solids and liquids (507)	92	104	85	128
Respiratory conditions due to other and unspecified external agents (508)	2	*188*	23	679

ICD - International Classification of Diseases, 9th Revision   LCL - lower confidence limit   UCL - upper confidence limit
NOTE: PMRs in **bold** are significantly different from 100 (p<0.05). PMRs in *italics* are based on fewer than five observed deaths. PMRs are based on underlying and contributing cause of death. Some values could not be calculated because the number of observed or expected deaths was zero; such values are indicated by ---. See appendices for source description, methods, ICD codes, and a list of selected states.
SOURCE: National Center for Health Statistics multiple-cause-of-death data

Pneumoconiosis and Other Lung Disease Mortality within and by Agricultural Group

Figure 2-62. Coal workers' pneumoconiosis: Proportionate mortality ratio (PMR) adjusted for age, sex, and race/ethnicity by agricultural group, U.S. residents age 15 and over, selected states, 1988–1998

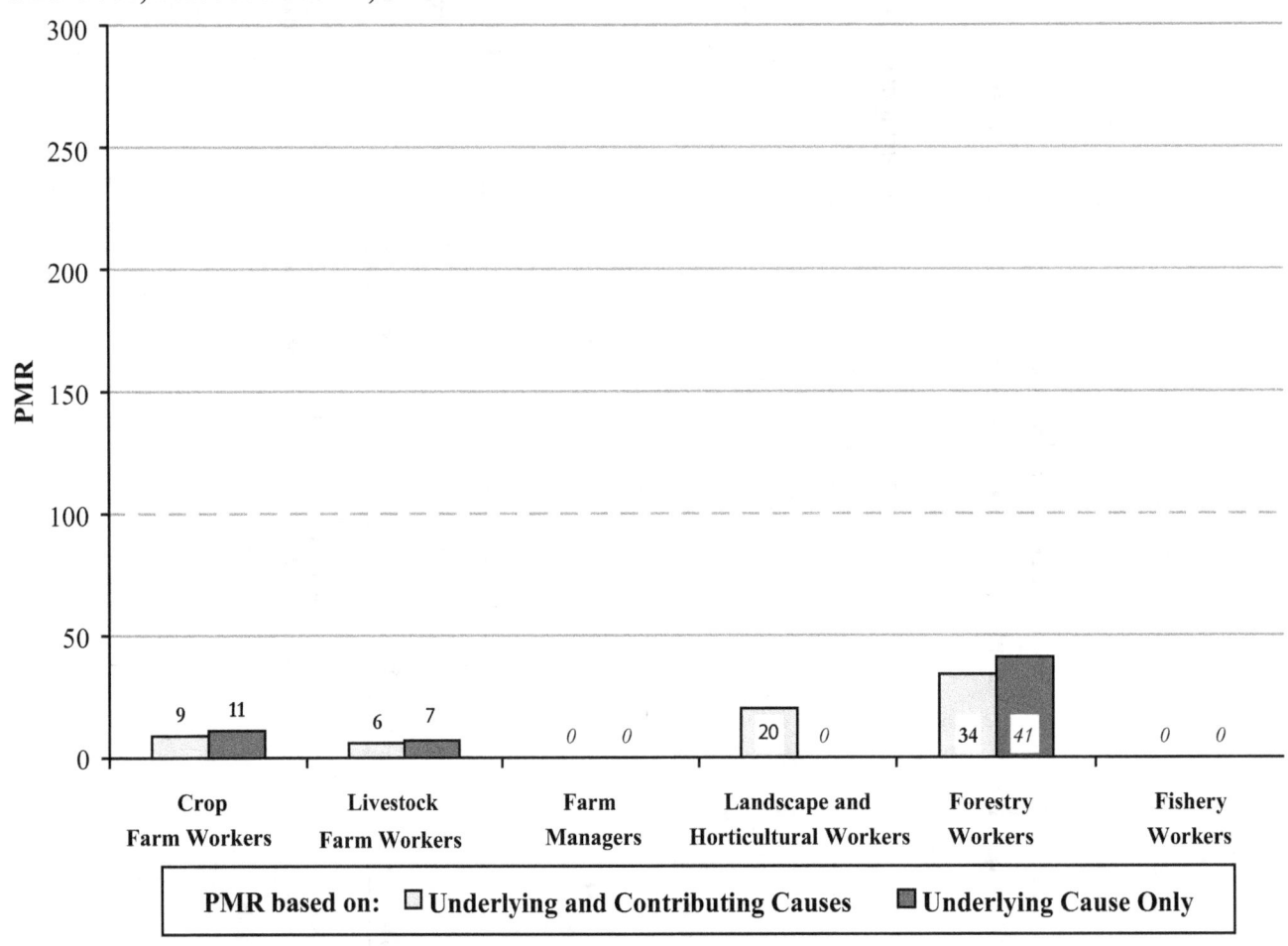

ICD - International Classification of Diseases, 9th Revision
NOTE: Coal workers' pneumoconiosis = ICD-9 code 500. PMRs in **bold** are significantly different from 100 ($p<0.05$). PMRs in *italics* are based on fewer than five observed deaths. PMRs are based on underlying and contributing cause of death. See appendices for source description, methods, ICD codes, and a list of selected states.
SOURCE: National Center for Health Statistics multiple-cause-of-death data

Figure 2-63. Asbestosis: Proportionate mortality ratio (PMR) adjusted for age, sex, and race/ethnicity by agricultural group, U.S. residents age 15 and over, selected states, 1988–1998

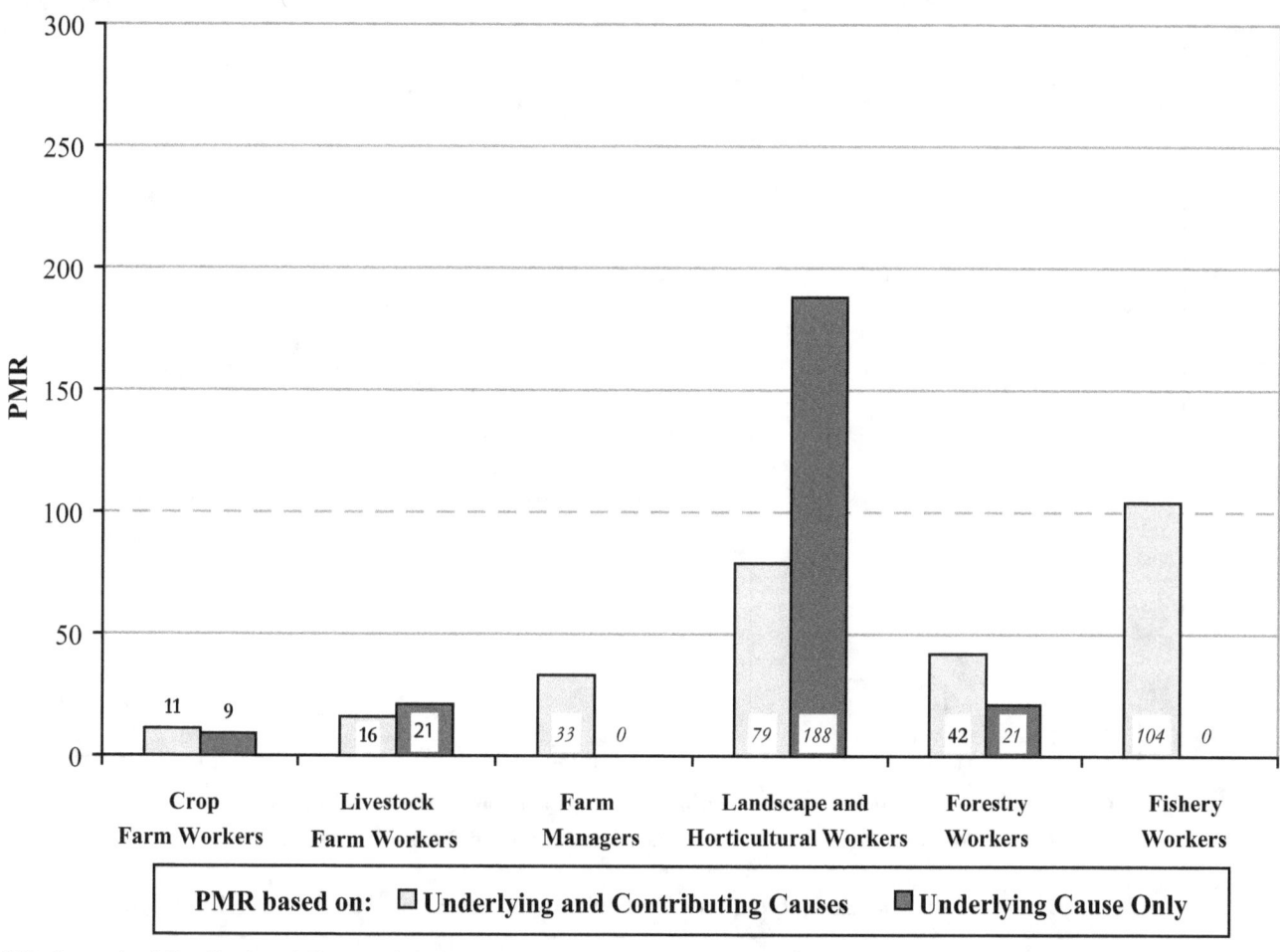

ICD - International Classification of Diseases, 9th Revision
NOTE: Asbestosis = ICD-9 code 501. PMRs in **bold** are significantly different from 100 (p<0.05). PMRs in *italics* are based on fewer than five observed deaths. PMRs are based on underlying and contributing cause of death. See appendices for source description, methods, ICD codes, and a list of selected states.
SOURCE: National Center for Health Statistics multiple-cause-of-death data

Pneumoconiosis and Other Lung Disease Mortality within and by Agricultural Group

Figure 2-64. Pneumoconiosis due to other silica or silicates: Proportionate mortality ratio (PMR) adjusted for age, sex, and race/ethnicity by agricultural group, U.S. residents age 15 and over, selected states, 1988–1998

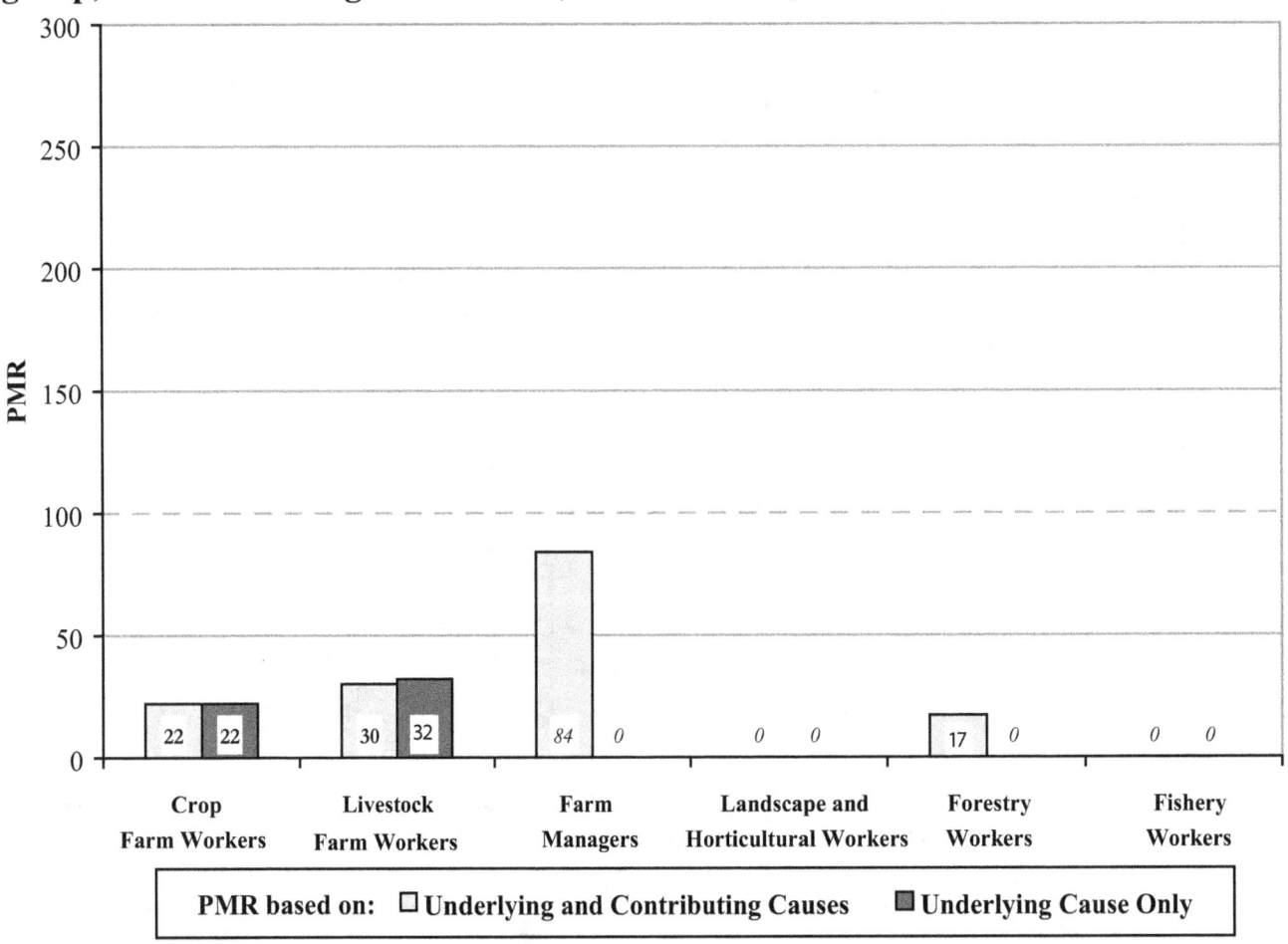

ICD - International Classification of Diseases, 9th Revision
NOTE: Pneumoconiosis due to other silica or silicates = ICD-9 code 502. PMRs in **bold** are significantly different from 100 (p<0.05). PMRs in *italics* are based on fewer than five observed deaths. PMRs are based on underlying and contributing cause of death. See appendices for source description, methods, ICD codes, and a list of selected states.
SOURCE: National Center for Health Statistics multiple-cause-of-death data

Figure 2-65. Pneumoconiosis, unspecified: Proportionate mortality ratio (PMR) adjusted for age, sex, and race/ethnicity by agricultural group, U.S. residents age 15 and over, selected states, 1988–1998

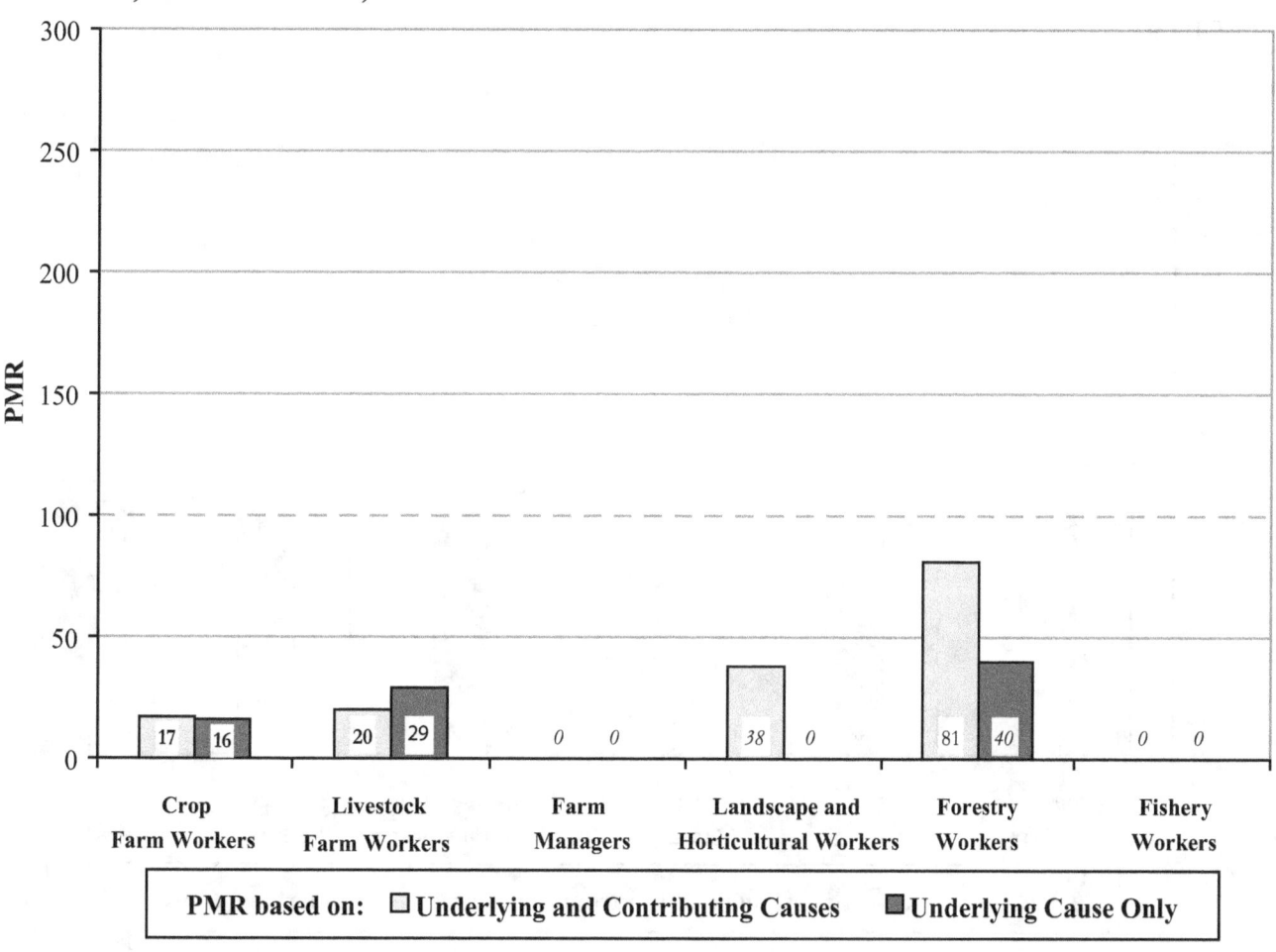

ICD - International Classification of Diseases, 9th Revision
NOTE: Pneumoconiosis, unspecified = ICD-9 code 505. PMRs in **bold** are significantly different from 100 (p<0.05). PMRs in *italics* are based on fewer than five observed deaths. PMRs are based on underlying and contributing cause of death. See appendices for source description, methods, ICD codes, and a list of selected states.
SOURCE: National Center for Health Statistics multiple-cause-of-death data

Pneumoconiosis and Other Lung Disease Mortality within and by Agricultural Group

Figure 2-66. Pneumonitis due to solids and liquids: Proportionate mortality ratio (PMR) adjusted for age, sex, and race/ethnicity by agricultural group, U.S. residents age 15 and over, selected states, 1988–1998

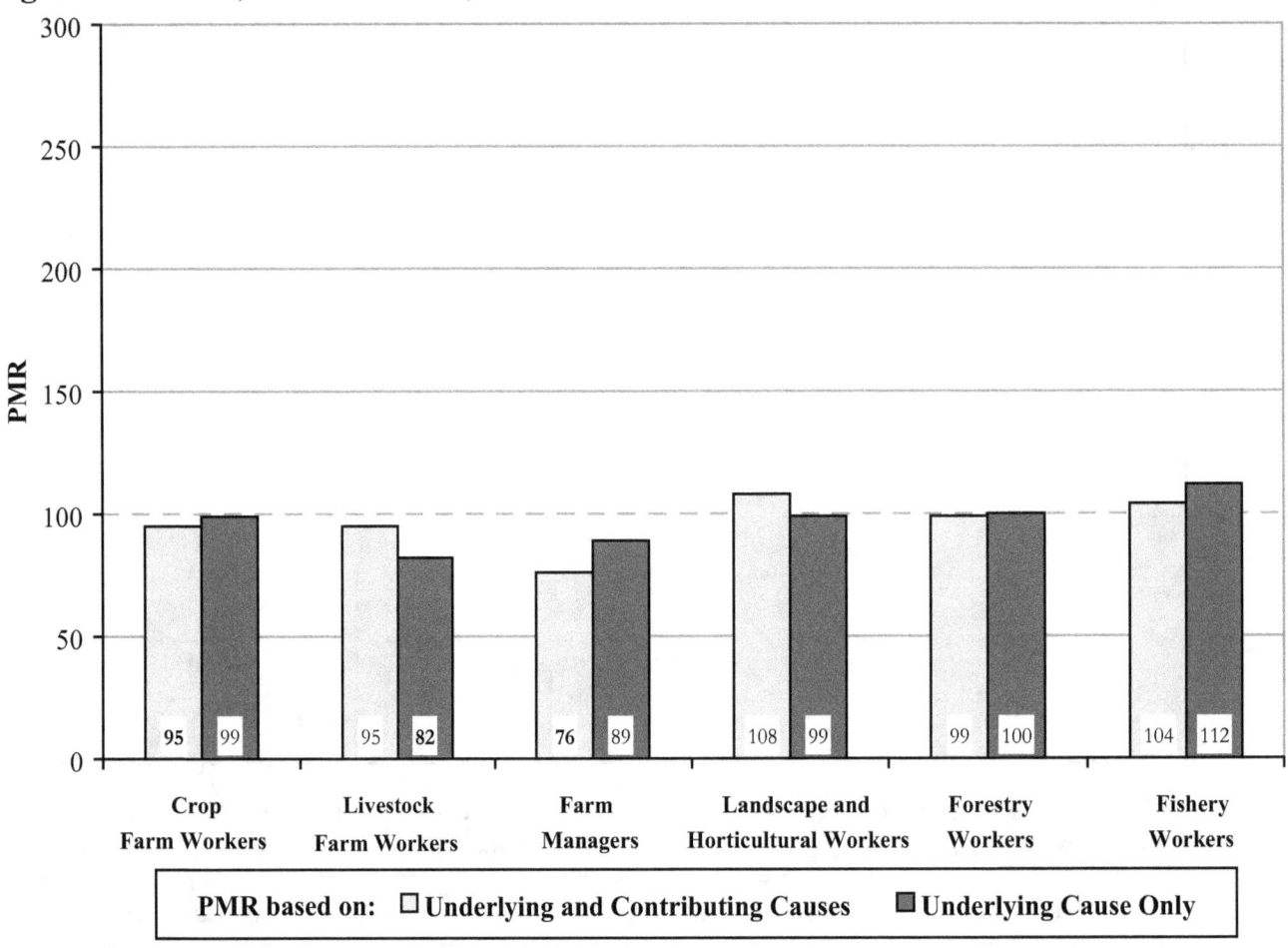

ICD - International Classification of Diseases, 9th Revision
NOTE: Pneumonitis due to solids and liquids = ICD-9 code 507. PMRs in **bold** are significantly different from 100 (p<0.05). PMRs in *italics* are based on fewer than five observed deaths. PMRs are based on underlying and contributing cause of death. See appendices for source description, methods, ICD codes, and a list of selected states.
SOURCE: National Center for Health Statistics multiple-cause-of-death data

Pneumoconiosis and Other Lung Disease Mortality within and by Agricultural Group

Figure 2-67. Respiratory conditions due to other and unspecified external agents: Proportionate mortality ratio (PMR) adjusted for age, sex, and race/ethnicity by agricultural group, U.S. residents age 15 and over, selected states, 1988–1998

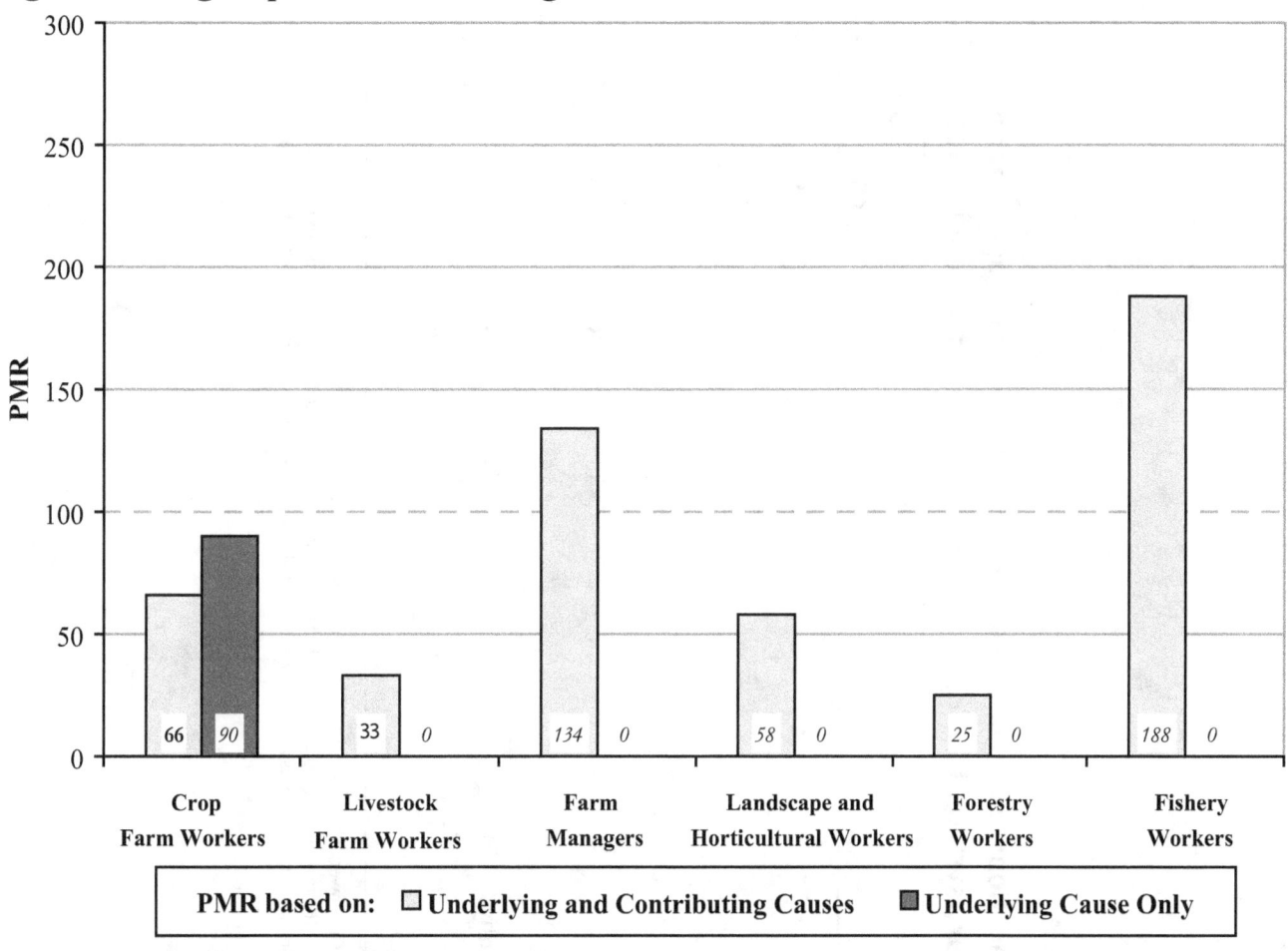

ICD - International Classification of Diseases, 9th Revision
NOTE: Respiratory conditions due to other and unspecified external agents = ICD-9 code 508. PMRs in **bold** are significantly different from 100 ($p<0.05$). PMRs in *italics* are based on fewer than five observed deaths. PMRs are based on underlying and contributing cause of death. See appendices for source description, methods, ICD codes, and a list of selected states.
SOURCE: National Center for Health Statistics multiple-cause-of-death data

Other Diseases of Respiratory System Mortality within and by Agricultural Group

Table 2-85. Crop farm workers: Proportionate mortality ratio (PMR) adjusted for age, sex, and race/ethnicity for other diseases of respiratory system, U.S. residents age 15 and over, selected states, 1988–1998

Disease Category (ICD Code)	Number of Deaths	PMR	95% Confidence Interval LCL	95% Confidence Interval UCL
Empyema (510)	134	90	76	107
Pleurisy (511)	952	**89**	84	95
Pneumothorax (512)	301	95	85	106
Abscess of lung and mediastinum (513)	153	**120**	102	141
Pulmonary congestion and hypostasis (514)	1,830	**113**	108	118
Postinflammatory pulmonary fibrosis (515)	1,165	**87**	82	92
Other alveolar and parietoalveolar pneumonopathy (516)	74	80	63	101
Other diseases of the lung (518)	2,701	**88**	85	91
Other diseases of respiratory system (519)	396	104	94	115

ICD - International Classification of Diseases, 9th Revision LCL - lower confidence limit UCL - upper confidence limit
NOTE: PMRs in **bold** are significantly different from 100 (p<0.05). PMRs in *italics* are based on fewer than five observed deaths. PMRs are based on underlying and contributing cause of death. Some values could not be calculated because the number of observed or expected deaths was zero; such values are indicated by ---. See appendices for source description, methods, ICD codes, and a list of selected states.
SOURCE: National Center for Health Statistics multiple-cause-of-death data

Other Diseases of Respiratory System Mortality within and by Agricultural Group

Table 2-86. Livestock farm workers: Proportionate mortality ratio (PMR) adjusted for age, sex, and race/ethnicity for other diseases of respiratory system, U.S. residents age 15 and over, selected states, 1988–1998

Disease Category (ICD Code)	Number of Deaths	PMR	95% Confidence Interval	
			LCL	UCL
Empyema (510)	47	119	88	158
Pleurisy (511)	203	73	64	84
Pneumothorax (512)	47	**58**	43	77
Abscess of lung and mediastinum (513)	29	93	62	134
Pulmonary congestion and hypostasis (514)	469	109	100	119
Postinflammatory pulmonary fibrosis (515)	322	**87**	78	97
Other alveolar and parietoalveolar pneumonopathy (516)	19	75	45	117
Other diseases of the lung (518)	676	**83**	77	89
Other diseases of respiratory system (519)	105	104	86	126

ICD - International Classification of Diseases, 9th Revision   LCL - lower confidence limit   UCL - upper confidence limit
NOTE: PMRs in **bold** are significantly different from 100 (p<0.05). PMRs in *italics* are based on fewer than five observed deaths. PMRs are based on underlying and contributing cause of death. Some values could not be calculated because the number of observed or expected deaths was zero; such values are indicated by ---. See appendices for source description, methods, ICD codes, and a list of selected states.
SOURCE: National Center for Health Statistics multiple-cause-of-death data

Other Diseases of Respiratory System Mortality within and by Agricultural Group

Table 2-87. Farm managers: Proportionate mortality ratio (PMR) adjusted for age, sex, and race/ethnicity for other diseases of respiratory system, U.S. residents age 15 and over, selected states, 1988–1998

Disease Category (ICD Code)	Number of Deaths	PMR	95% Confidence Interval LCL	95% Confidence Interval UCL
Empyema (510)	2	*80*	10	289
Pleurisy (511)	15	*90*	50	149
Pneumothorax (512)	8	155	67	305
Abscess of lung and mediastinum (513)	1	*50*	1	278
Pulmonary congestion and hypostasis (514)	20	80	49	124
Postinflammatory pulmonary fibrosis (515)	22	97	61	147
Other alveolar and parietoalveolar pneumonopathy (516)	1	*62*	2	344
Other diseases of the lung (518)	46	91	67	121
Other diseases of respiratory system (519)	2	*32*	4	116

ICD - International Classification of Diseases, 9th Revision   LCL - lower confidence limit   UCL - upper confidence limit
NOTE: PMRs in **bold** are significantly different from 100 (p<0.05). PMRs in *italics* are based on fewer than five observed deaths. PMRs are based on underlying and contributing cause of death. Some values could not be calculated because the number of observed or expected deaths was zero; such values are indicated by ---. See appendices for source description, methods, ICD codes, and a list of selected states.
SOURCE: National Center for Health Statistics multiple-cause-of-death data

Other Diseases of Respiratory System Mortality within and by Agricultural Group

Table 2-88. Landscape and horticultural workers: Proportionate mortality ratio (PMR) adjusted for age, sex, and race/ethnicity for other diseases of respiratory system, U.S. residents age 15 and over, selected states, 1988–1998

Disease Category (ICD Code)	Number of Deaths	PMR	95% Confidence Interval LCL	95% Confidence Interval UCL
Empyema (510)	9	126	58	239
Pleurisy (511)	30	86	58	123
Pneumothorax (512)	10	72	35	132
Abscess of lung and mediastinum (513)	13	**190**	101	325
Pulmonary congestion and hypostasis (514)	48	85	63	113
Postinflammatory pulmonary fibrosis (515)	31	79	54	112
Other alveolar and parietoalveolar pneumonopathy (516)	2	*43*	5	155
Other diseases of the lung (518)	98	**80**	65	98
Other diseases of respiratory system (519)	11	71	36	127

ICD - International Classification of Diseases, 9th Revision   LCL - lower confidence limit   UCL - upper confidence limit
NOTE: PMRs in **bold** are significantly different from 100 (p<0.05). PMRs in *italics* are based on fewer than five observed deaths. PMRs are based on underlying and contributing cause of death. Some values could not be calculated because the number of observed or expected deaths was zero; such values are indicated by ---. See appendices for source description, methods, ICD codes, and a list of selected states.
SOURCE: National Center for Health Statistics multiple-cause-of-death data

Other Diseases of Respiratory System Mortality within and by Agricultural Group

Table 2-89. Forestry workers: Proportionate mortality ratio (PMR) adjusted for age, sex, and race/ethnicity for other diseases of respiratory system, U.S. residents age 15 and over, selected states, 1988–1998

Disease Category (ICD Code)	Number of Deaths	PMR	95% Confidence Interval	
			LCL	UCL
Empyema (510)	9	65	30	123
Pleurisy (511)	57	**74**	56	96
Pneumothorax (512)	24	89	57	132
Abscess of lung and mediastinum (513)	13	99	53	169
Pulmonary congestion and hypostasis (514)	118	100	83	120
Postinflammatory pulmonary fibrosis (515)	64	**69**	54	88
Other alveolar and parietoalveolar pneumonopathy (516)	5	62	20	145
Other diseases of the lung (518)	223	91	80	104
Other diseases of respiratory system (519)	36	117	84	162

ICD - International Classification of Diseases, 9th Revision LCL - lower confidence limit   UCL - upper confidence limit
NOTE: PMRs in **bold** are significantly different from 100 (p<0.05). PMRs in *italics* are based on fewer than five observed deaths. PMRs are based on underlying and contributing cause of death. Some values could not be calculated because the number of observed or expected deaths was zero; such values are indicated by ---. See appendices for source description, methods, ICD codes, and a list of selected states.
SOURCE: National Center for Health Statistics multiple-cause-of-death data

Table 2-90. Fishery workers: Proportionate mortality ratio (PMR) adjusted for age, sex, and race/ethnicity for other diseases of respiratory system, U.S. residents age 15 and over, selected states, 1988-1998

Disease Category (ICD Code)	Number of Deaths	PMR	95% Confidence Interval LCL	95% Confidence Interval UCL
Empyema (510)	3	*79*	16	231
Pleurisy (511)	13	64	34	109
Pneumothorax (512)	6	84	31	183
Abscess of lung and mediastinum (513)	6	**183**	67	399
Pulmonary congestion and hypostasis (514)	32	101	69	103
Postinflammatory pulmonary fibrosis (515)	24	94	60	140
Other alveolar and parietoalveolar pneumonopathy (516)	2	*87*	11	314
Other diseases of the lung (518)	60	89	68	115
Other diseases of respiratory system (519)	4	48	13	123

ICD - International Classification of Diseases, 9th Revision   LCL - lower confidence limit   UCL - upper confidence limit
NOTE: PMRs in **bold** are significantly different from 100 (p<0.05). PMRs in *italics* are based on fewer than five observed deaths. PMRs are based on underlying and contributing cause of death. Some values could not be calculated because the number of observed or expected deaths was zero, such values are indicated by ---. See appendices for source description, methods, ICD codes, and a list of selected states.
SOURCE: National Center for Health Statistics multiple-cause-of-death data

Other Diseases of Respiratory System Mortality within and by Agricultural Group

Figure 2-68. Empyema: Proportionate mortality ratio (PMR) adjusted for age, sex, and race/ethnicity by agricultural group, U.S. residents age 15 and over, selected states, 1988–1998

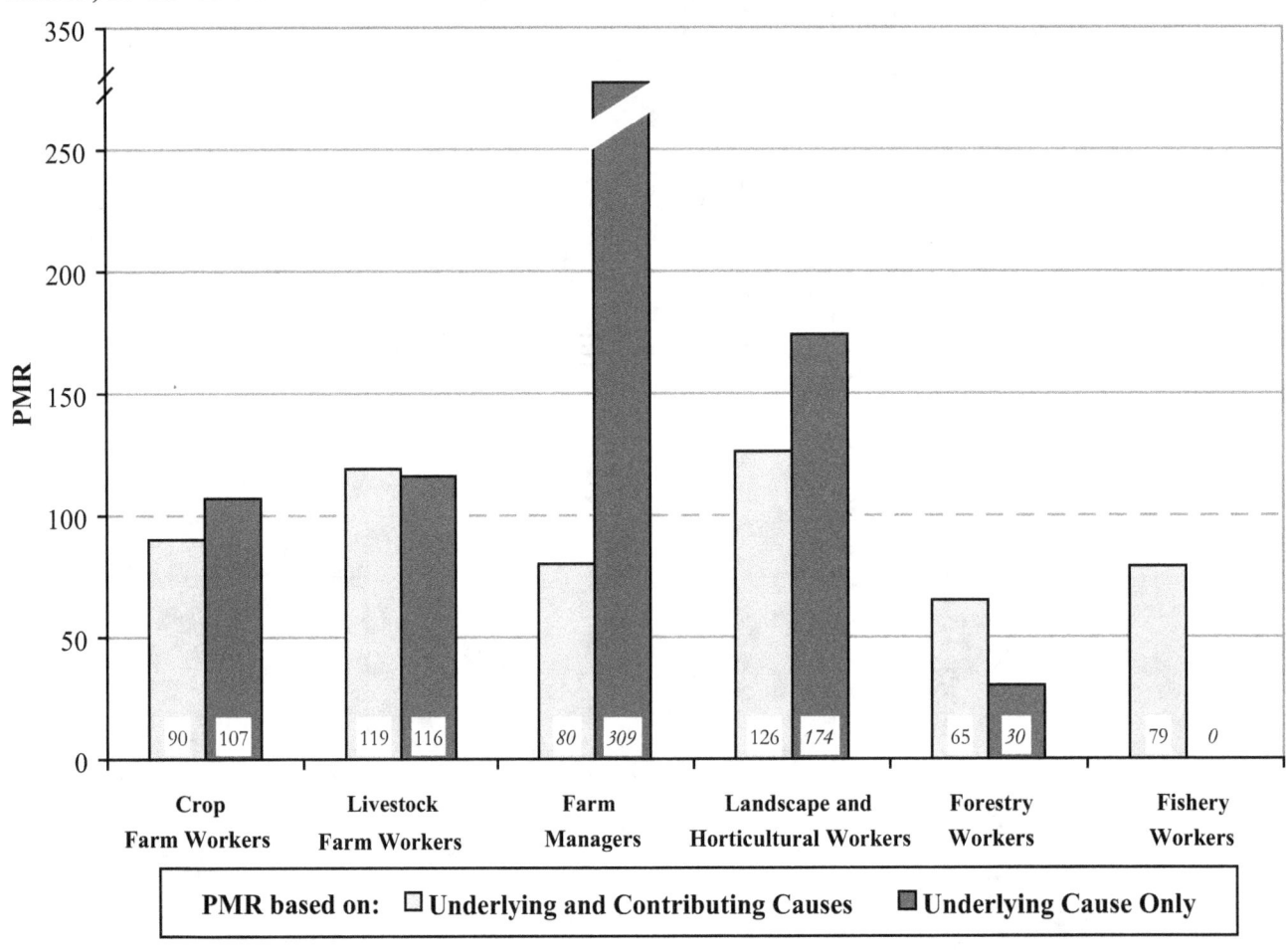

ICD - International Classification of Diseases, 9th Revision
NOTE: Empyema = ICD-9 code 510. PMRs in **bold** are significantly different from 100 (p<0.05). PMRs in *italics* are based on fewer than five observed deaths. PMRs are based on underlying and contributing cause of death. See appendices for source description, methods, ICD codes, and a list of selected states.
SOURCE: National Center for Health Statistics multiple-cause-of-death data

Other Diseases of Respiratory System Mortality within and by Agricultural Group

Figure 2-69. Pleurisy: Proportionate mortality ratio (PMR) adjusted for age, sex, and race/ethnicity by agricultural group, U.S. residents age 15 and over, selected states, 1988–1998

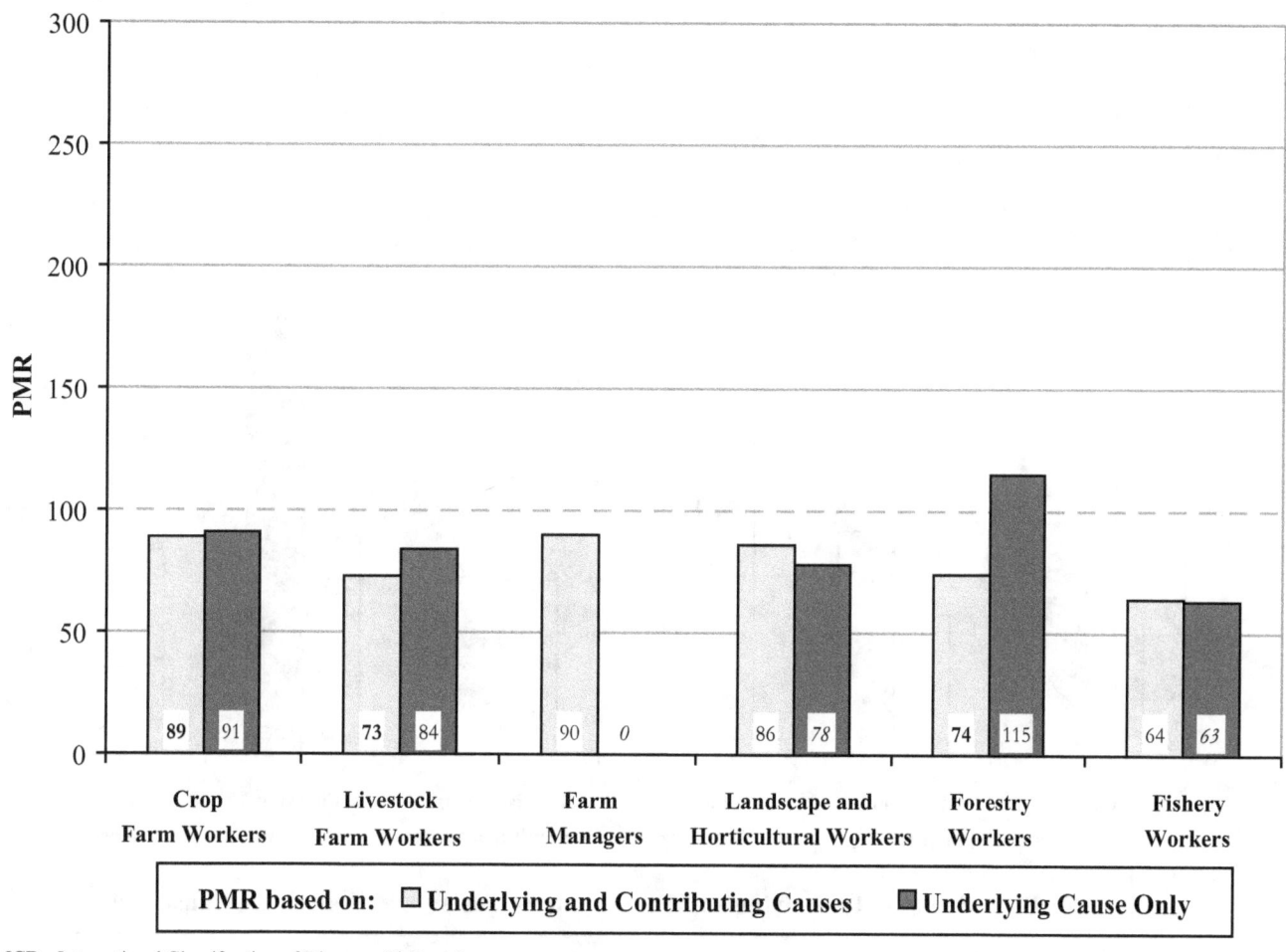

ICD - International Classification of Diseases, 9th Revision
NOTE: Pleurisy = ICD-9 code 511. PMRs in **bold** are significantly different from 100 (p<0.05). PMRs in *italics* are based on fewer than five observed deaths. PMRs are based on underlying and contributing cause of death. See appendices for source description, methods, ICD codes, and a list of selected states.
SOURCE: National Center for Health Statistics multiple-cause-of-death data

Other Diseases of Respiratory System Mortality within and by Agricultural Group

Figure 2-70. Pneumothorax: Proportionate mortality ratio (PMR) adjusted for age, sex, and race/ethnicity by agricultural group, U.S. residents age 15 and over, selected states, 1988–1998

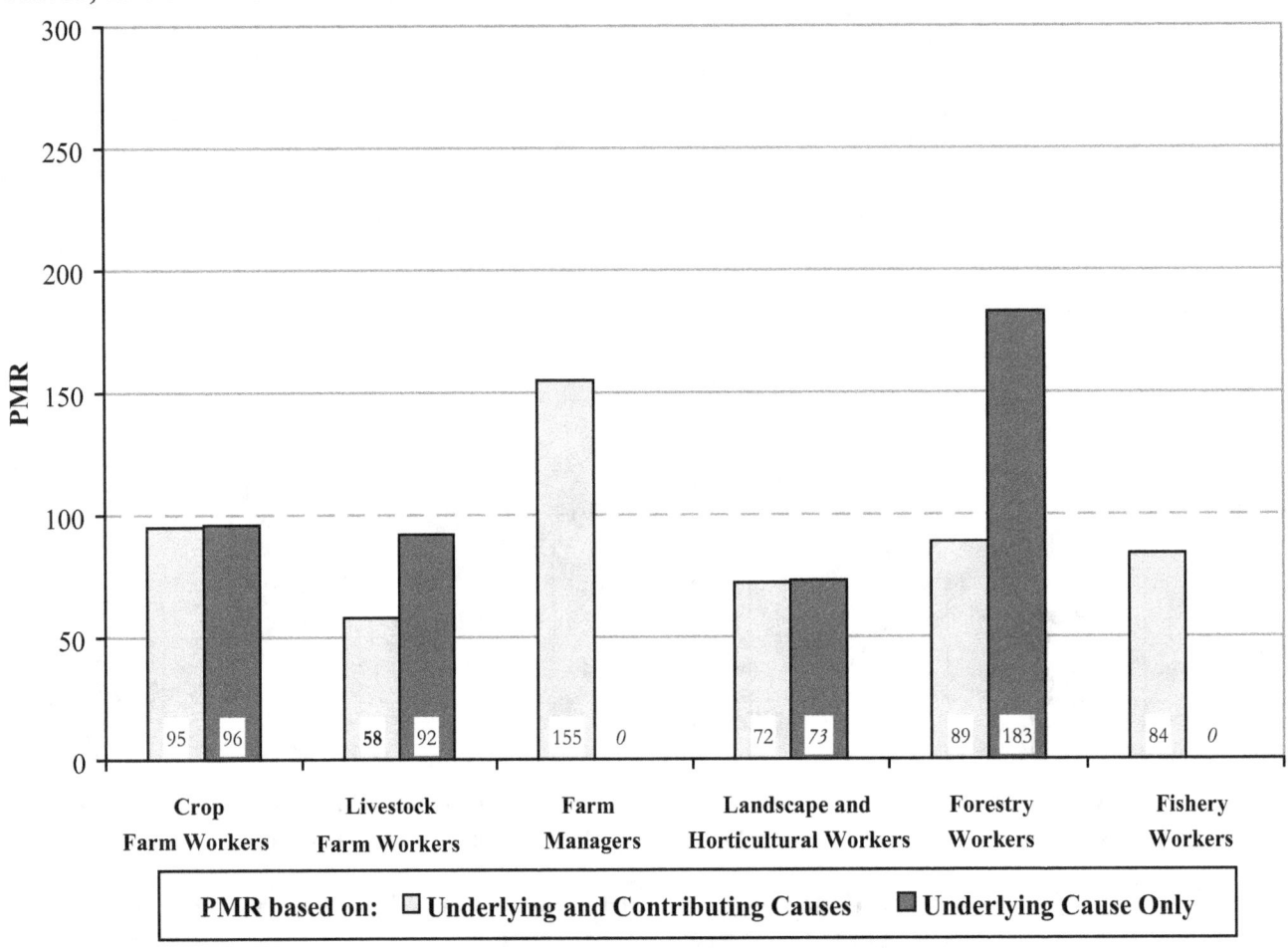

ICD - International Classification of Diseases, 9th Revision
NOTE: Pneumothorax = ICD-9 code 512. PMRs in **bold** are significantly different from 100 (p<0.05). PMRs in *italics* are based on fewer than five observed deaths. PMRs are based on underlying and contributing cause of death. See appendices for source description, methods, ICD codes, and a list of selected states.
SOURCE: National Center for Health Statistics multiple-cause-of-death data

Other Diseases of Respiratory System Mortality within and by Agricultural Group

Figure 2-71. Abscess of lung and mediastinum: Proportionate mortality ratio (PMR) adjusted for age, sex, and race/ethnicity by agricultural group, U.S. residents age 15 and over, selected states, 1988–1998

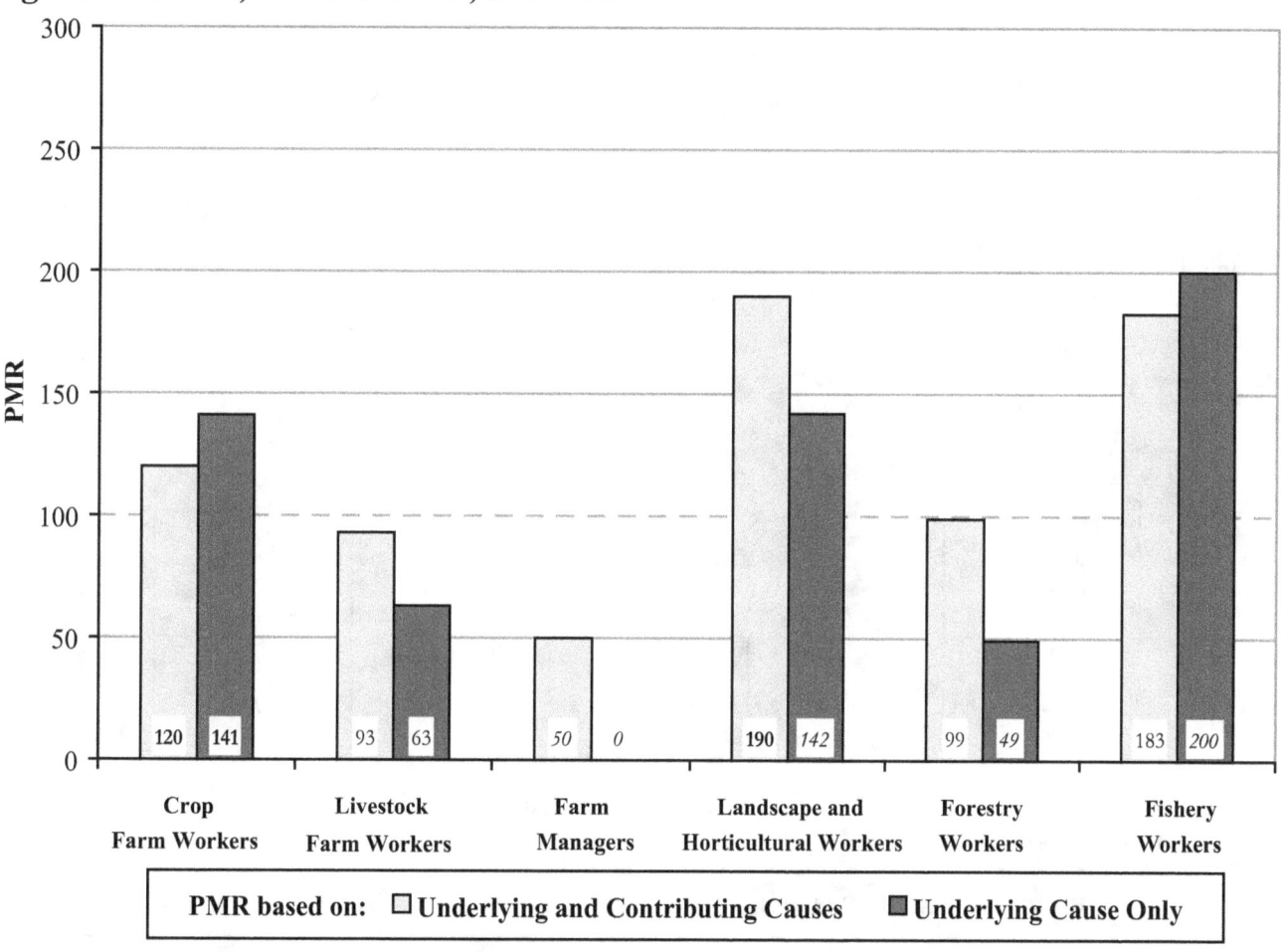

ICD - International Classification of Diseases, 9th Revision
NOTE: Abscess of lung and mediastinum = ICD-9 code 513. PMRs in **bold** are significantly different from 100 (p<0.05). PMRs in *italics* are based on fewer than five observed deaths. PMRs are based on underlying and contributing cause of death. See appendices for source description, methods, ICD codes, and a list of selected states.
SOURCE: National Center for Health Statistics multiple-cause-of-death data

Other Diseases of Respiratory System Mortality within and by Agricultural Group

Figure 2-72. Pulmonary congestion and hypostasis: Proportionate mortality ratio (PMR) adjusted for age, sex, and race/ethnicity by agricultural group, U.S. residents age 15 and over, selected states, 1988–1998

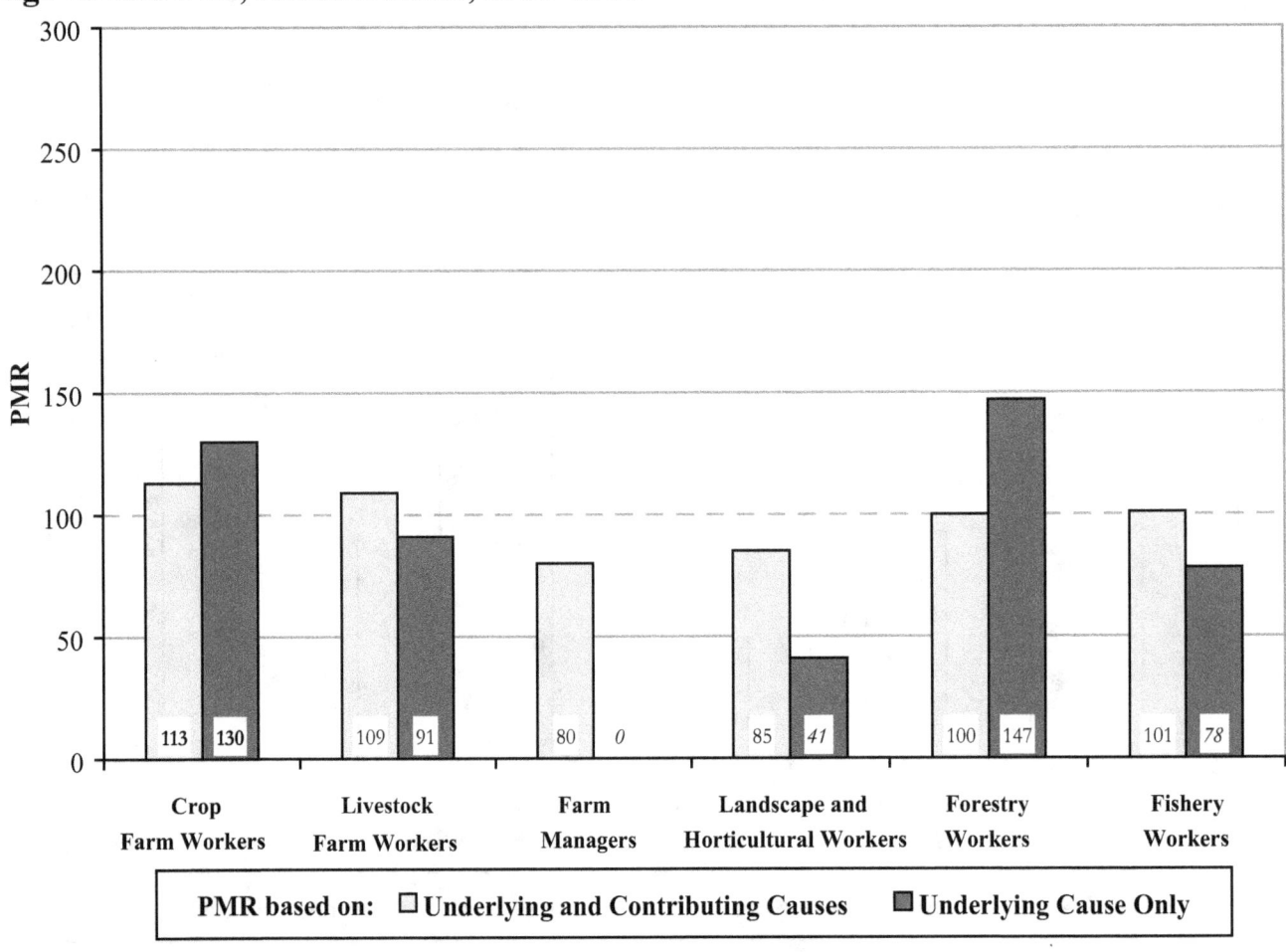

ICD - International Classification of Diseases, 9th Revision
NOTE: Pulmonary congestion and hypostasis = ICD-9 code 514. PMRs in **bold** are significantly different from 100 (p<0.05). PMRs in *italics* are based on fewer than five observed deaths. PMRs are based on underlying and contributing cause of death. See appendices for source description, methods, ICD codes, and a list of selected states.
SOURCE: National Center for Health Statistics multiple-cause-of-death data

Other Diseases of Respiratory System Mortality within and by Agricultural Group

Figure 2-73. Postinflammatory pulmonary fibrosis: Proportionate mortality ratio (PMR) adjusted for age, sex, and race/ethnicity by agricultural group, U.S. residents age 15 and over, selected states, 1988–1998

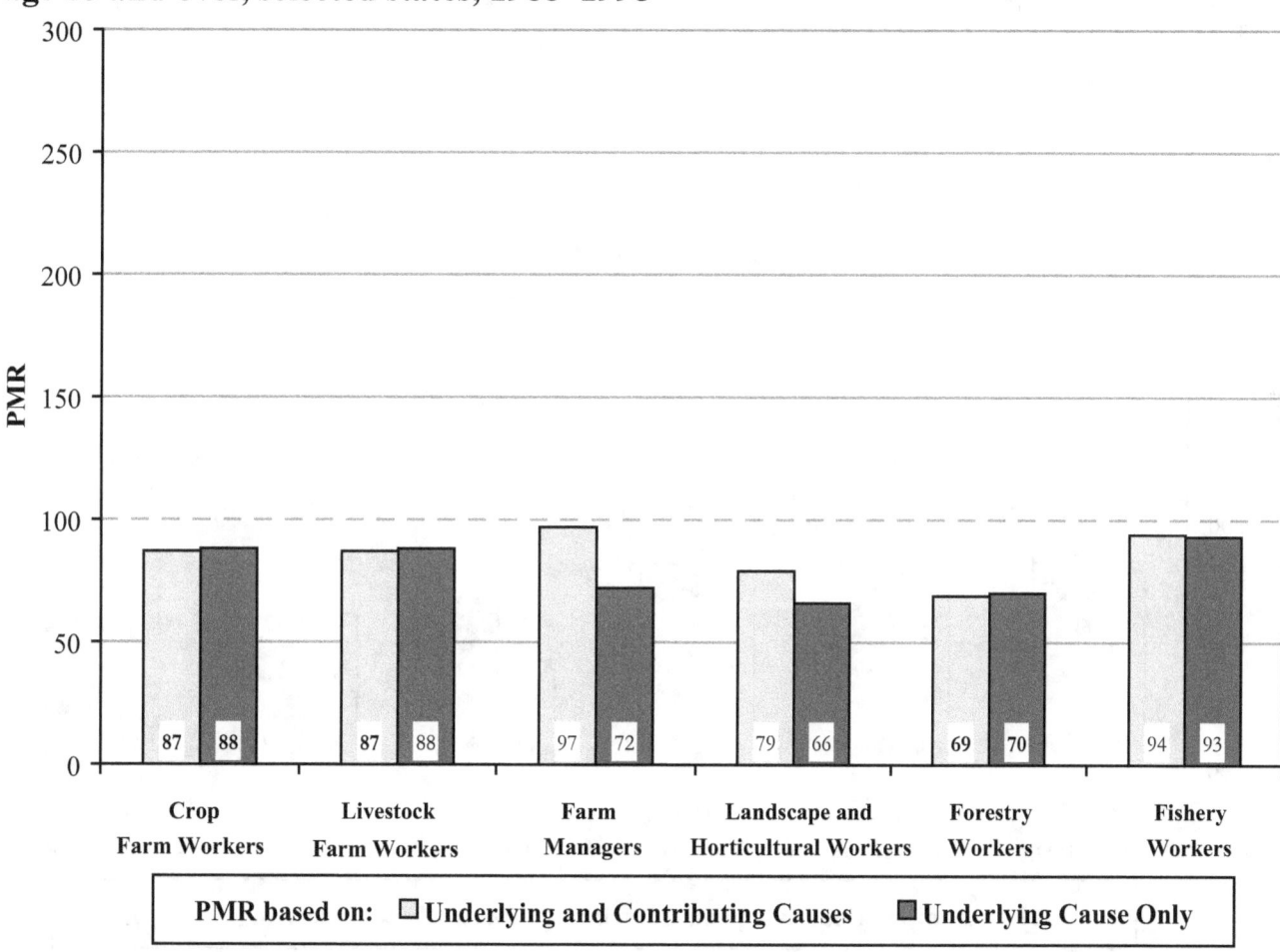

ICD - International Classification of Diseases, 9th Revision
NOTE: Postinflammatory pulmonary fibrosis = ICD-9 code 515. PMRs in **bold** are significantly different from 100 (p<0.05). PMRs in *italics* are based on fewer than five observed deaths. PMRs are based on underlying and contributing cause of death. See appendices for source description, methods, ICD codes, and a list of selected states.
SOURCE: National Center for Health Statistics multiple-cause-of-death data

Other Diseases of Respiratory System Mortality within and by Agricultural Group

Figure 2-74. Other alveolar and parietoalveolar pneumonopathy: Proportionate mortality ratio (PMR) adjusted for age, sex, and race/ethnicity by agricultural group, U.S. residents age 15 and over, selected states, 1988–1998

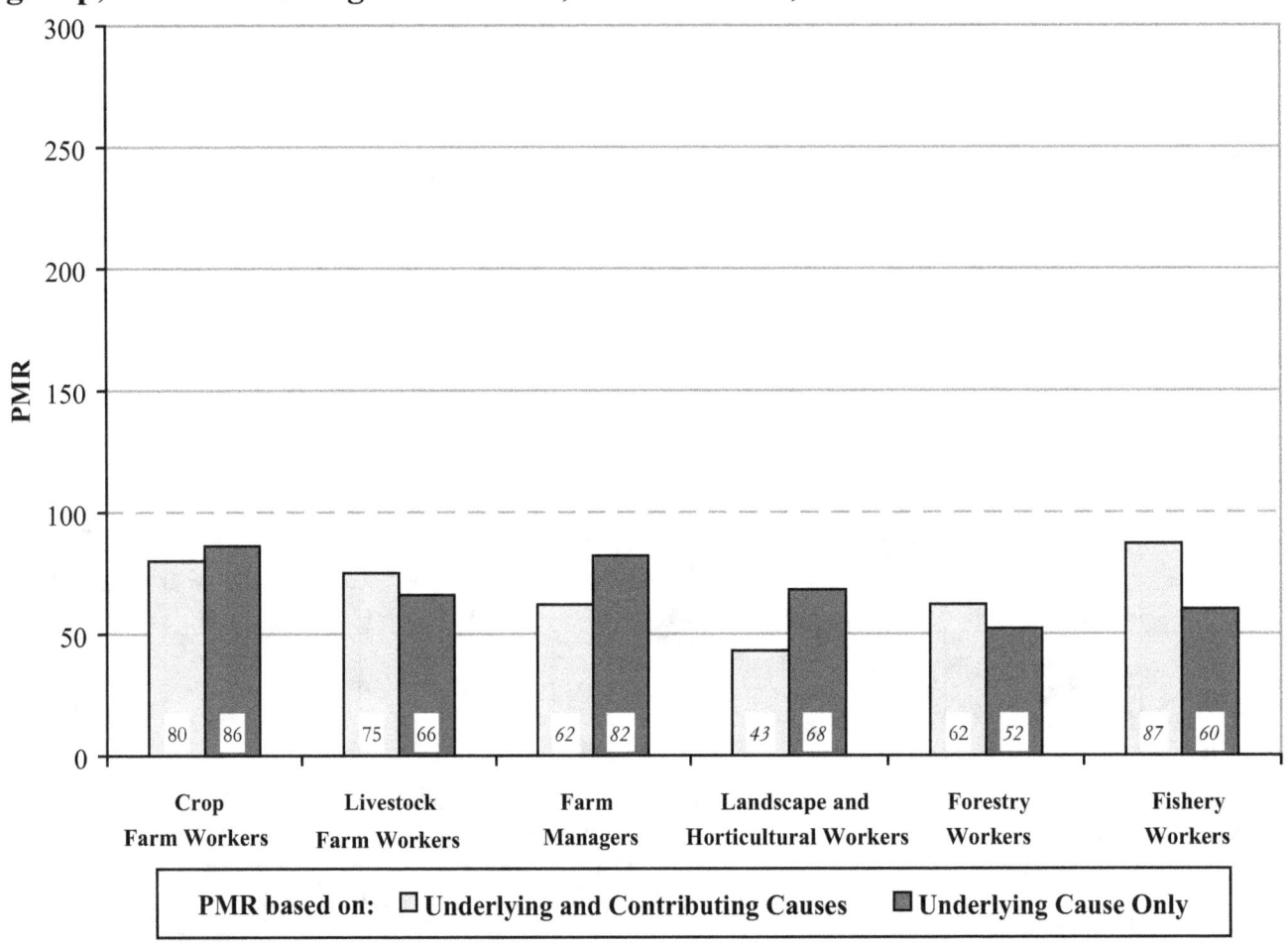

ICD - International Classification of Diseases, 9th Revision
NOTE: Other alveolar and parietoalveolar pneumonopathy = ICD-9 code 516. PMRs in **bold** are significantly different from 100 (p<0.05). PMRs in *italics* are based on fewer than five observed deaths. PMRs are based on underlying and contributing cause of death. See appendices for source description, methods, ICD codes, and a list of selected states.
SOURCE: National Center for Health Statistics multiple-cause-of-death data

Figure 2-75. Other diseases of the lung: Proportionate mortality ratio (PMR) adjusted for age, sex, and race/ethnicity by agricultural group, U.S. residents age 15 and over, selected states, 1988–1998

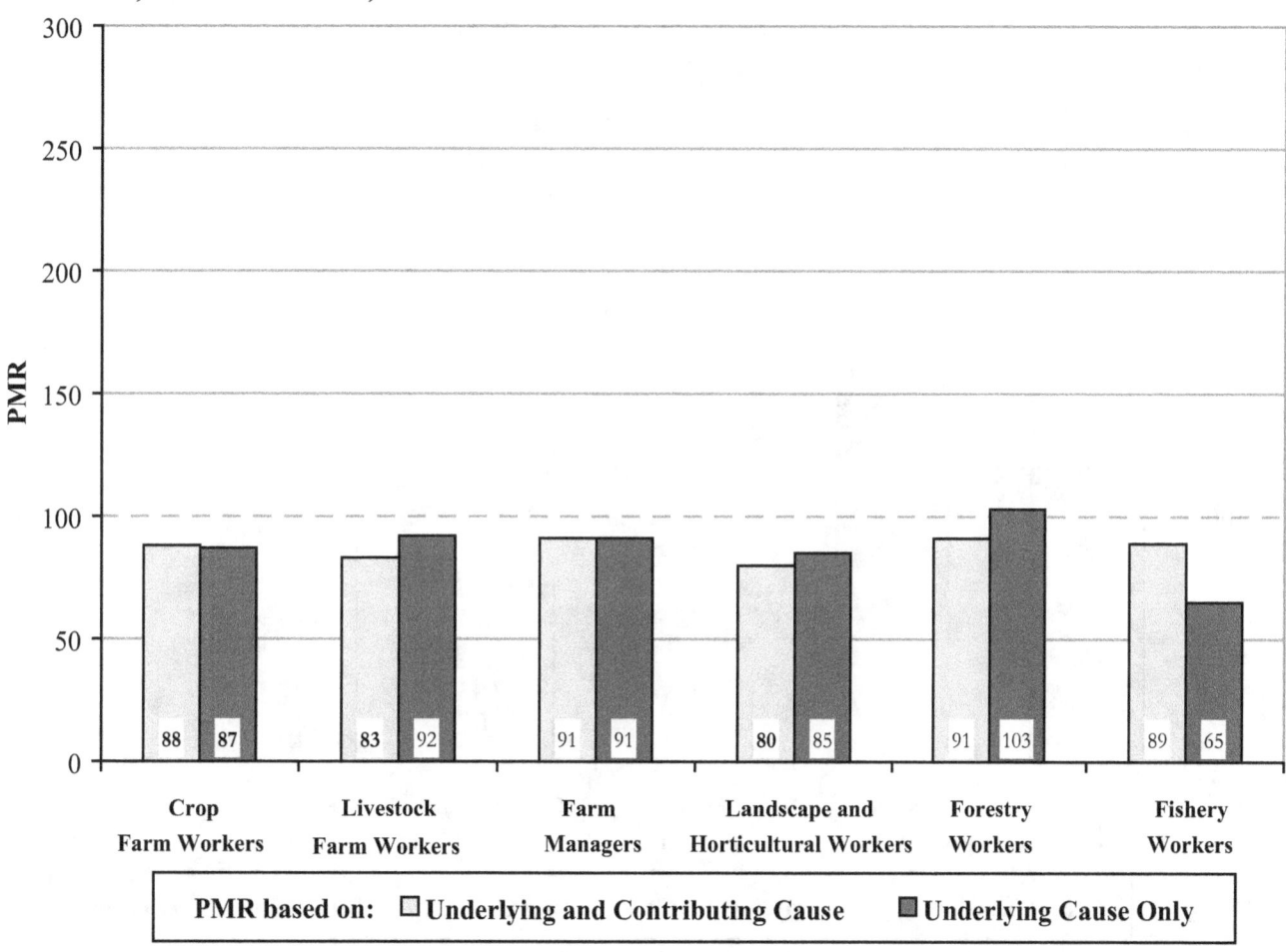

ICD - International Classification of Diseases, 9th Revision
NOTE: Other diseases of the lung = ICD-9 code 518. PMRs in **bold** are significantly different from 100 (p<0.05). PMRs in *italics* are based on fewer than five observed deaths. PMRs are based on underlying and contributing cause of death. See appendices for source description, methods, ICD codes, and a list of selected states.
SOURCE: National Center for Health Statistics multiple-cause-of-death data

Other Diseases of Respiratory System Mortality within and by Agricultural Group

Figure 2-76. Other diseases of respiratory system: Proportionate mortality ratio (PMR) adjusted for age, sex, and race/ethnicity by agricultural group, U.S. residents age 15 and over, selected states, 1988–1998

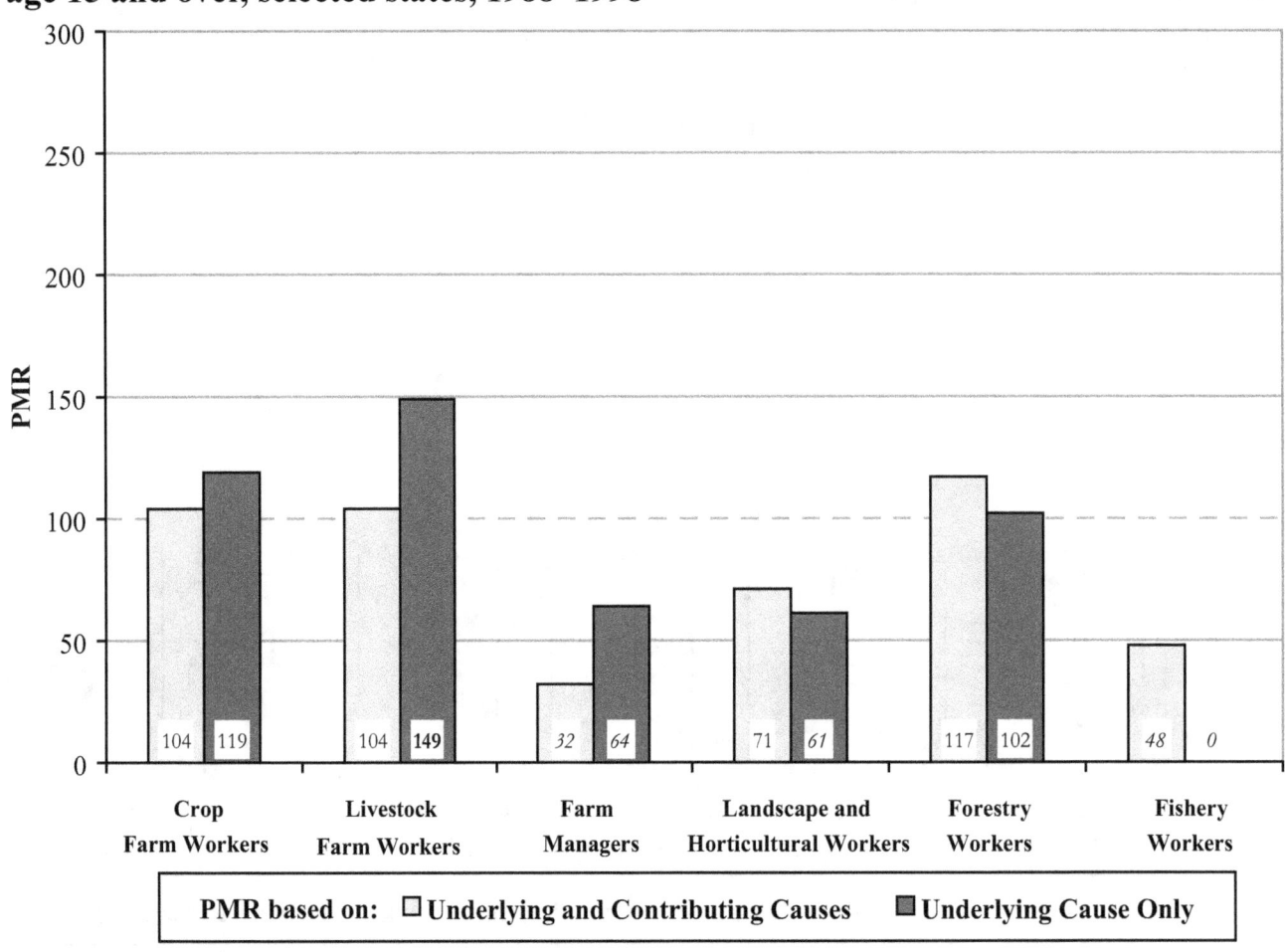

ICD - International Classification of Diseases, 9th Revision
NOTE: Other diseases of respiratory system = ICD-9 code 519. PMRs in **bold** are significantly different from 100 (p<0.05). PMRs in *italics* are based on fewer than five observed deaths. PMRs are based on underlying and contributing cause of death. See appendices for source description, methods, ICD codes, and a list of selected states.
SOURCE: National Center for Health Statistics multiple-cause-of-death data

Section 3

Morbidity

Morbidity by Agricultural Group within Respiratory Condition–NHIS

Table 3-1. Hayfever (past year): Estimated prevalence and prevalence ratio (PR) adjusted for age, sex, race/ethnicity, and smoking status by agricultural group and survey year, U.S. residents age 18 and over, 1997–1999

Worker Group	Survey Year	Number Observed	Estimated Prevalence of Condition in U.S. n	(%)	PR	95% Confidence Interval LCL	UCL
Farm Workers	1997	14	118,596	8.2			
	1998	15	103,577	7.9			
	1999	17	134,222	8.5			
	1997–1999			8.2	103	76	137
Farm Managers	1997	8	67,341	6.2			
	1998	7	50,950	5.2			
	1999	12	82,875	9.6			
	1997–1999			6.9	75	49	109
Forestry/Fishery Workers	1997	1	2,666	2.5			
	1998	2	12,347	9.3			
	1999	0	0	0.0			
	1997–1999			4.1	45	9	132
All Non-agricultural Workers	1997	3,270	17,855,843	9.3			
	1998	2,893	17,571,918	9.1			
	1999	2,630	17,528,032	9.0			
	1997–1999			9.1	100		

n - estimated number   LCL - lower confidence limit   UCL - upper confidence limit
NOTE: Based on responses to the question "During the past 12 months, have you been told by a doctor or other health professional that you had hayfever?" Estimated number in U.S., estimated percent with condition, and PR are based on weighted sample results. PRs in **bold** are significantly different from 100 (p<0.05). PRs in *italics* are based on fewer than five observed cases. See appendices for source description and methods.
SOURCE: National Center for Health Statistics, National Health Interview Survey

Morbidity by Agricultural Group within Respiratory Condition–NHIS

Table 3-2. Sinusitis (past year): Estimated prevalence and prevalence ratio (PR) adjusted for age, sex, race/ethnicity, and smoking status by agricultural group and survey year, U.S. residents age 18 and over, 1997–1999

Worker Group	Survey Year	Number Observed	Estimated Prevalence of Condition in U.S. n	(%)	PR	95% Confidence Interval LCL	UCL
Farm Workers	1997	21	151,233	10.4			
	1998	14	124,020	9.4			
	1999	17	116,580	7.4			
	1997–1999			9.0	77	59	101
Farm Managers	1997	16	116,148	10.9			
	1998	10	76,391	7.7			
	1999	9	81,861	9.5			
	1997–1999			9.4	**69**	48	96
Forestry/Fishery Workers	1997	2	9,338	8.7			
	1998	2	13,714	10.3			
	1999	1	9,114	7.1			
	1997–1999			8.7	68	22	159
All Non-agricultural Workers	1997	5,859	31,367,704	16.4			
	1998	5,185	31,438,933	16.3			
	1999	4,688	30,496,524	15.6			
	1997–1999			16.0	100		

n - estimated number   LCL - lower confidence limit   UCL - upper confidence limit
NOTE: Based on responses to the question "During the past 12 months, have you been told by a doctor or other health professional that you had sinusitis?" Estimated number in U.S., estimated percent with condition, and PR are based on weighted sample results. PRs in **bold** are significantly different from 100 (p<0.05). PRs in *italics* are based on fewer than five observed cases. See appendices for source description and methods.
SOURCE: National Center for Health Statistics, National Health Interview Survey

Morbidity by Agricultural Group within Respiratory Condition–NHIS

Table 3-3. Chronic bronchitis (past year): Estimated prevalence and prevalence ratio (PR) adjusted for age, sex, race/ethnicity, and smoking status by agricultural group and survey year, U.S. residents age 18 and over, 1997–1999

Worker Group	Survey Year	Number Observed	Estimated Prevalence of Condition in U.S. n	(%)	PR	95% Confidence Interval LCL	UCL
Farm Workers	1997	6	39,014	2.7			
	1998	8	69,040	5.2			
	1999	7	40,345	2.6			
	1997–1999			3.4	108	67	165
Farm Managers	1997	4	26,239	2.5			
	1998	1	6,507	0.7			
	1999	3	20,890	2.4			
	1997–1999			1.8	52	22	102
Forestry/Fishery Workers	1997	0	0	0.0			
	1998	1	9,448	7.1			
	1999	1	9,114	7.1			
	1997–1999			5.0	*143*	*17*	*516*
All Non-agricultural Workers	1997	1,857	9,593,307	5.0			
	1998	1,554	8,863,177	4.6			
	1999	1,409	8,751,682	4.5			
	1997–1999			4.7	100		

n - estimated number   LCL - lower confidence limit   UCL - upper confidence limit
NOTE: Based on responses to the question "During the past 12 months, have you been told by a doctor or other health professional that you had chronic bronchitis?" Estimated number in U.S., estimated percent with condition, and PR are based on weighted sample results. PRs in **bold** are significantly different from 100 (p<0.05). PRs in *italics* are based on fewer than five observed cases. See appendices for source description and methods.
SOURCE: National Center for Health Statistics, National Health Interview Survey

Table 3-4. Emphysema (ever): Estimated prevalence and prevalence ratio (PR) adjusted for age, sex, race/ethnicity, and smoking status by agricultural group and survey year, U.S. residents age 18 and over, 1997–1999

Worker Group	Survey Year	Number Observed	Estimated Prevalence of Condition in U.S. n	Estimated Prevalence of Condition in U.S. (%)	PR	95% Confidence Interval LCL	95% Confidence Interval UCL
Farm Workers	1997	0	0	0.0			
	1998	3	20,760	1.6			
	1999	1	1,969	0.1			
	1997–1999			0.5	71	19	182
Farm Managers	1997	1	10,486	1.0			
	1998	0	0	0.0			
	1999	0	0	0.0			
	1997–1999			0.4	20	1	111
Forestry/Fishery Workers	1997	0	0	0.0			
	1998	0	0	0.0			
	1999	0	0	0.0			
	1997–1999			0.0	0	---	---
All Non-agricultural Workers	1997	612	3,205,415	1.7			
	1998	542	2,934,923	1.5			
	1999	460	2,783,974	1.4			
	1997–1999			1.5	100		

n - estimated number   LCL - lower confidence limit   UCL - upper confidence limit
NOTE: Based on responses to the question "Have you EVER been told by a doctor or other health professional that you had emphysema?" Estimated number in U.S., estimated percent with condition, and PR are based on weighted sample results. PRs in **bold** are significantly different from 100 (p<0.05). PRs in *italics* are based on fewer than five observed cases. See appendices for source description and methods.
SOURCE: National Center for Health Statistics, National Health Interview Survey

Table 3-5. Asthma (ever): Estimated prevalence and prevalence ratio (PR) adjusted for age, sex, race/ethnicity, and smoking status by agricultural group and survey year, U.S. residents age 18 and over, 1997–1999

Worker Group	Survey Year	Number Observed	Estimated Prevalence of Condition in U.S. n	Estimated Prevalence of Condition in U.S. (%)	PR	95% Confidence Interval LCL	95% Confidence Interval UCL
Farm Workers	1997	17	114,894	7.9			
	1998	16	132,838	10.1			
	1999	16	142,603	9.1			
	1997–1999			9.0	109	81	144
Farm Managers	1997	5	29,316	2.7			
	1998	9	63,150	6.3			
	1999	7	59,530	6.9			
	1997–1999			5.2	68	42	104
Forestry/Fishery Workers	1997	0	0	0.0			
	1998	3	15,828	11.9			
	1999	1	10,946	8.5			
	1997–1999			7.2	89	24	228
All Non-agricultural Workers	1997	3,232	17,442,022	9.1			
	1998	2,925	17,457,507	9.0			
	1999	2,631	16,660,804	8.5			
	1997–1999			8.9	100		

n - estimated number   LCL - lower confidence limit   UCL - upper confidence limit

NOTE: Based on responses to the question "Have you EVER been told by a doctor or other health professional that you had asthma?" Estimated number in U.S., estimated percent with condition, and PR are based on weighted sample results. PRs in **bold** are significantly different from 100 (p<0.05). PRs in *italics* are based on fewer than five observed cases. See appendices for source description and methods.

SOURCE: National Center for Health Statistics, National Health Interview Survey

Morbidity by Agricultural Group within Respiratory Condition–NHIS

Table 3-6. Lung cancer (ever): Estimated prevalence and prevalence ratio (PR) adjusted for age, sex, race/ethnicity, and smoking status by agricultural group and survey year, U.S. residents age 18 and over, 1997–1999

Worker Group	Survey Year	Number Observed	Estimated Prevalence of Condition in U.S. n	(%)	PR	95% Confidence Interval LCL	UCL
Farm Workers	1997	0	0	0.0			
	1998	0	0	0.0			
	1999	0	0	0.0			
	1997–1999			0.0	*0*	---	---
Farm Managers	1997	0	0	0.0			
	1998	0	0	0.0			
	1999	1	13,927	1.6			
	1997–1999			0.5	223	6	1,239
Forestry/Fishery Workers	1997	0	0	0.0			
	1998	0	0	0.0			
	1999	0	0	0.0			
	1997–1999			0.0	*0*	---	---
All Non-agricultural Workers	1997	78	388,060	0.2			
	1998	54	339,485	0.2			
	1999	70	439,496	0.2			
	1997–1999			0.2	100		

n - estimated number   LCL - lower confidence limit   UCL - upper confidence limit
NOTE: Based on responses to the question "Have you EVER been told by a doctor or other health professional that you had cancer or a malignancy of any kind? What kind of cancer was it? ... lung?" Estimated number in U.S., estimated percent with condition, and PR are based on weighted sample results. PRs in **bold** are significantly different from 100 (p<0.05). PRs in *italics* are based on fewer than five observed cases. See appendices for source description and methods.
SOURCE: National Center for Health Statistics, National Health Interview Survey

Figure 3-1. Respiratory conditions: Prevalence ratio (PR) adjusted for age, sex, race/ethnicity, and smoking status by agricultural group, U.S. residents age 18 and over, 1997–1999

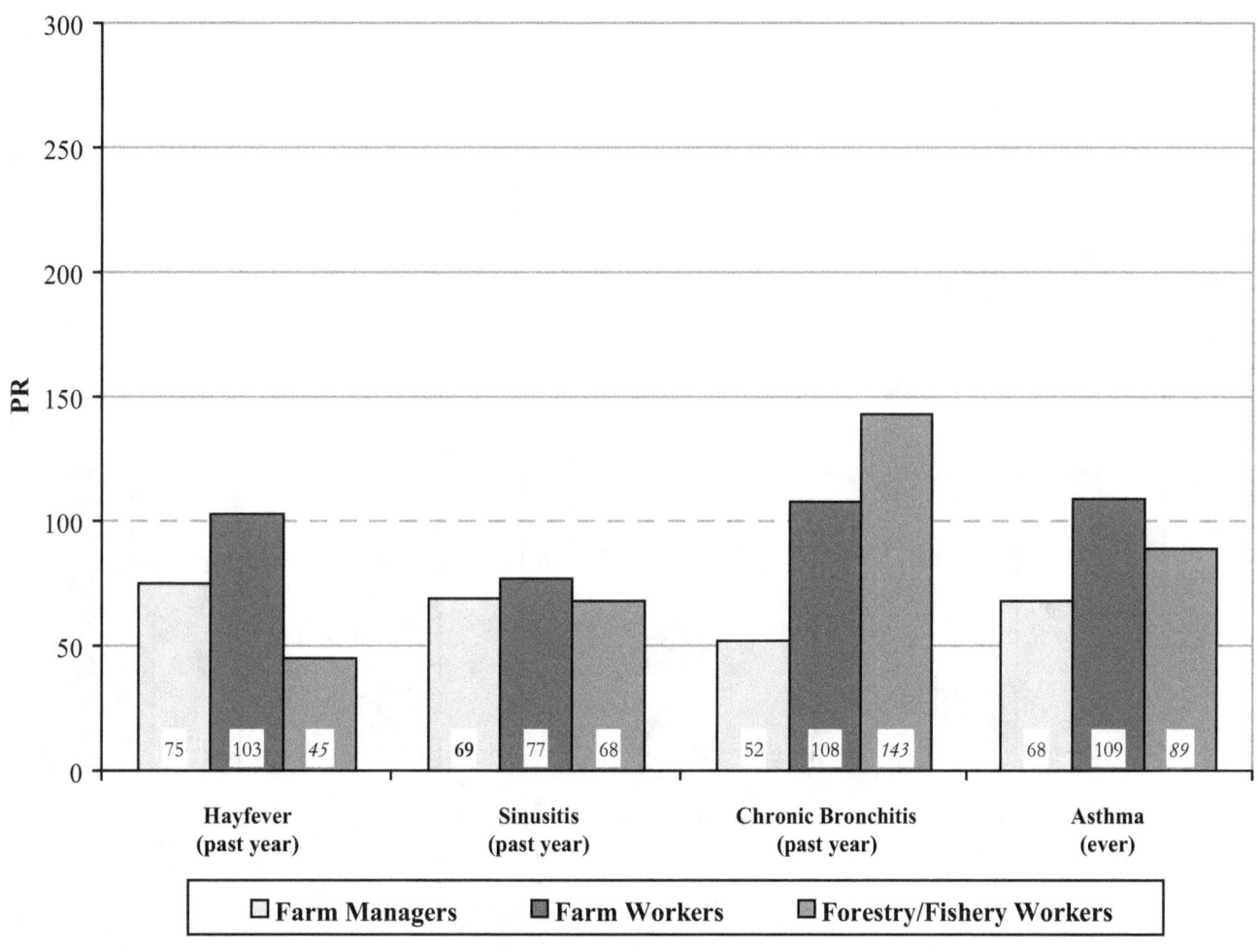

NOTE: Based on responses to the following questions:
  "During the past 12 months, have you been told by a doctor or other health professional that you had hayfever?"
  "During the past 12 months, have you been told by a doctor or other health professional that you had sinusitis?"
  "During the past 12 months, have you been told by a doctor or other health professional that you had chronic bronchitis?"
  "Have you EVER been told by a doctor or other health professional that you had asthma?"
PRs in **bold** are significantly different from 100 (p<0.05). PRs in *italics* are based on fewer than five observed cases. See appendices for source description and methods.
SOURCE: National Center for Health Statistics, National Health Interview Survey

Morbidity by Respiratory Condition and Sex within Agricultural Group–NHIS

Figure 3-2. Farm workers: Prevalence ratio (PR) adjusted for age, race/ethnicity, and smoking status by respiratory condition and sex, U.S. residents age 18 and over, 1997–1999

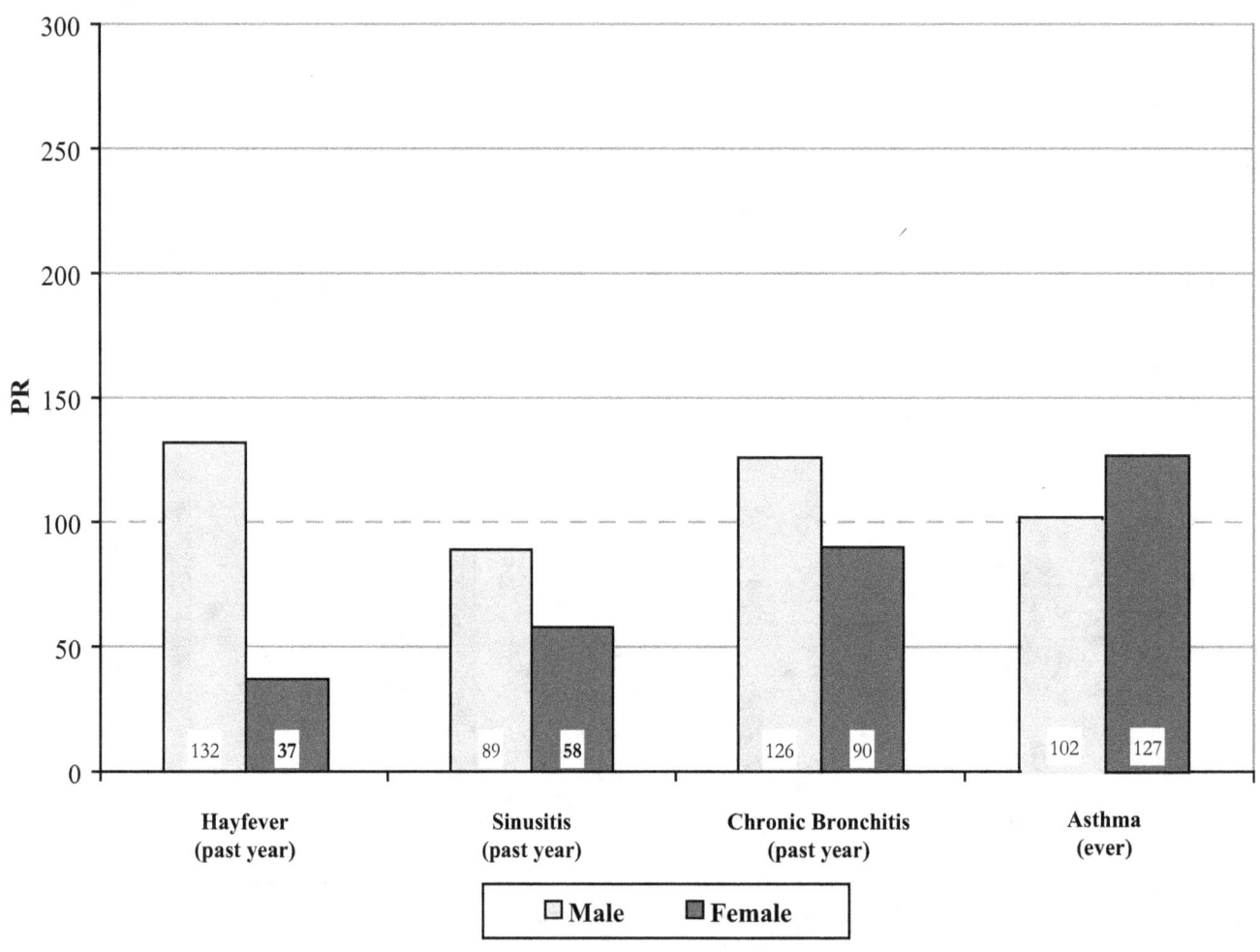

NOTE: Based on responses to the following questions:
"During the past 12 months, have you been told by a doctor or other health professional that you had hayfever?"
"During the past 12 months, have you been told by a doctor or other health professional that you had sinusitis?"
"During the past 12 months, have you been told by a doctor or other health professional that you had chronic bronchitis?"
"Have you EVER been told by a doctor or other health professional that you had asthma?"
PRs in **bold** are significantly different from 100 (p<0.05). PRs in *italics* are based on fewer than five observed cases. See appendices for source description and methods.
SOURCE: National Center for Health Statistics, National Health Interview Survey

Morbidity by Respiratory Condition and Sex within Agricultural Group–NHIS

Figure 3-3. Farm managers: Prevalence ratio (PR) adjusted for age, race/ethnicity, and smoking status by respiratory condition and sex, U.S. residents age 18 and over, 1997–1999

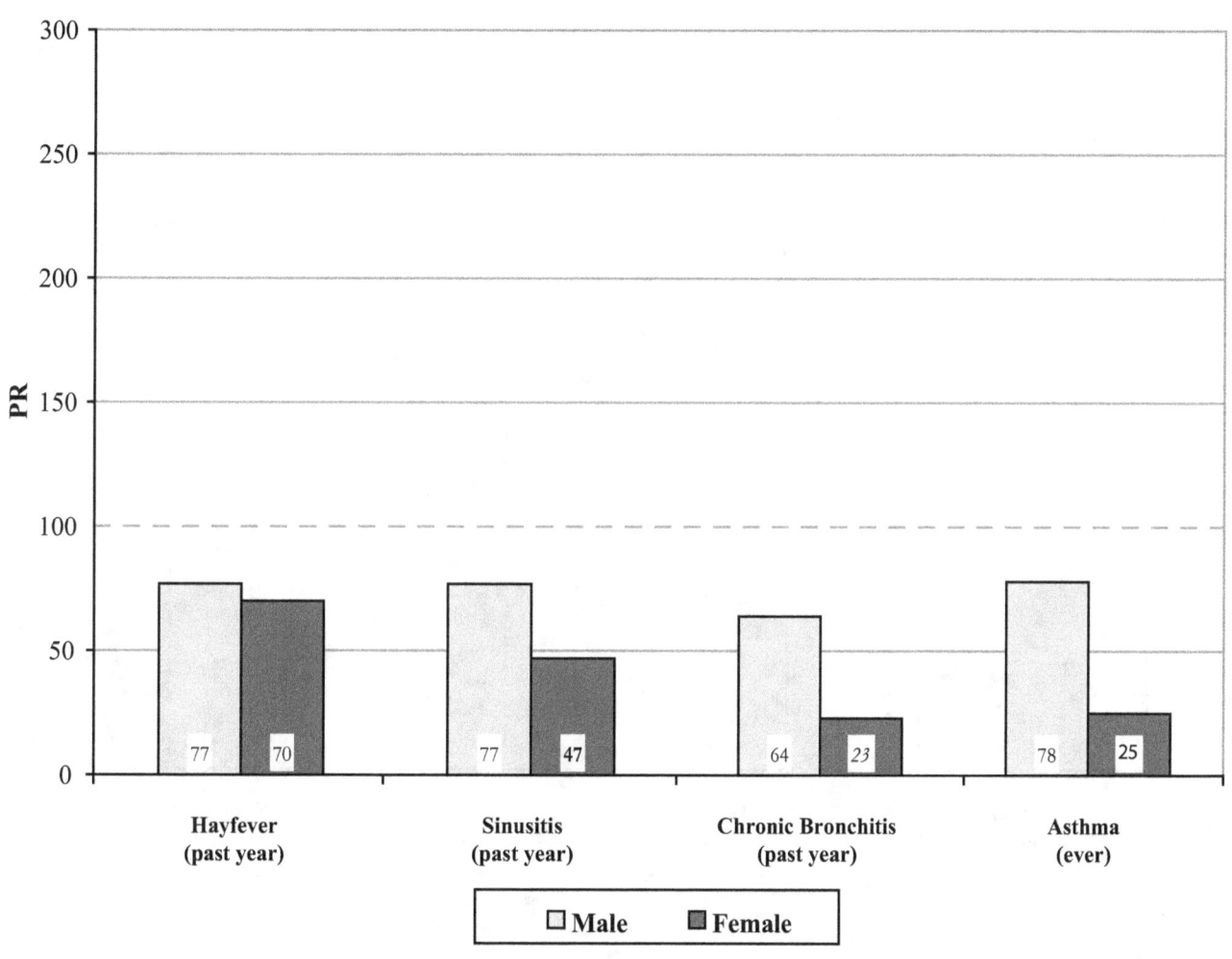

NOTE: Based on responses to the following questions:
  "During the past 12 months, have you been told by a doctor or other health professional that you had hayfever?"
  "During the past 12 months, have you been told by a doctor or other health professional that you had sinusitis?"
  "During the past 12 months, have you been told by a doctor or other health professional that you had chronic bronchitis?"
  "Have you EVER been told by a doctor or other health professional that you had asthma?"
PRs in **bold** are significantly different from 100 (p<0.05). PRs in *italics* are based on fewer than five observed cases. See appendices for source description and methods.
SOURCE: National Center for Health Statistics, National Health Interview Survey

Morbidity by Respiratory Condition and Sex within Agricultural Group–NHIS

Figure 3-4. Forestry/fishery workers: Prevalence ratio (PR) adjusted for age, race/ethnicity, and smoking status by respiratory condition and sex, U.S. residents age 18 and over, 1997–1999

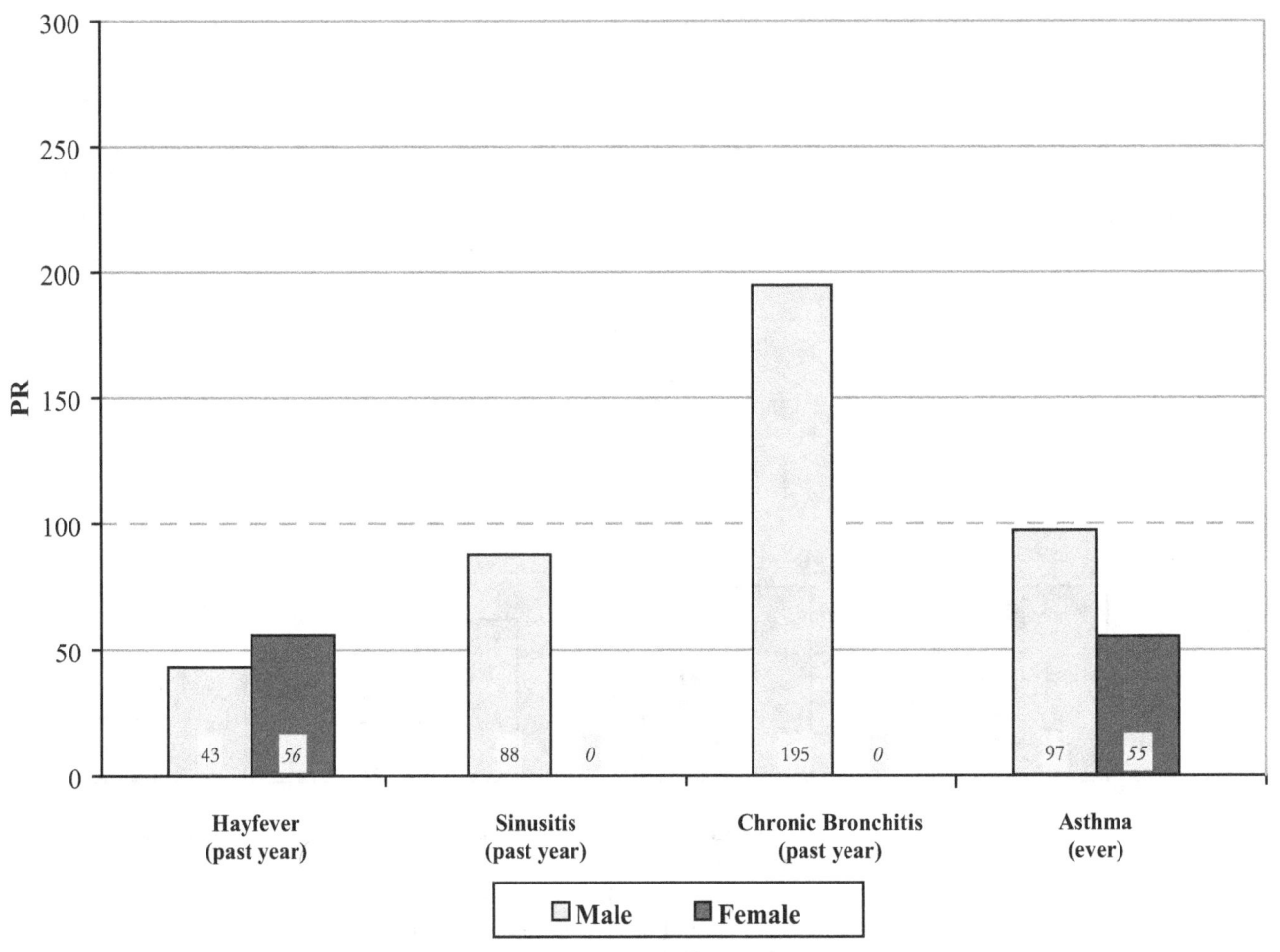

NOTE: Based on responses to the following questions:
  "During the past 12 months, have you been told by a doctor or other health professional that you had hayfever?"
  "During the past 12 months, have you been told by a doctor or other health professional that you had sinusitis?"
  "During the past 12 months, have you been told by a doctor or other health professional that you had chronic bronchitis?"
  "Have you EVER been told by a doctor or other health professional that you had asthma?"
PRs in **bold** are significantly different from 100 (p<0.05). PRs in *italics* are based on fewer than five observed cases. See appendices for source description and methods.
SOURCE: National Center for Health Statistics, National Health Interview Survey

Morbidity by Respiratory Condition and Race/Ethnicity within Argricultural Group–NHIS

Figure 3-5. Farm workers: Prevalence ratio (PR) adjusted for age, sex, and smoking status by respiratory condition and race/ethnicity, U.S. residents age 18 and over, 1997–1999

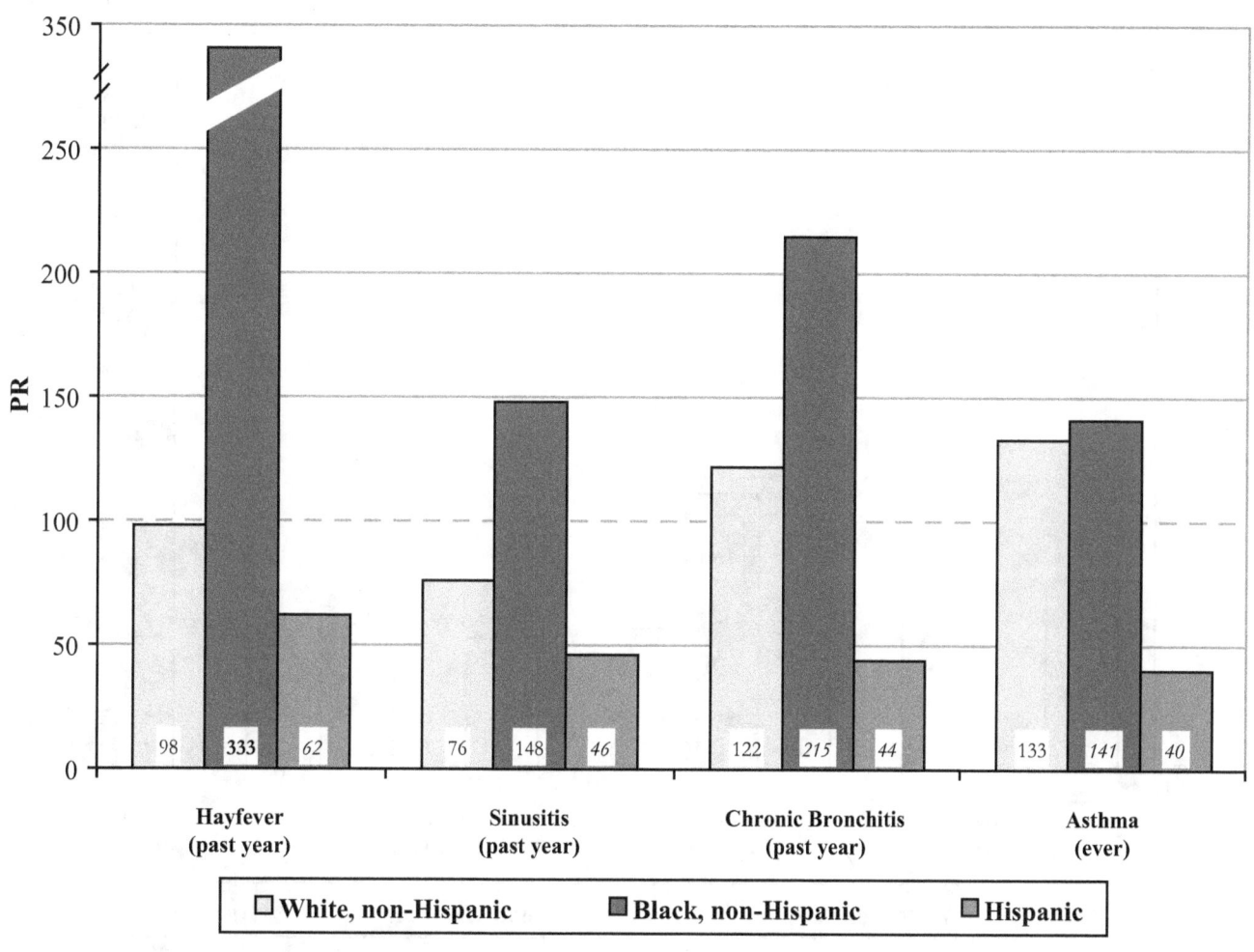

NOTE: Based on responses to the following questions:
 "During the past 12 months, have you been told by a doctor or other health professional that you had hayfever?"
 "During the past 12 months, have you been told by a doctor or other health professional that you had sinusitis?"
 "During the past 12 months, have you been told by a doctor or other health professional that you had chronic bronchitis?"
 "Have you EVER been told by a doctor or other health professional that you had asthma?"
PRs in **bold** are significantly different from 100 ($p<0.05$). PRs in *italics* are based on fewer than five observed cases. See appendices for source description and methods.
SOURCE: National Center for Health Statistics, National Health Interview Survey

Figure 3-6. Farm managers: Prevalence ratio (PR) adjusted for age, sex, and smoking status by respiratory condition and race/ethnicity, U.S. residents age 18 and over, 1997–1999

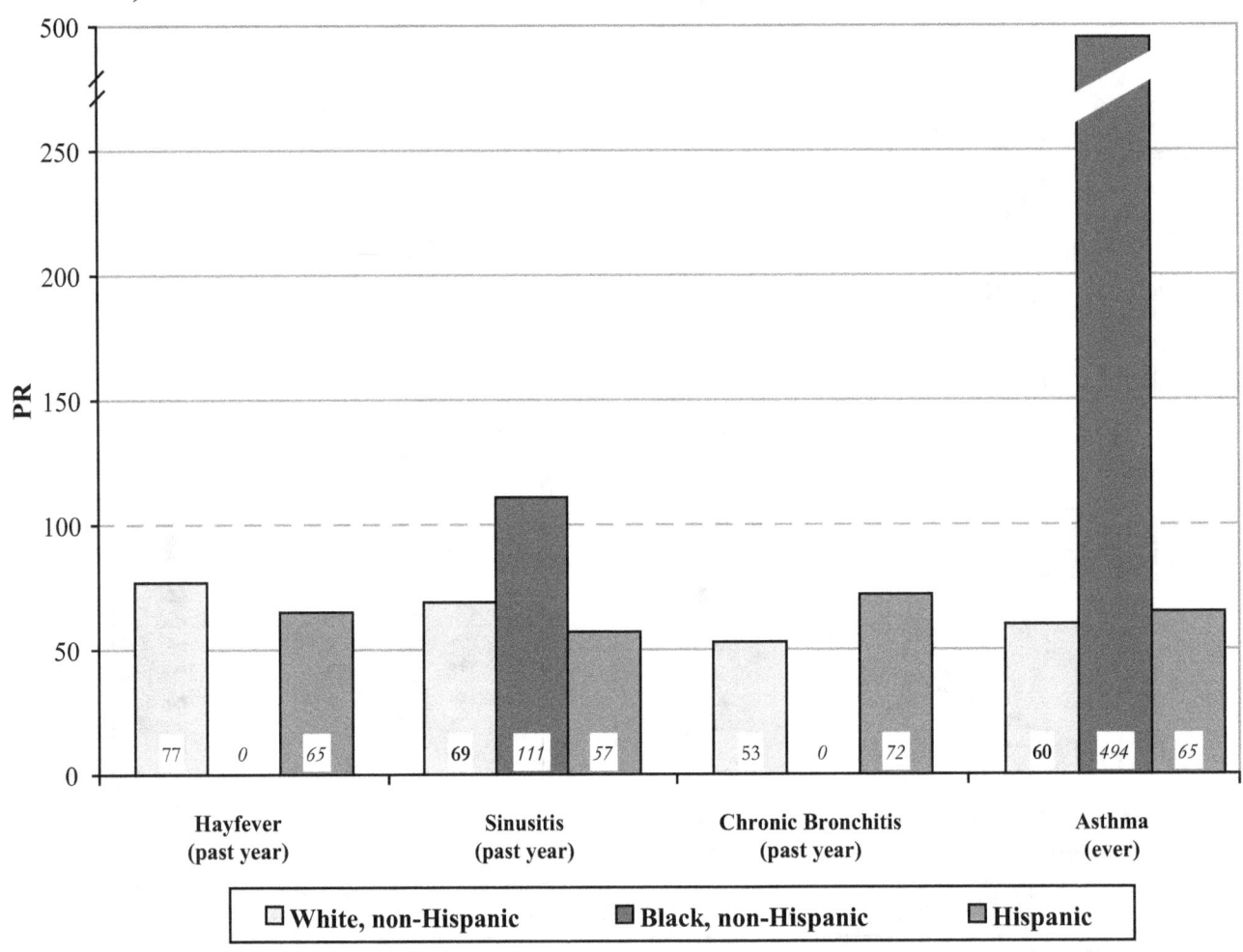

NOTE: Based on responses to the following questions:
  "During the past 12 months, have you been told by a doctor or other health professional that you had hayfever?"
  "During the past 12 months, have you been told by a doctor or other health professional that you had sinusitis?"
  "During the past 12 months, have you been told by a doctor or other health professional that you had chronic bronchitis?"
  "Have you EVER been told by a doctor or other health professional that you had asthma?"
PRs in **bold** are significantly different from 100 (p<0.05). PRs in *italics* are based on fewer than five observed cases. See appendices for source description and methods.
SOURCE: National Center for Health Statistics, National Health Interview Survey

Figure 3-7. Forestry/fishery workers: Prevalence ratio (PR) adjusted for age, sex, and smoking status by respiratory condition and race/ethnicity, U.S. residents age 18 and over, 1997–1999

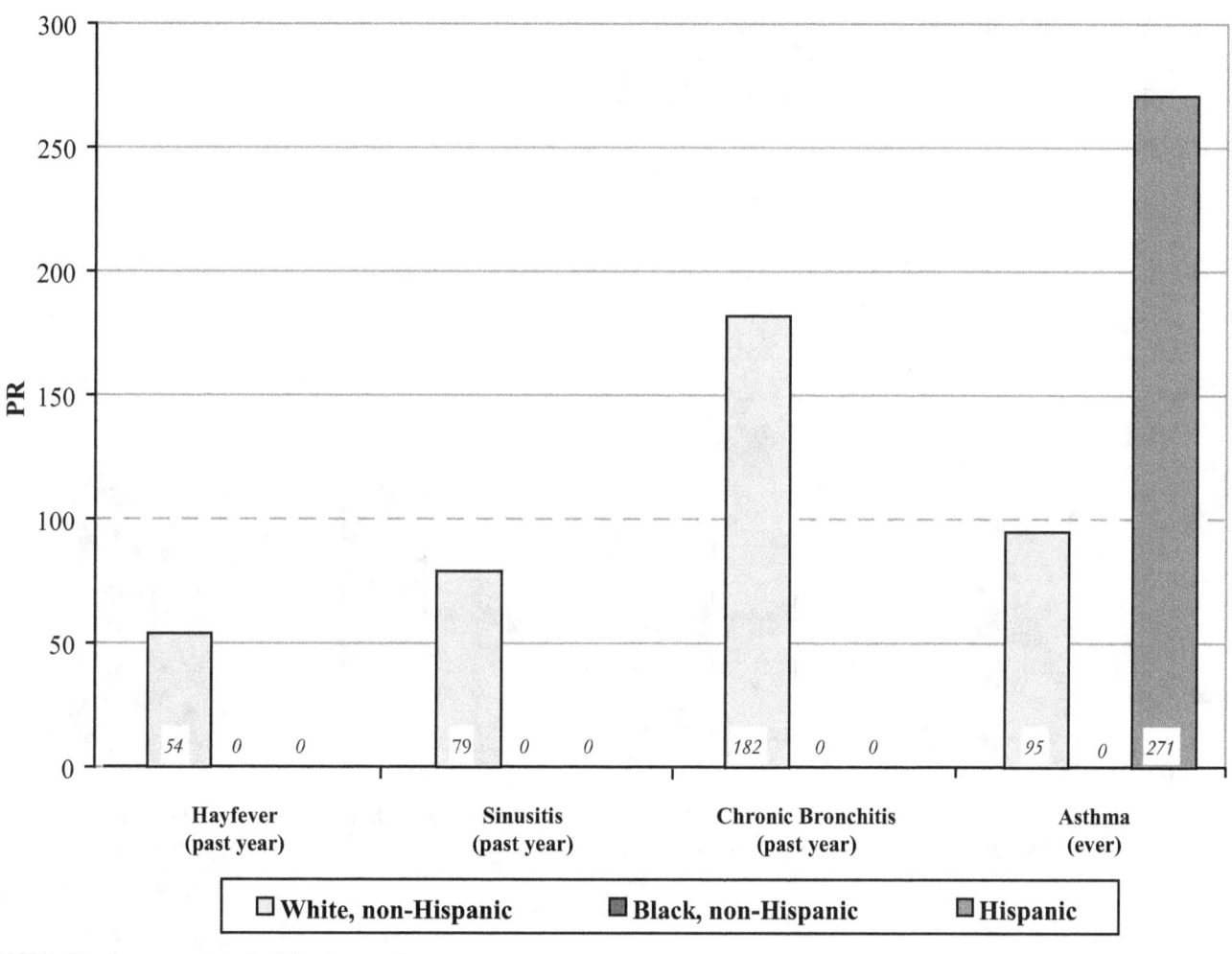

NOTE: Based on responses to the following questions:
 "During the past 12 months, have you been told by a doctor or other health professional that you had hayfever?"
 "During the past 12 months, have you been told by a doctor or other health professional that you had sinusitis?"
 "During the past 12 months, have you been told by a doctor or other health professional that you had chronic bronchitis?"
 "Have you EVER been told by a doctor or other health professional that you had asthma?"
PRs in **bold** are significantly different from 100 ($p<0.05$). PRs in *italics* are based on fewer than five observed cases. See appendices for source description and methods.
SOURCE: National Center for Health Statistics, National Health Interview Survey

Morbidity by Respiratory Condition and Smoking Status within Agricultural Group–NHIS

Figure 3-8. Farm workers: Prevalence ratio (PR) adjusted for age, sex, and race/ethnicity by respiratory condition and smoking status, U.S. residents age 18 and over, 1997–1999

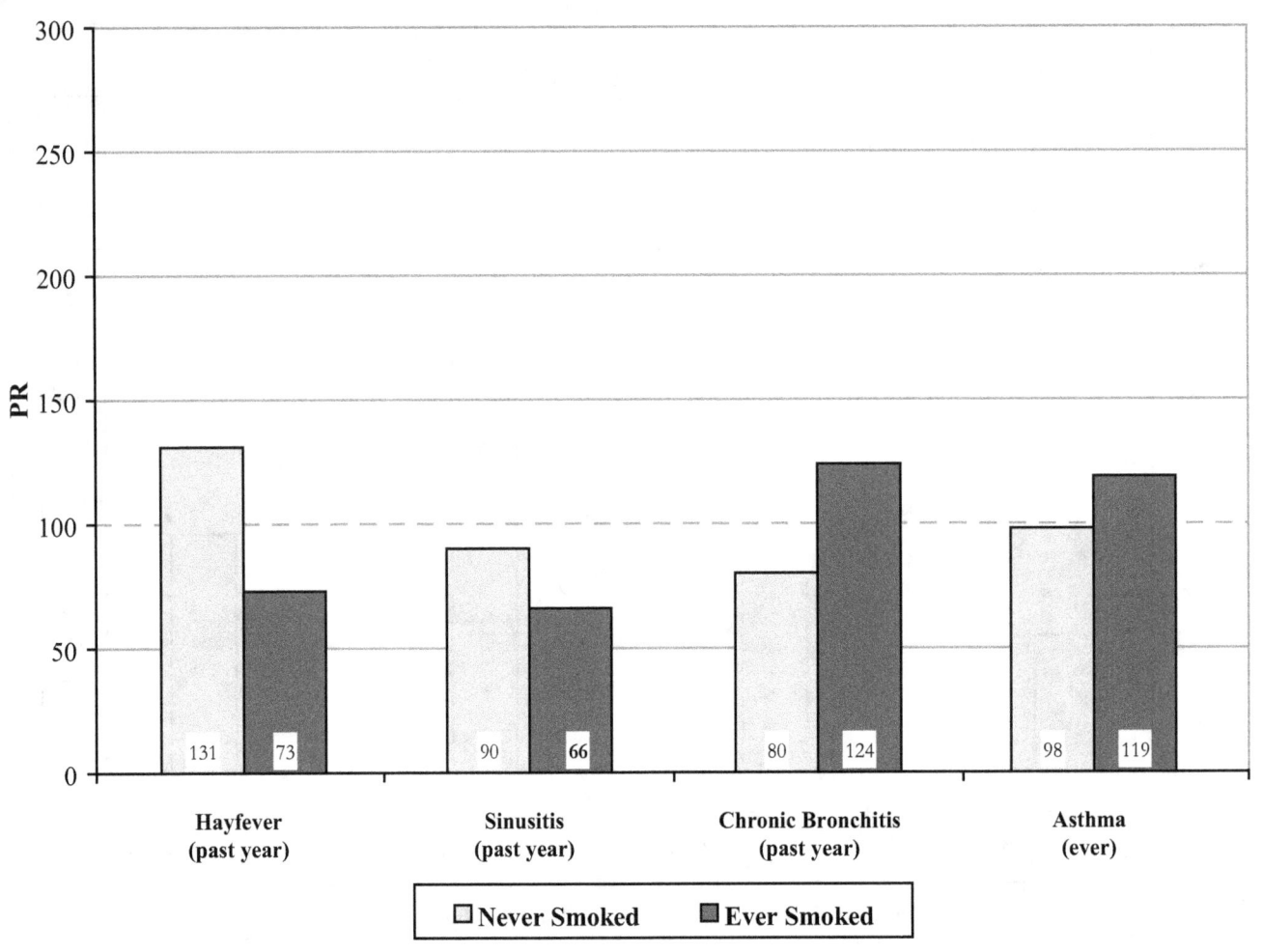

NOTE: Based on responses to the following questions:
 "During the past 12 months, have you been told by a doctor or other health professional that you had hayfever?"
 "During the past 12 months, have you been told by a doctor or other health professional that you had sinusitis?"
 "During the past 12 months, have you been told by a doctor or other health professional that you had chronic bronchitis?"
 "Have you EVER been told by a doctor or other health professional that you had asthma?"
PRs in **bold** are significantly different from 100 ($p<0.05$). PRs in *italics* are based on fewer than five observed cases. See appendices for source description and methods.
SOURCE: National Center for Health Statistics, National Health Interview Survey

Morbidity by Respiratory Condition and Smoking Status within Agricultural Group–NHIS

Figure 3-9. Farm managers: Prevalence ratio (PR) adjusted for age, sex, and race/ethnicity by respiratory condition and smoking status, U.S. residents age 18 and over, 1997–1999

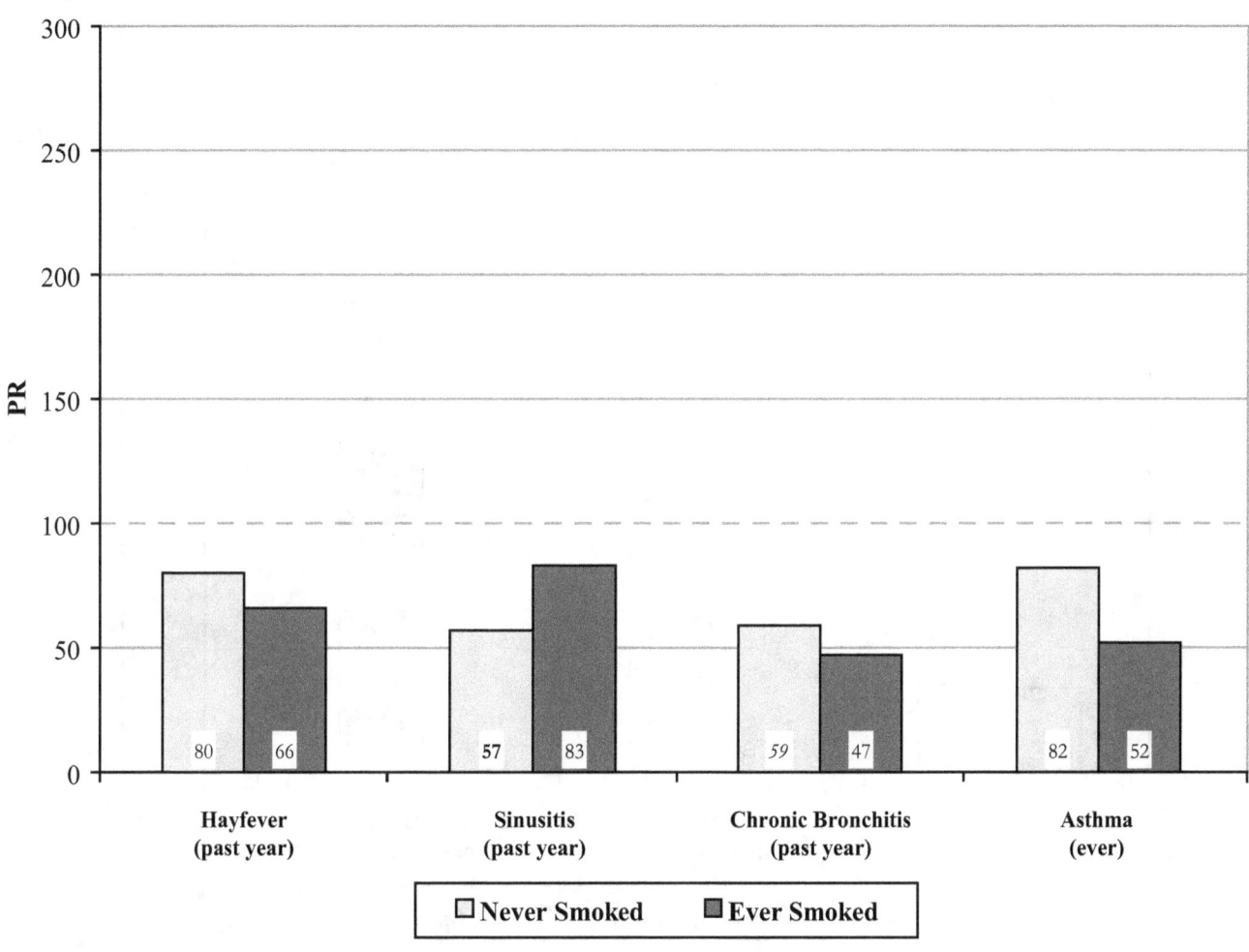

NOTE: Based on responses to the following questions:
   "During the past 12 months, have you been told by a doctor or other health professional that you had hayfever?"
   "During the past 12 months, have you been told by a doctor or other health professional that you had sinusitis?"
   "During the past 12 months, have you been told by a doctor or other health professional that you had chronic bronchitis?"
   "Have you EVER been told by a doctor or other health professional that you had asthma?"
PRs in **bold** are significantly different from 100 ($p<0.05$). PRs in *italics* are based on fewer than five observed cases. See appendices for source description and methods.
SOURCE: National Center for Health Statistics, National Health Interview Survey

Morbidity by Respiratory Condition and Smoking Status within Agricultural Group–NHIS

Figure 3-10. Forestry/fishery workers: Prevalence ratio (PR) adjusted for age, sex, and race/ethnicity by respiratory condition and smoking status, U.S. residents age 18 and over, 1997–1999

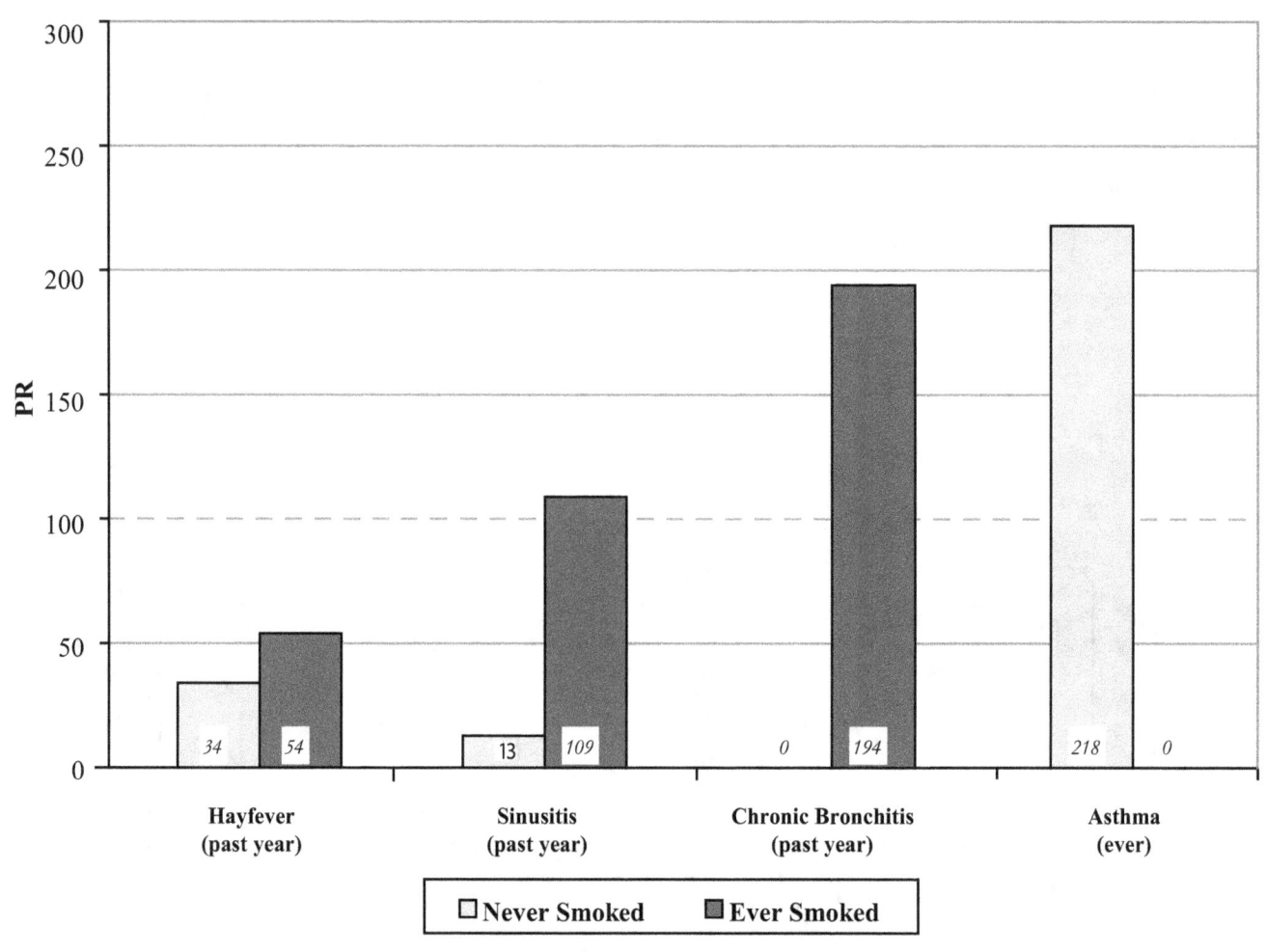

NOTE: Based on responses to the following questions:
"During the past 12 months, have you been told by a doctor or other health professional that you had hayfever?"
"During the past 12 months, have you been told by a doctor or other health professional that you had sinusitis?"
"During the past 12 months, have you been told by a doctor or other health professional that you had chronic bronchitis?"
"Have you EVER been told by a doctor or other health professional that you had asthma?"
PRs in **bold** are significantly different from 100 (p<0.05). PRs in *italics* are based on fewer than five observed cases. See appendices for source description and methods.
SOURCE: National Center for Health Statistics, National Health Interview Survey

Morbidity by Agricultural Group within Respiratory Condition–NHANES III

Table 3-7. Wheezing, apart from a cold (current): Estimated prevalence and prevalence ratio (PR) adjusted for age, sex, race/ethnicity, and smoking status by agricultural group, U.S. residents age 17 and over, 1988–1994

Worker Group	Number Observed	Estimated Prevalence of Condition in U.S. n	Estimated Prevalence of Condition in U.S. (%)	PR	95% Confidence Interval LCL	95% Confidence Interval UCL
Farm Workers	95	436,107	12.0	111	90	136
Farm Managers	53	370,558	14.1	115	88	151
Other Agricultural Workers	19	138,713	10.7	100	60	156
All Non-agricultural Workers	1,905	21,380,987	12.0	100		

n - estimated number LCL - lower confidence limit UCL - upper confidence limit
NOTE: Based on responses to the question "Apart from when you have a cold, does your chest ever sound wheezy or whistling?" Estimated number in U.S., estimated percent with condition, and PR are based on weighted sample results. PRs in **bold** are significantly different from 100 (p<0.05). PRs in *italics* are based on fewer than five observed cases. See appendices for source description and methods.
SOURCE: National Center for Health Statistics, Third National Health and Nutrition Examination Survey (NHANES III)

Table 3-8. Cough (current): Estimated prevalence and prevalence ratio (PR) adjusted for age, sex, race/ethnicity, and smoking status by agricultural group, U.S. residents age 17 and over, 1988–1994

Worker Group	Number Observed	Estimated Prevalence of Condition in U.S.		PR	95% Confidence Interval	
		n	(%)		LCL	UCL
Farm Workers	56	330,763	9.1	108	82	140
Farm Managers	48	231,149	8.8	90	67	119
Other Agricultural Workers	15	84,140	6.5	74	41	122
All Non-agricultural Workers	1,493	16,332,052	9.2	100		

n - estimated number   LCL - lower confidence limit   UCL - upper confidence limit
NOTE: Based on responses to the question "Do you usually cough on most days for 3 consecutive months or more during the year?" Estimated number in U.S., estimated percent with condition, and PR are based on weighted sample results. PRs in **bold** are significantly different from 100 (p<0.05). PRs in *italics* are based on fewer than five observed cases. See appendices for source description and methods.
SOURCE: National Center for Health Statistics, Third National Health and Nutrition Examination Survey (NHANES III)

Table 3-9. Phlegm (current): Estimated prevalence and prevalence ratio (PR) adjusted for age, sex, race/ethnicity, and smoking status by agricultural group, U.S. residents age 17 and over, 1988–1994

Worker Group	Number Observed	Estimated Prevalence of Condition in U.S. n	Estimated Prevalence of Condition in U.S. (%)	PR	95% Confidence Interval LCL	95% Confidence Interval UCL
Farm Workers	75	425,288	11.7	**133**	106	167
Farm Managers	58	266,299	10.1	94	72	122
Other Agricultural Workers	17	101,183	7.8	86	50	138
All Non-agricultural Workers	1,554	15,218,500	8.5	100		

n - estimated number   LCL - lower confidence limit   UCL - upper confidence limit
NOTE: Based on responses to the question "Do you bring up phlegm on most days for 3 consecutive months or more during the year?" Estimated number in U.S., estimated percent with condition, and PR are based on weighted sample results. PRs in **bold** are significantly different from 100 (p<0.05). PRs in *italics* are based on fewer than five observed cases. See appendices for source description and methods.
SOURCE: National Center for Health Statistics, Third National Health and Nutrition Examination Survey (NHANES III)

Table 3-10. Shortness of breath (current): Estimated prevalence and prevalence ratio (PR) adjusted for age, sex, race/ethnicity, and smoking status by agricultural group, U.S. residents age 17 and over, 1988–1994

Worker Group	Number Observed	Estimated Prevalence of Condition in U.S.		PR	95% Confidence Interval	
		n	(%)		LCL	UCL
Farm Workers	218	915,335	25.2	110	97	126
Farm Managers	125	615,034	23.5	94	79	112
Other Agricultural Workers	41	249,033	19.2	101	74	137
All Non-agricultural Workers	4,642	41,389,383	23.3	100		

n - estimated number   LCL - lower confidence limit   UCL - upper confidence limit
NOTE: Based on responses to the question "Are you troubled by shortness of breath when hurrying on level ground or walking up a slight hill?" Estimated number in U.S., estimated percent with condition, and PR are based on weighted sample results. PRs in **bold** are significantly different from 100 (p<0.05). PRs in *italics* are based on fewer than five observed cases. See appendices for source description and methods.
SOURCE: National Center for Health Statistics, Third National Health and Nutrition Examination Survey (NHANES III)

Morbidity by Agricultural Group within Respiratory Condition–NHANES III

Table 3-11. Stuffy, itchy, runny nose (past year): Estimated prevalence and prevalence ratio (PR) adjusted for age, sex, race/ethnicity, and smoking status by agricultural group, U.S. residents age 17 and over, 1988–1994

Worker Group	Number Observed	Estimated Prevalence of Condition in U.S. n	Estimated Prevalence of Condition in U.S. (%)	PR	95% Confidence Interval LCL	95% Confidence Interval UCL
Farm Workers	273	1,563,707	43.1	**87**	77	98
Farm Managers	176	1,284,201	48.9	97	77	122
Other Agricultural Workers	66	401,689	31.0	**59**	46	75
All Non-agricultural Workers	8,726	98,356,204	55.3	100		

n - estimated number   LCL - lower confidence limit   UCL - upper confidence limit
NOTE: Based on responses to the question "During the past 12 months, have you had any episodes of stuffy, itchy, or runny nose?" Estimated number in U.S., estimated percent with condition, and PR are based on weighted sample results. PRs in **bold** are significantly different from 100 (p<0.05). PRs in *italics* are based on fewer than five observed cases. See appendices for source description and methods.
SOURCE: National Center for Health Statistics, Third National Health and Nutrition Examination Survey (NHANES III)

Table 3-12. Cold or flu (past year): Estimated prevalence and prevalence ratio (PR) adjusted for age, sex, race/ethnicity, and smoking status by agricultural group, U.S. residents age 17 and over, 1988–1994

Worker Group	Number Observed	Estimated Prevalence of Condition in U.S. n	Estimated Prevalence of Condition in U.S. (%)	PR	95% Confidence Interval LCL	95% Confidence Interval UCL
Farm Workers	502	2,373,949	65.4	96	88	105
Farm Managers	203	1,767,314	67.2	108	95	124
Other Agricultural Workers	117	770,838	59.4	83	69	100
All Non-agricultural Workers	12,029	123,171,528	69.2	100		

n - estimated number   LCL - lower confidence limit   UCL - upper confidence limit
NOTE: Based on responses to the question "During the past 12 months, have you had a cold or the flu?" Estimated number in U.S., estimated percent with condition, and PR are based on weighted sample results. PRs in **bold** are significantly different from 100 (p<0.05). PRs in *italics* are based on fewer than five observed cases. See appendices for source description and methods.
SOURCE: National Center for Health Statistics, Third National Health and Nutrition Examination Survey (NHANES III)

Morbidity by Agricultural Group within Respiratory Condition–NHANES III

Table 3-13. Sinusitis (past year): Estimated prevalence and prevalence ratio (PR) adjusted for age, sex, race/ethnicity, and smoking status by agricultural group, U.S. residents age 17 and over, 1988–1994

Worker Group	Number Observed	Estimated Prevalence of Condition in U.S. n	Estimated Prevalence of Condition in U.S. (%)	PR	95% Confidence Interval LCL	95% Confidence Interval UCL
Farm Workers	142	851,416	23.5	**80**	68	94
Farm Managers	95	792,658	30.5	97	79	119
Other Agricultural Workers	36	264,361	20.4	**69**	49	96
All Non-agricultural Workers	5,572	64,524,553	36.3	100		

n - estimated number   LCL - lower confidence limit   UCL - upper confidence limit
NOTE: Based on responses to the question "During the past 12 months, have you had sinusitis or sinus problems?" Estimated number in U.S., estimated percent with condition, and PR are based on weighted sample results. PRs in **bold** are significantly different from 100 (p<0.05). PRs in *italics* are based on fewer than five observed cases. See appendices for source description and methods.
SOURCE: National Center for Health Statistics, Third National Health and Nutrition Examination Survey (NHANES III)

Table 3-14. Pneumonia (past year): Estimated prevalence and prevalence ratio (PR) adjusted for age, sex, race/ethnicity, and smoking status by agricultural group, U.S. residents age 17 and over, 1988–1994

Worker Group	Number Observed	Estimated Prevalence of Condition in U.S. n	Estimated Prevalence of Condition in U.S. (%)	PR	95% Confidence Interval LCL	95% Confidence Interval UCL
Farm Workers	17	52,159	1.4	78	45	125
Farm Managers	12	68,095	2.6	136	70	237
Other Agricultural Workers	6	14,449	1.1	78	29	170
All Non-agricultural Workers	374	3,851,288	2.2	100		

n - estimated number LCL - lower confidence limit UCL - upper confidence limit
NOTE: Based on responses to the question "During the past 12 months, have you had pneumonia?" Estimated number in U.S., estimated percent with condition, and PR are based on weighted sample results. PRs in **bold** are significantly different from 100 (p<0.05). PRs in *italics* are based on fewer than five observed cases. See appendices for source description and methods.
SOURCE: National Center for Health Statistics, Third National Health and Nutrition Examination Survey (NHANES III)

Morbidity by Agricultural Group within Respiratory Condition–NHANES III

Table 3-15. Wheezing (past year): Estimated prevalence and prevalence ratio (PR) adjusted for age, sex, race/ethnicity, and smoking status by agricultural group, U.S. residents age 17 and over, 1988–1994

Worker Group	Number Observed	Estimated Prevalence of Condition in U.S. n	Estimated Prevalence of Condition in U.S. (%)	PR	95% Confidence Interval LCL	95% Confidence Interval UCL
Farm Workers	120	553,404	15.2	97	80	116
Farm Managers	61	419,760	16.0	98	77	126
Other Agricultural Workers	22	166,787	12.8	77	48	117
All Non-agricultural Workers	2,737	31,130,384	17.5	100		

n - estimated number   LCL - lower confidence limit   UCL - upper confidence limit
NOTE: Based on responses to the question "Have you had wheezing or whistling in your chest at any time in the past 12 months?" Estimated number in U.S., estimated percent with condition, and PR are based on weighted sample results. PRs in **bold** are significantly different from 100 (p<0.05). PRs in *italics* are based on fewer than five observed cases. See appendices for source description and methods.
SOURCE: National Center for Health Statistics, Third National Health and Nutrition Examination Survey (NHANES III)

203

Morbidity by Agricultural Group within Respiratory Condition–NHANES III

Table 3-16. Asthma (ever): Estimated prevalence and prevalence ratio (PR) adjusted for age, sex, race/ethnicity, and smoking status by agricultural group, U.S. residents age 17 and over, 1988–1994

Worker Group	Number Observed	Estimated Prevalence of Condition in U.S. n	Estimated Prevalence of Condition in U.S. (%)	PR	95% Confidence Interval LCL	95% Confidence Interval UCL
Farm Workers	53	224,856	6.2	92	70	120
Farm Managers	21	118,254	4.5	**63**	39	96
Other Agricultural Workers	13	139,067	10.7	164	87	280
All Non-agricultural Workers	1,289	14,103,066	7.9	100		

n - estimated number   LCL - lower confidence limit   UCL - upper confidence limit
NOTE: Based on responses to the question "Has a doctor ever told you that you had asthma?" Estimated number in U.S., estimated percent with condition, and PR are based on weighted sample results. PRs in **bold** are significantly different from 100 (p<0.05). PRs in *italics* are based on fewer than five observed cases. See appendices for source description and methods.
SOURCE: National Center for Health Statistics, Third National Health and Nutrition Examination Survey (NHANES III)

Table 3-17. Chronic bronchitis (ever): Estimated prevalence and prevalence ratio (PR) adjusted for age, sex, race/ethnicity, and smoking status by agricultural group, U.S. residents age 17 and over, 1988-1994

Worker Group	Number Observed	Estimated Prevalence of Condition in U.S. n	Estimated Prevalence of Condition in U.S. (%)	PR	95% Confidence Interval LCL	95% Confidence Interval UCL
Farm Workers	43	170,254	4.7	97	71	131
Farm Managers	20	113,407	4.3	75	46	116
Other Agricultural Workers	8	67,224	5.2	120	52	236
All Non-agricultural Workers	1,062	11,141,371	6.3	100		

n - estimated number   LCL - lower confidence limit   UCL - upper confidence limit
NOTE: Based on responses to the question "Has a doctor ever told you that you had chronic bronchitis?" Estimated number in U.S., estimated percent with condition, and PR are based on weighted sample results. PRs in **bold** are significantly different from 100 (p<0.05). PRs in *italics* are based on fewer than five observed cases. See appendices for source description and methods.
SOURCE: National Center for Health Statistics, Third National Health and Nutrition Examination Survey (NHANES III)

Table 3-18. Emphysema (ever): Estimated prevalence and prevalence ratio (PR) adjusted for age, sex, race/ethnicity, and smoking status by agricultural group, U.S. residents age 17 and over, 1988–1994

Worker Group	Number Observed	Estimated Prevalence of Condition in U.S. n	Estimated Prevalence of Condition in U.S. (%)	PR	95% Confidence Interval LCL	95% Confidence Interval UCL
Farm Workers	20	104,616	2.9	134	82	207
Farm Managers	28	112,971	4.3	124	83	179
Other Agricultural Workers	5	27,826	2.1	128	41	299
All Non-agricultural Workers	375	3,401,032	1.9	100		

n - estimated number   LCL - lower confidence limit   UCL - upper confidence limit
NOTE: Based on responses to the question "Has a doctor ever told you that you had emphysema?" Estimated number in U.S., estimated percent with condition, and PR are based on weighted sample results. PRs in **bold** are significantly different from 100 (p<0.05). PRs in *italics* are based on fewer than five observed cases. See appendices for source description and methods.
SOURCE: National Center for Health Statistics, Third National Health and Nutrition Examination Survey (NHANES III)

Morbidity by Agricultural Group within Respiratory Condition–NHANES III

Table 3-19. Hayfever (ever): Estimated prevalence and prevalence ratio (PR) adjusted for age, sex, race/ethnicity, and smoking status by agricultural group, U.S. residents age 17 and over, 1988–1994

Worker Group	Number Observed	Estimated Prevalence of Condition in U.S. n	Estimated Prevalence of Condition in U.S. (%)	PR	95% Confidence Interval LCL	95% Confidence Interval UCL
Farm Workers	30	316,422	8.7	92	62	131
Farm Managers	26	147,088	5.6	**49**	32	72
Other Agricultural Workers	6	32,288	2.5	**26**	10	57
All Non-agricultural Workers	1,710	21,672,453	12.2	100		

n - estimated number   LCL - lower confidence limit   UCL - upper confidence limit
NOTE: Based on responses to the question "Has a doctor ever told you that you had hay fever?" Estimated number in U.S., estimated percent with condition, and PR are based on weighted sample results. PRs in **bold** are significantly different from 100 (p<0.05). PRs in *italics* are based on fewer than five observed cases. See appendices for source description and methods.
SOURCE: National Center for Health Statistics, Third National Health and Nutrition Examination Survey (NHANES III)

Morbidity by Agricultural Group within Respiratory Condition–NHANES III

Figure 3-11. Respiratory conditions (current): Prevalence ratio (PR) adjusted for age, sex, race/ethnicity, and smoking status by agricultural group, U.S. residents age 17 and over, 1988–1994

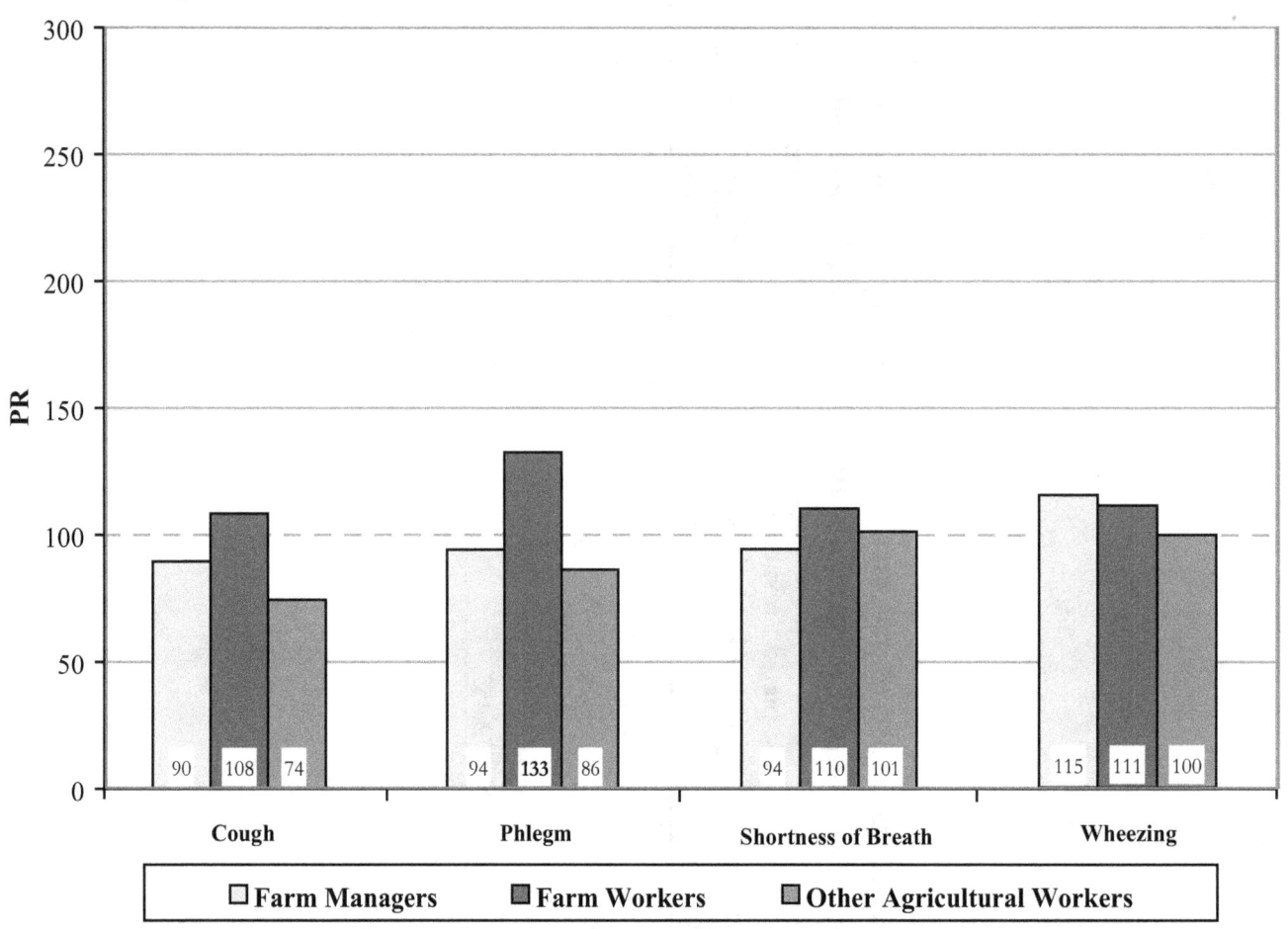

NOTE: Based on responses to the following questions:
  "Do you usually cough on most days for 3 consecutive months or more during the year?"
  "Do you bring up phlegm on most days for 3 consecutive months or more during the year?"
  "Are you troubled by shortness of breath when hurrying on level ground or walking up a slight hill?"
  "Apart from when you have a cold, does your chest ever sound wheezy or whistling?"
PRs in **bold** are significantly different from 100 ($p<0.05$). PRs in *italics* are based on fewer than five observed cases. See appendices for source description and methods.
SOURCE: National Center for Health Statistics, Third National Health and Nutrition Examination Survey (NHANES III)

Morbidity by Agricultural Group within Respiratory Condition–NHANES III

Figure 3-12. Respiratory conditions (past year): Prevalence ratio (PR) adjusted for age, sex, race/ethnicity, and smoking status by agricultural group, U.S. residents age 17 and over, 1988–1994

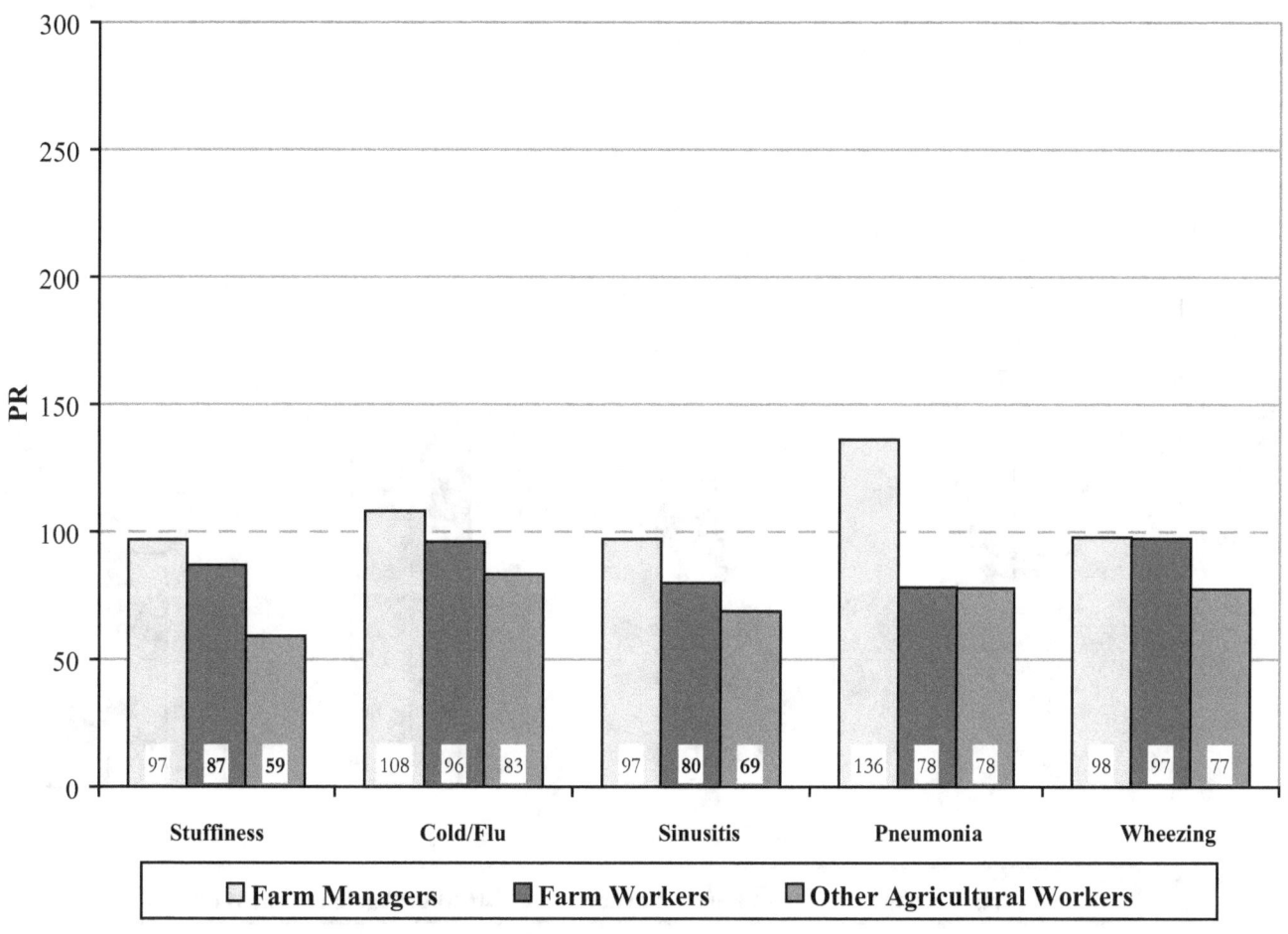

NOTE: Based on responses to the following questions:
"During the past 12 months, have you had any episodes of stuffy, itchy, or runny nose?"
"During the past 12 months, have you had a cold or the flu?"
"During the past 12 months, have you had sinusitis or sinus problems?"
"During the past 12 months, have you had pneumonia?"
"Have you had wheezing or whistling in your chest at any time in the past 12 months?"
PRs in **bold** are significantly different from 100 ($p<0.05$). PRs in *italics* are based on fewer than five observed cases. See appendices for source description and methods.
SOURCE: National Center for Health Statistics, Third National Health and Nutrition Examination Survey (NHANES III)

Morbidity by Agricultural Group within Respiratory Condition–NHANES III

Figure 3-13. Respiratory conditions (ever): Prevalence ratio (PR) adjusted for age, sex, race/ethnicity, and smoking status by agricultural group, U.S. residents age 17 and over, 1988–1994

NOTE: Based on responses to the following questions:
  "Has a doctor ever told you that you had asthma?"
  "Has a doctor ever told you that you had chronic bronchitis?"
  "Has a doctor ever told you that you had emphysema?"
  "Has a doctor ever told you that you had hay fever?"
PRs in **bold** are significantly different from 100 (p<0.05). PRs in *italics* are based on fewer than five observed cases. See appendices for source description and methods.
SOURCE: National Center for Health Statistics, Third National Health and Nutrition Examination Survey (NHANES III)

Morbidity by Sex within Respiratory Condition and Agricultural Group–NHANES III

Figure 3-14. Respiratory conditions (current), farm workers: Prevalence ratio (PR) adjusted for age, race/ethnicity, and smoking status by sex, U.S. residents age 17 and over, 1988–1994

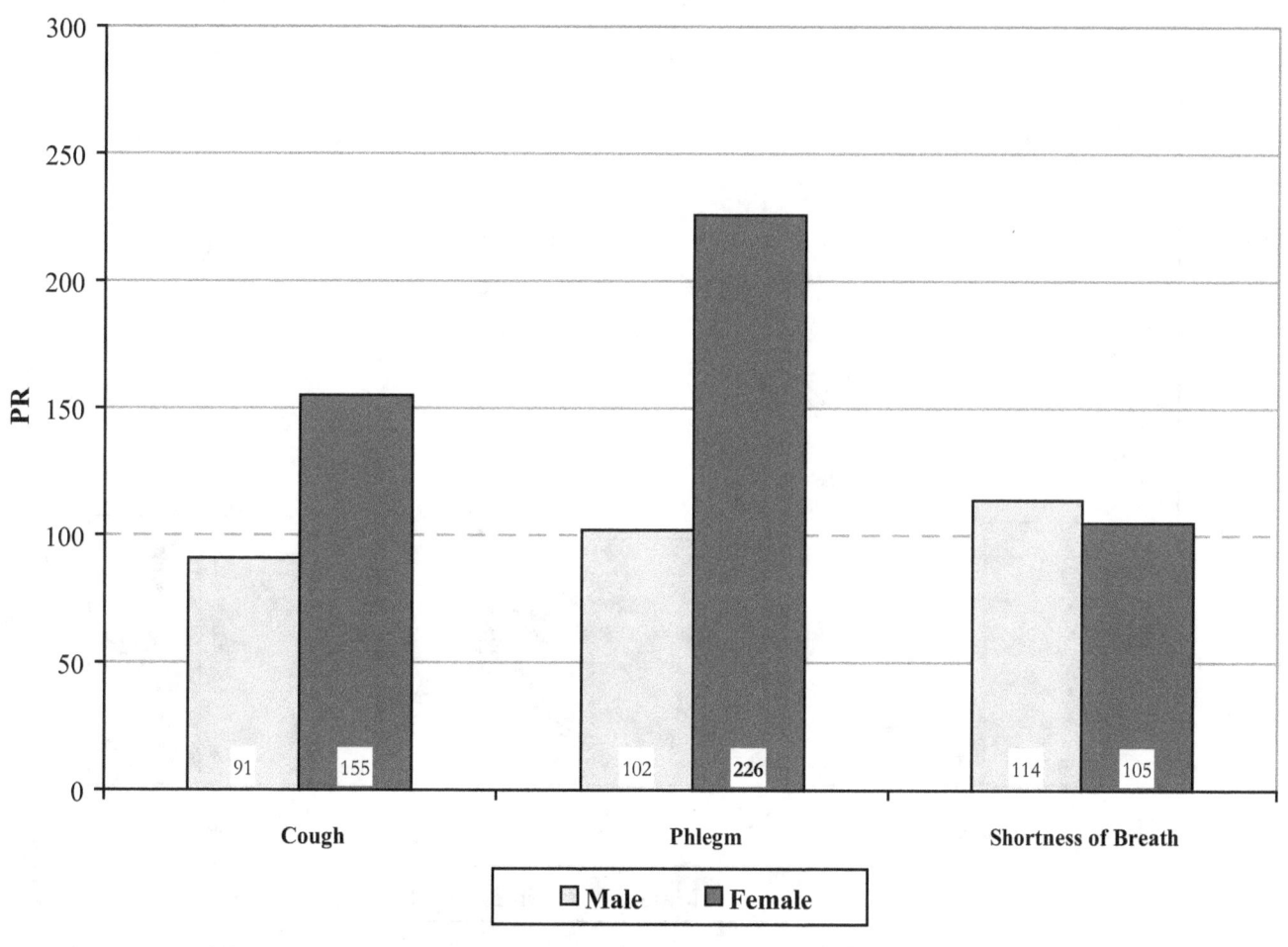

NOTE: Based on responses to the following questions:
 "Do you usually cough on most days for 3 consecutive months or more during the year?"
 "Do you bring up phlegm on most days for 3 consecutive months or more during the year?"
 "Are you troubled by shortness of breath when hurrying on level ground or walking up a slight hill?"
PRs in **bold** are significantly different from 100 (p<0.05). PRs in *italics* are based on fewer than five observed cases. See appendices for source description and methods.
SOURCE: National Center for Health Statistics, Third National Health and Nutrition Examination Survey (NHANES III)

Morbidity by Sex within Respiratory Condition and Agricultural Group–NHANES III

Figure 3-15. Respiratory conditions (current), farm managers: Prevalence ratio (PR) adjusted for age, race/ethnicity, and smoking status by sex, U.S. residents age 17 and over, 1988–1994

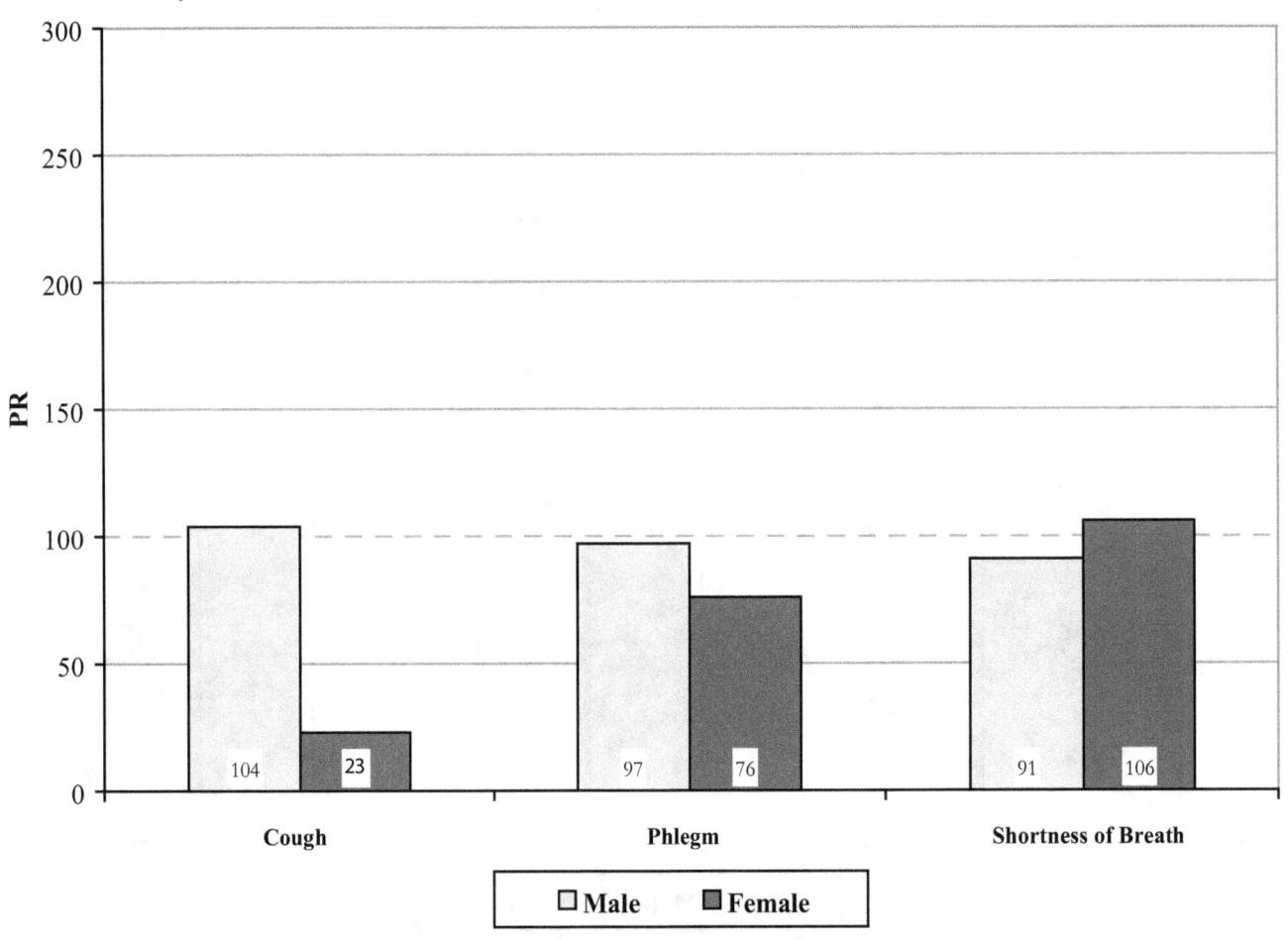

NOTE: Based on responses to the following questions:
 "Do you usually cough on most days for 3 consecutive months or more during the year?"
 "Do you bring up phlegm on most days for 3 consecutive months or more during the year?"
 "Are you troubled by shortness of breath when hurrying on level ground or walking up a slight hill?"
PRs in **bold** are significantly different from 100 (p<0.05). PRs in *italics* are based on fewer than five observed cases. See appendices for source description and methods.
SOURCE: National Center for Health Statistics, Third National Health and Nutrition Examination Survey (NHANES III)

Figure 3-16. Respiratory conditions (current), other agricultural workers: Prevalence ratio (PR) adjusted for age, race/ethnicity, and smoking status by sex, U.S. residents age 17 and over, 1988–1994

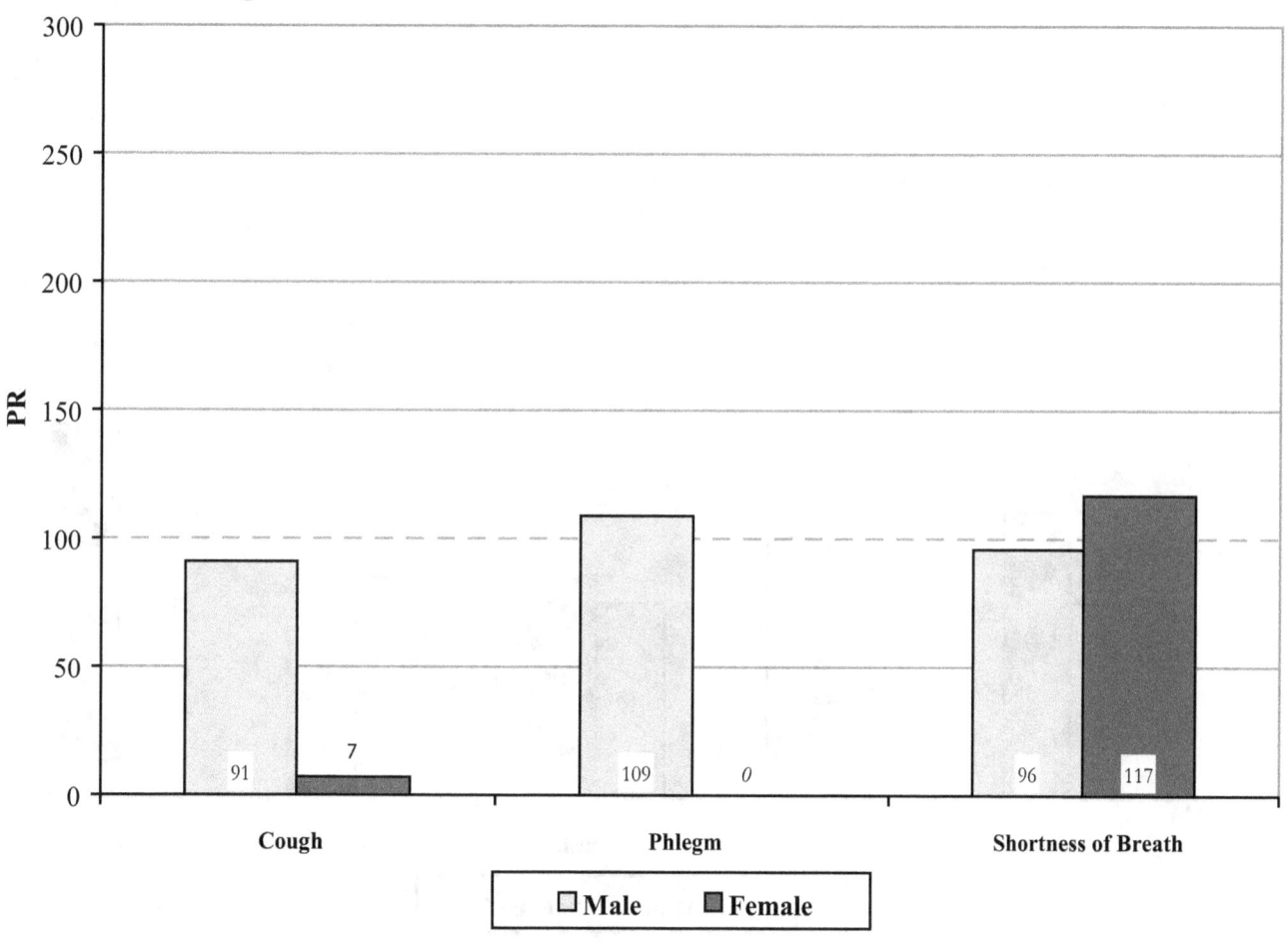

NOTE: Based on responses to the following questions:
 "Do you usually cough on most days for 3 consecutive months or more during the year?"
 "Do you bring up phlegm on most days for 3 consecutive months or more during the year?"
 "Are you troubled by shortness of breath when hurrying on level ground or walking up a slight hill?"
PRs in **bold** are significantly different from 100 (p<0.05). PRs in *italics* are based on fewer than five observed cases. See appendices for source description and methods.
SOURCE: National Center for Health Statistics, Third National Health and Nutrition Examination Survey (NHANES III)

Morbidity by Sex within Respiratory Condition and Agricultural Group–NHANES III

Figure 3-17. Respiratory conditions (past year), farm workers: Prevalence ratio (PR) adjusted for age, race/ethnicity, and smoking status by sex, U.S. residents age 17 and over, 1988–1994

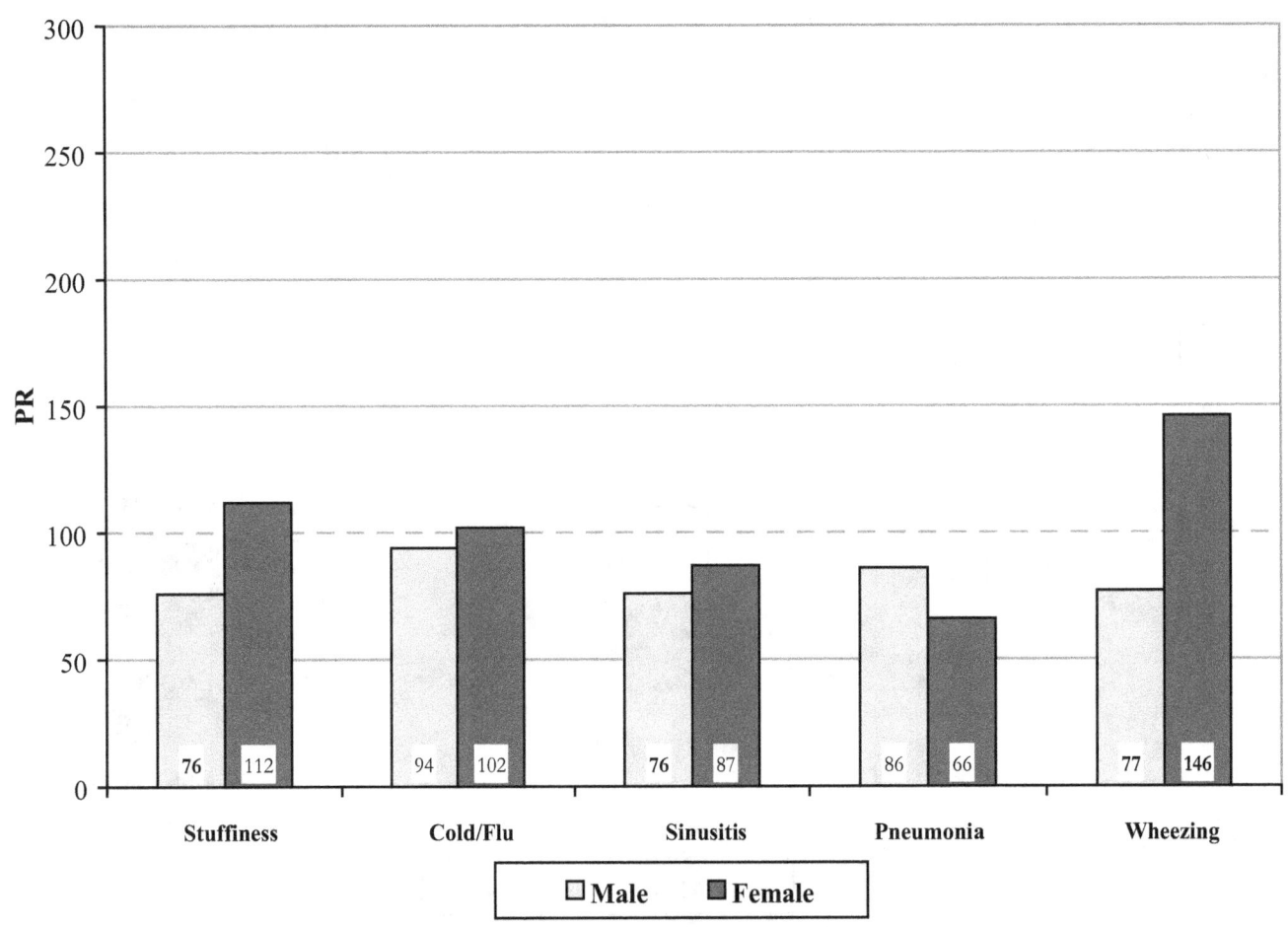

NOTE: Based on responses to the following questions:
  "During the past 12 months, have you had any episodes of stuffy, itchy, or runny nose?"
  "During the past 12 months, have you had a cold or the flu?"
  "During the past 12 months, have you had sinusitis or sinus problems?"
  "During the past 12 months, have you had pneumonia?"
  "Have you had wheezing or whistling in your chest at any time in the past 12 months?"
PRs in **bold** are significantly different from 100 (p<0.05). PRs in *italics* are based on fewer than five observed cases. See appendices for source description and methods.
SOURCE: National Center for Health Statistics, Third National Health and Nutrition Examination Survey (NHANES III)

Morbidity by Sex within Respiratory Condition and Agricultural Group–NHANES III

Figure 3-18. Respiratory conditions (past year), farm managers: Prevalence ratio (PR) adjusted for age, race/ethnicity, and smoking status by sex, U.S. residents age 17 and over, 1988–1994

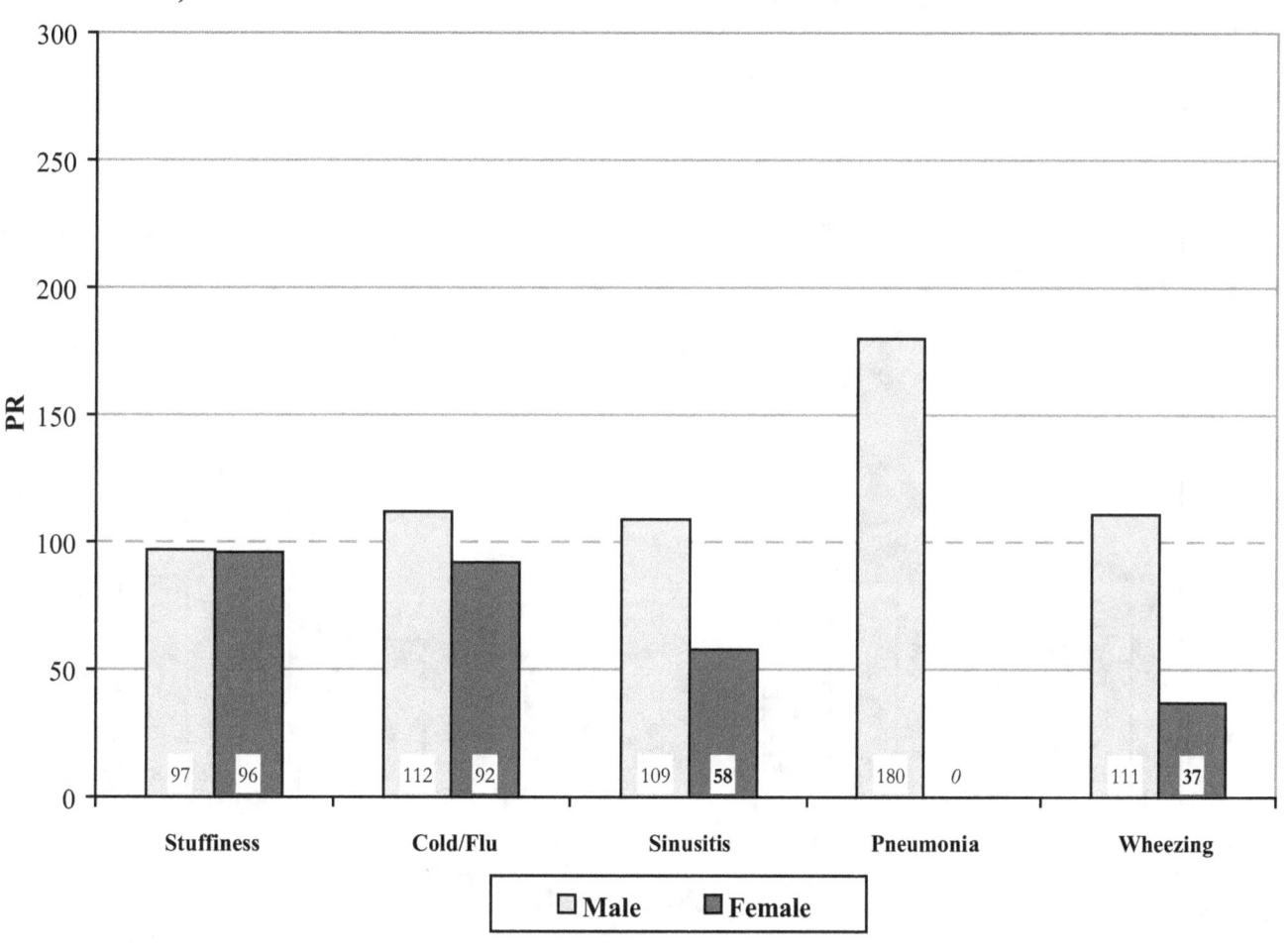

NOTE: Based on responses to the following questions:
  "During the past 12 months, have you had any episodes of stuffy, itchy, or runny nose?"
  "During the past 12 months, have you had a cold or the flu?"
  "During the past 12 months, have you had sinusitis or sinus problems?"
  "During the past 12 months, have you had pneumonia?"
  "Have you had wheezing or whistling in your chest at any time in the past 12 months?"
PRs in **bold** are significantly different from 100 (p<0.05). PRs in *italics* are based on fewer than five observed cases. See appendices for source description and methods.
SOURCE: National Center for Health Statistics, Third National Health and Nutrition Examination Survey (NHANES III)

Morbidity by Sex within Respiratory Condition and Agricultural Group–NHANES III

Figure 3-19. Respiratory conditions (past year), other agricultural workers: Prevalence ratio (PR) adjusted for age, race/ethnicity, and smoking status by sex, U.S. residents age 17 and over, 1988–1994

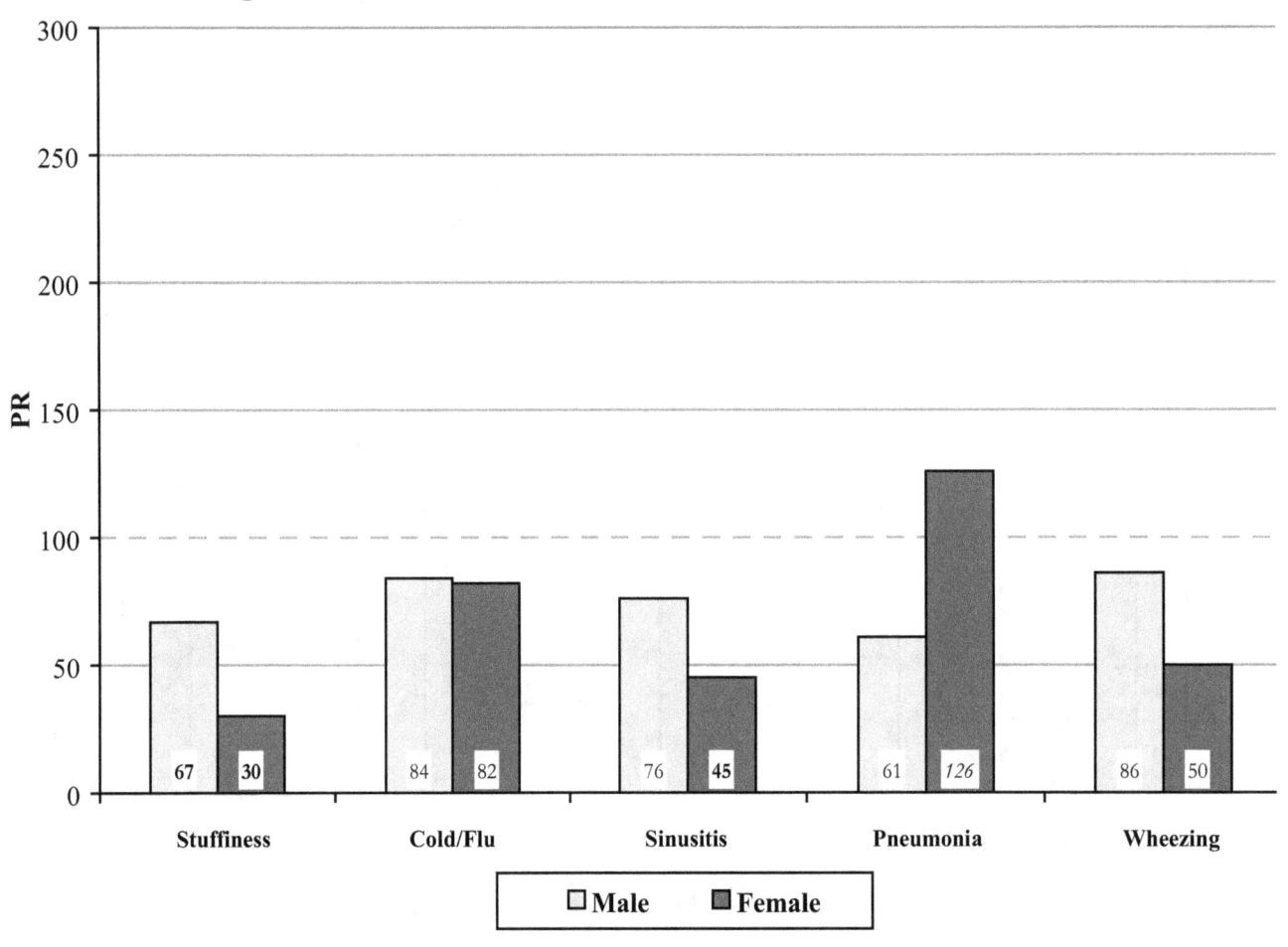

NOTE: Based on responses to the following questions:
 "During the past 12 months, have you had any episodes of stuffy, itchy, or runny nose?"
 "During the past 12 months, have you had a cold or the flu?"
 "During the past 12 months, have you had sinusitis or sinus problems?"
 "During the past 12 months, have you had pneumonia?"
 "Have you had wheezing or whistling in your chest at any time in the past 12 months?"
PRs in **bold** are significantly different from 100 ($p<0.05$). PRs in *italics* are based on fewer than five observed cases. See appendices for source description and methods.
SOURCE: National Center for Health Statistics, Third National Health and Nutrition Examination Survey (NHANES III)

216

Morbidity by Sex within Respiratory Condition and Agricultural Group–NHANES III

Figure 3-20. Respiratory conditions (ever), farm workers: Prevalence ratio (PR) adjusted for age, race/ethnicity, and smoking status by sex, U.S. residents age 17 and over, 1988–1994

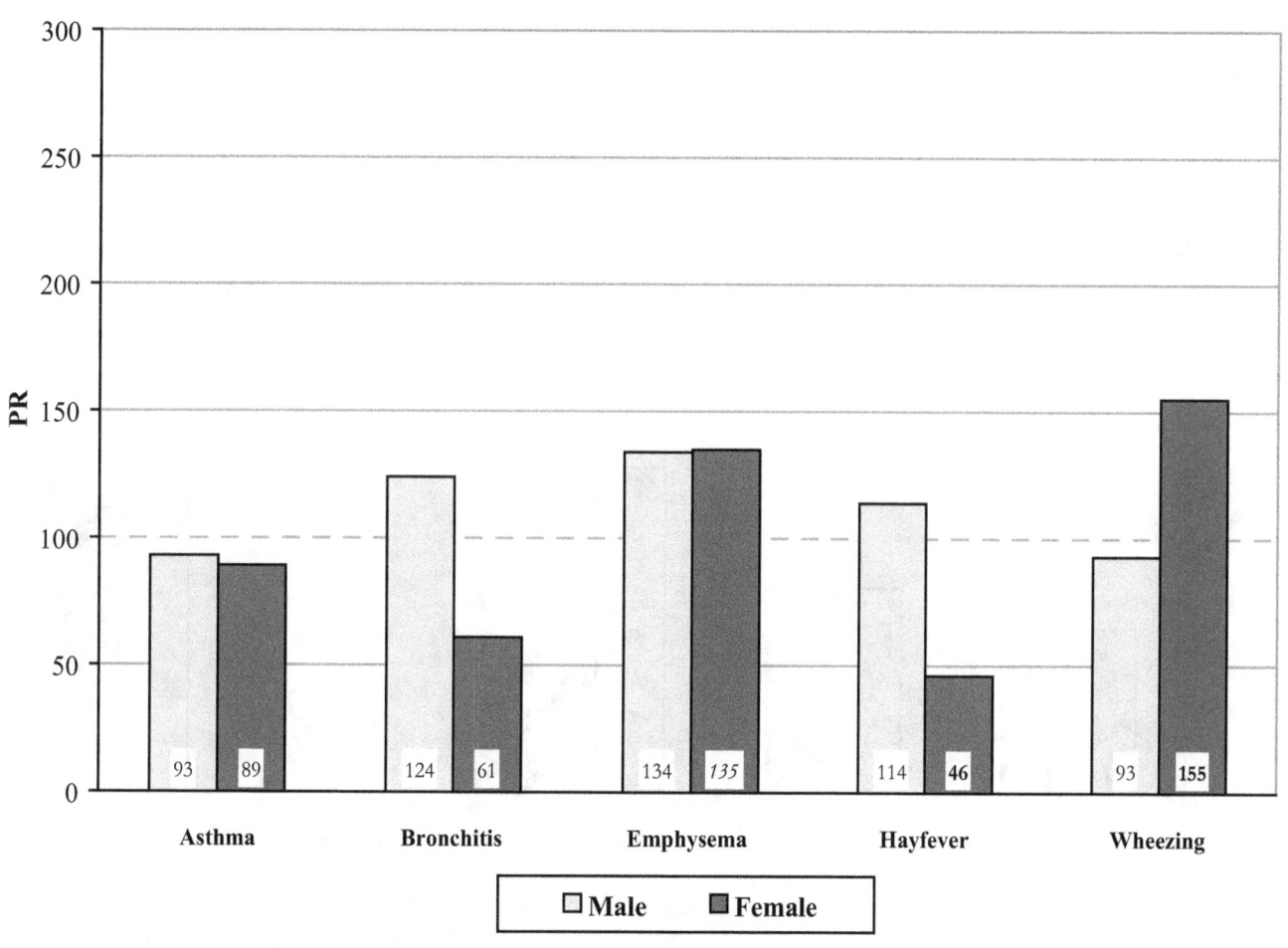

NOTE: Based on responses to the following questions:
  "Has a doctor ever told you that you had asthma?"
  "Has a doctor ever told you that you had chronic bronchitis?"
  "Has a doctor ever told you that you had emphysema?"
  "Has a doctor ever told you that you had hay fever?"
  "Apart from when you have a cold, does your chest ever sound wheezy or whistling?"
PRs in **bold** are significantly different from 100 ($p<0.05$). PRs in *italics* are based on fewer than five observed cases. See appendices for source description and methods.
SOURCE: National Center for Health Statistics, Third National Health and Nutrition Examination Survey (NHANES III)

Morbidity by Sex within Respiratory Condition and Agricultural Group–NHANES III

Figure 3-21. Respiratory conditions (ever), farm managers: Prevalence ratio (PR) adjusted for age, race/ethnicity, and smoking status by sex, U.S. residents age 17 and over, 1988–1994

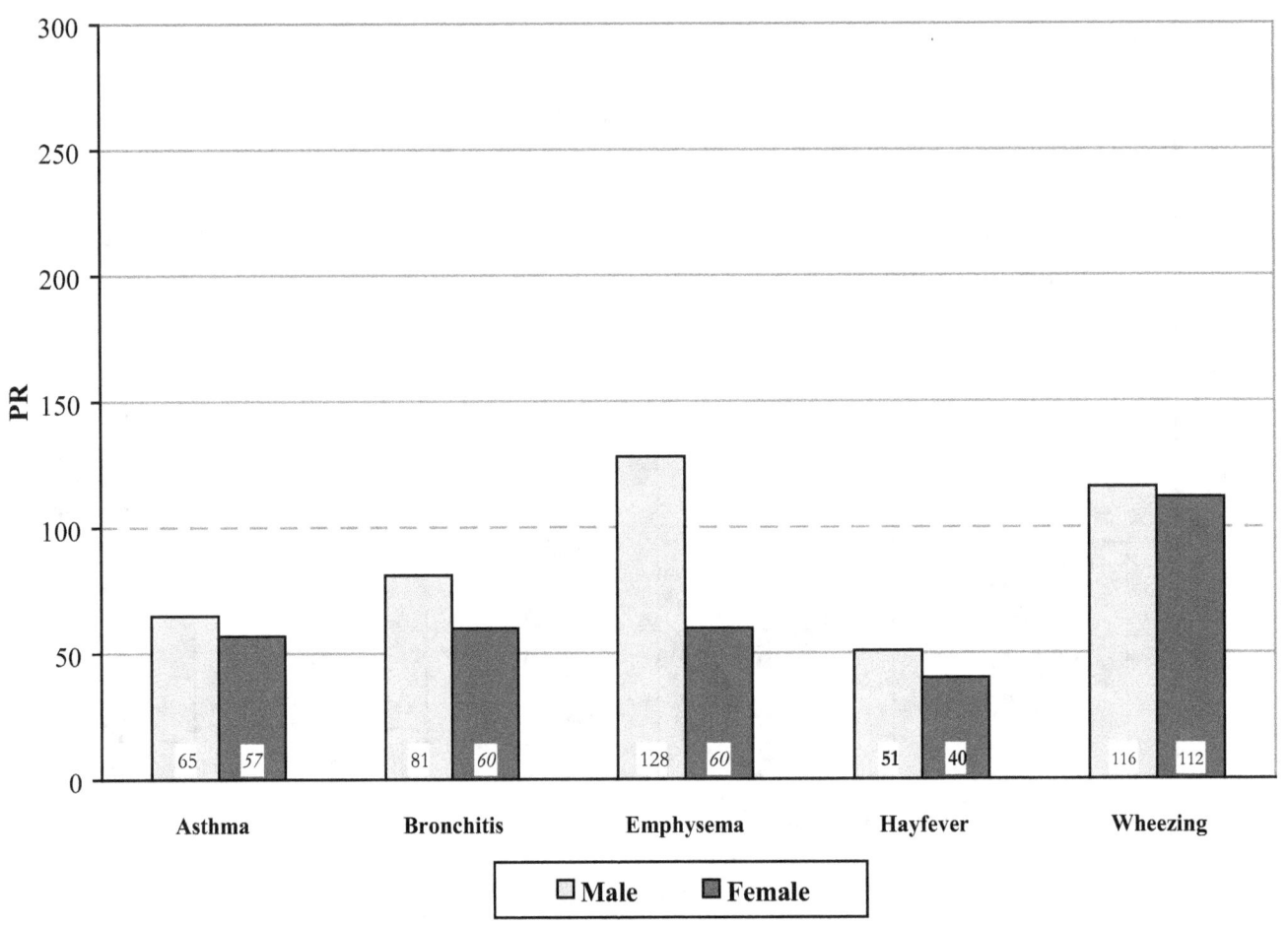

NOTE: Based on responses to the following questions:
 "Has a doctor ever told you that you had asthma?"
 "Has a doctor ever told you that you had chronic bronchitis?"
 "Has a doctor ever told you that you had emphysema?"
 "Has a doctor ever told you that you had hay fever?"
 "Apart from when you have a cold, does your chest ever sound wheezy or whistling?"
PRs in **bold** are significantly different from 100 (p<0.05). PRs in *italics* are based on fewer than five observed cases. See appendices for source description and methods.
SOURCE: National Center for Health Statistics, Third National Health and Nutrition Examination Survey (NHANES III)

Morbidity by Sex within Respiratory Condition and Agricultural Group–NHANES III

Figure 3-22. Respiratory conditions (ever), other agricultural workers: Prevalence ratio (PR) adjusted for age, race/ethnicity, and smoking status by sex, U.S. residents age 17 and over, 1988–1994

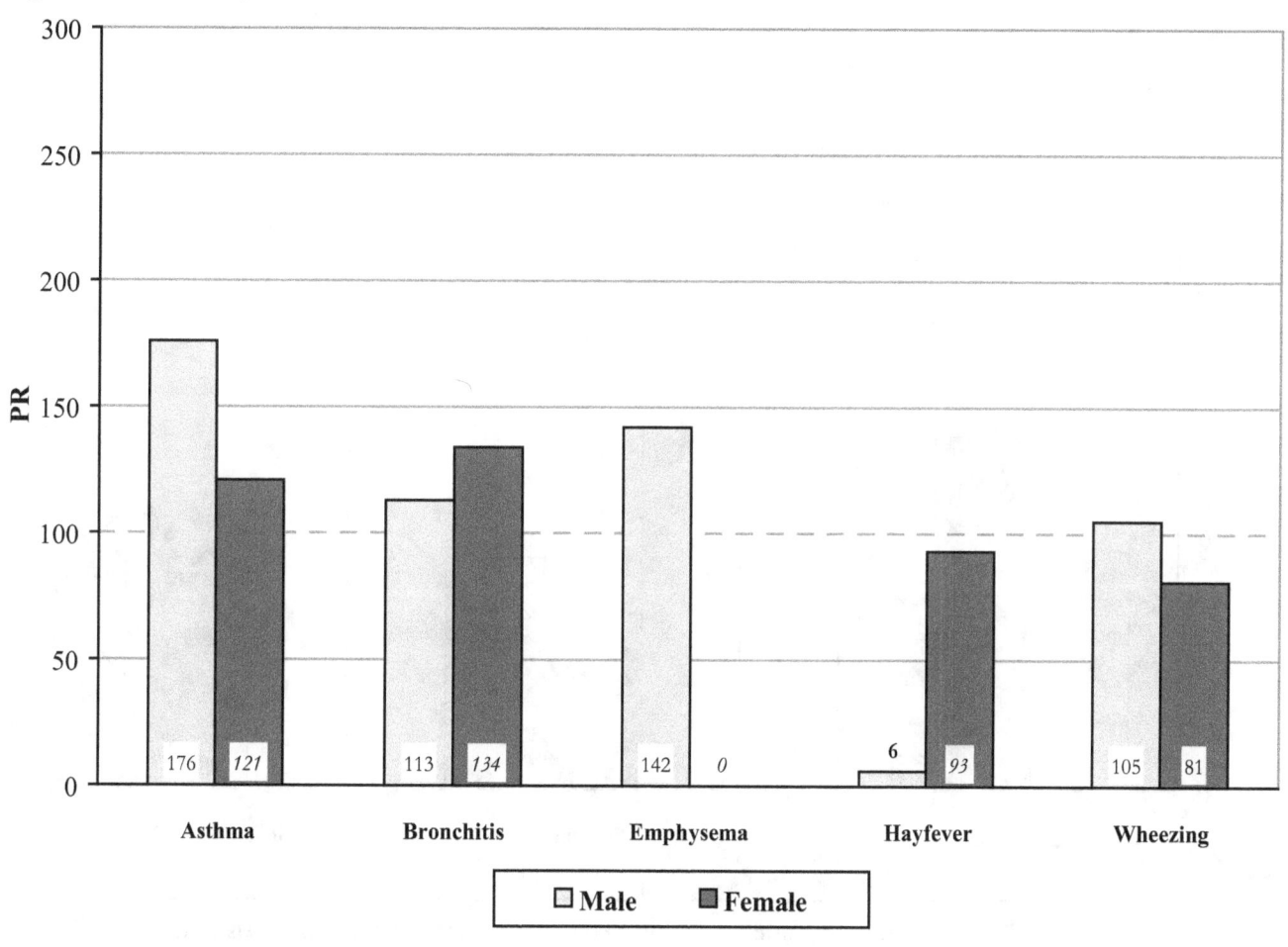

NOTE: Based on responses to the following questions:
 "Has a doctor ever told you that you had asthma?"
 "Has a doctor ever told you that you had chronic bronchitis?"
 "Has a doctor ever told you that you had emphysema?"
 "Has a doctor ever told you that you had hay fever?"
 "Apart from when you have a cold, does your chest ever sound wheezy or whistling?"
PRs in **bold** are significantly different from 100 (p<0.05). PRs in *italics* are based on fewer than five observed cases. See appendices for source description and methods.
SOURCE: National Center for Health Statistics, Third National Health and Nutrition Examination Survey (NHANES III)

*Morbidity by Race/Ethnicity within Respiratory Condition and Agricultural Group–
NHANES III*

Figure 3-23. Respiratory conditions (current), farm workers: Prevalence ratio (PR) adjusted for age, sex, and smoking status by race/ethnicity, U.S. residents age 17 and over, 1988–1994

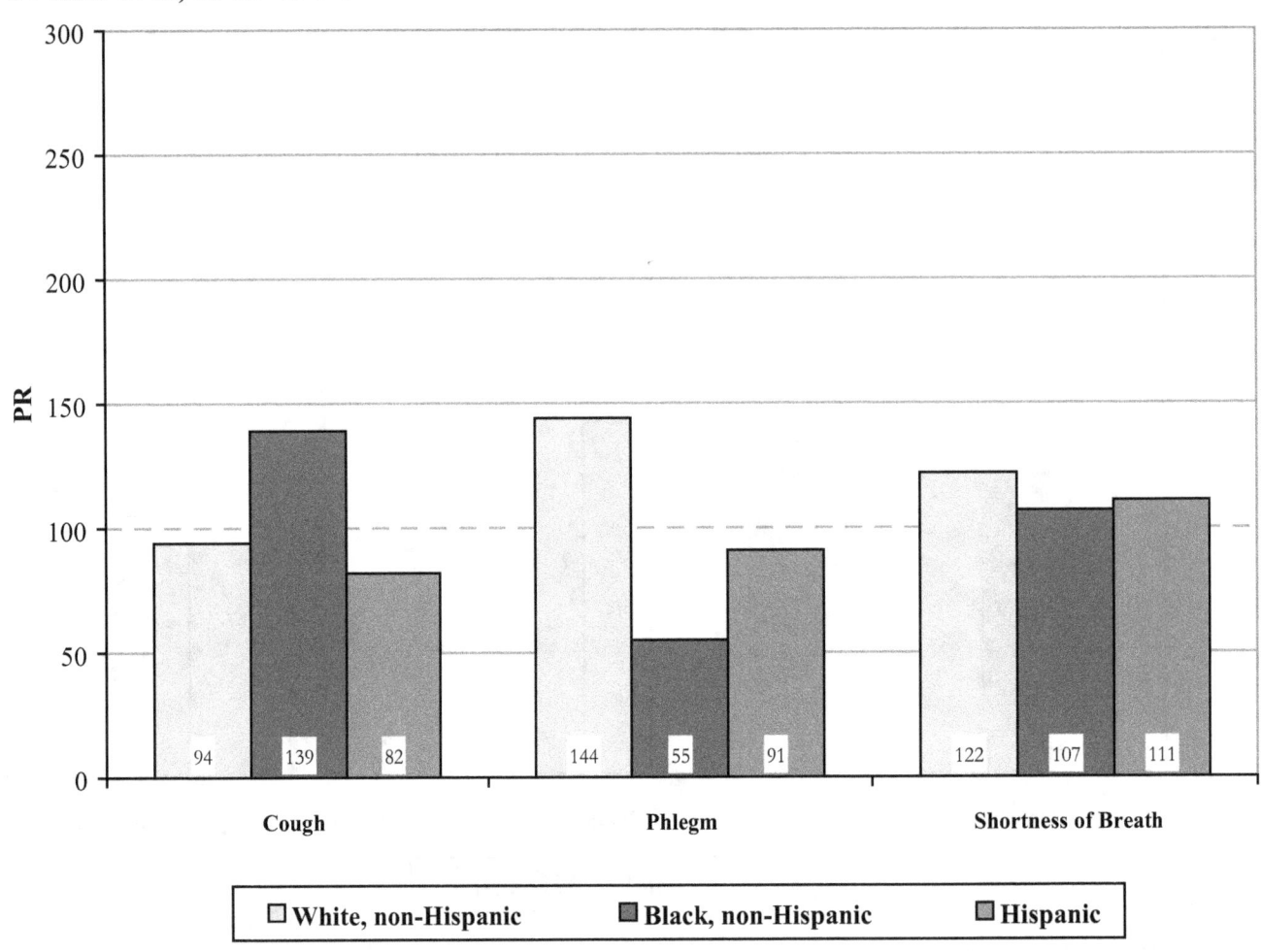

NOTE: Based on responses to the following questions:
 "Do you usually cough on most days for 3 consecutive months or more during the year?"
 "Do you bring up phlegm on most days for 3 consecutive months or more during the year?"
 "Are you troubled by shortness of breath when hurrying on level ground or walking up a slight hill?"
PRs in **bold** are significantly different from 100 ($p<0.05$). PRs in *italics* are based on fewer than five observed cases. See appendices for source description and methods.
SOURCE: National Center for Health Statistics, Third National Health and Nutrition Examination Survey (NHANES III)

Morbidity by Race/Ethnicity within Respiratory Condition and Agricultural Group–
NHANES III

Figure 3-24. Respiratory conditions (current), farm managers: Prevalence ratio (PR) adjusted for age, sex, and smoking status by race/ethnicity, U.S. residents age 17 and over, 1988–1994

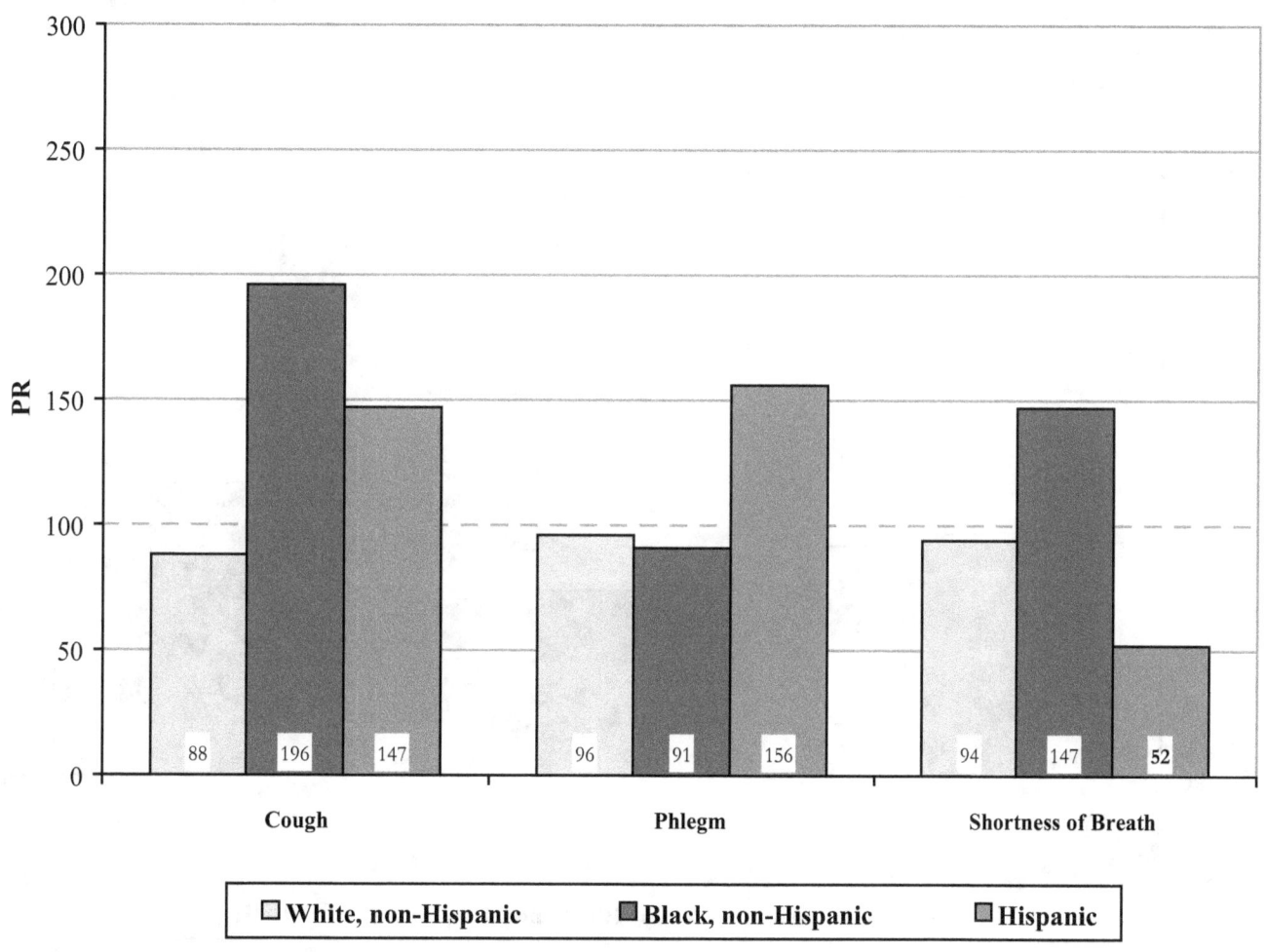

NOTE: Based on responses to the following questions:
 "Do you usually cough on most days for 3 consecutive months or more during the year?"
 "Do you bring up phlegm on most days for 3 consecutive months or more during the year?"
 "Are you troubled by shortness of breath when hurrying on level ground or walking up a slight hill?"
PRs in **bold** are significantly different from 100 (p<0.05). PRs in *italics* are based on fewer than five observed cases. See appendices for source description and methods.
SOURCE: National Center for Health Statistics, Third National Health and Nutrition Examination Survey (NHANES III)

Morbidity by Race/Ethnicity within Respiratory Condition and Agricultural Group–NHANES III

Figure 3-25. Respiratory conditions (current), other agricultural workers: Prevalence ratio (PR) adjusted for age, sex, and smoking status by race/ethnicity, U.S. residents age 17 and over, 1988–1994

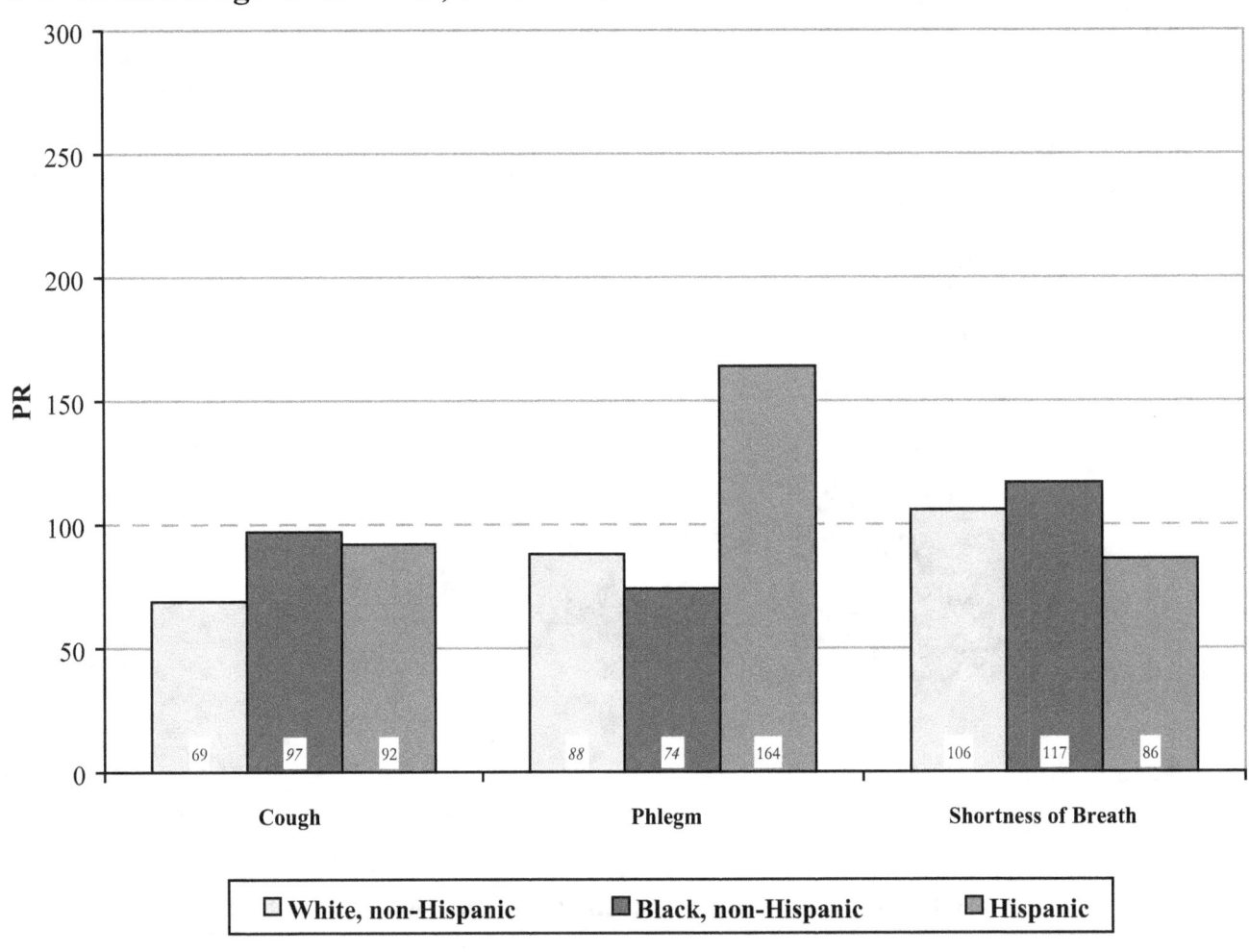

NOTE: Based on responses to the following questions:
 "Do you usually cough on most days for 3 consecutive months or more during the year?"
 "Do you bring up phlegm on most days for 3 consecutive months or more during the year?"
 "Are you troubled by shortness of breath when hurrying on level ground or walking up a slight hill?"
PRs in **bold** are significantly different from 100 (p<0.05). PRs in *italics* are based on fewer than five observed cases. See appendices for source description and methods.
SOURCE: National Center for Health Statistics, Third National Health and Nutrition Examination Survey (NHANES III)

Morbidity by Race/Ethnicity within Respiratory Condition and Agricultural Group–
NHANES III

Figure 3-26. Respiratory conditions (past year), farm workers: Prevalence ratio (PR) adjusted for age, sex, and smoking status by race/ethnicity, U.S. residents age 17 and over, 1988–1994

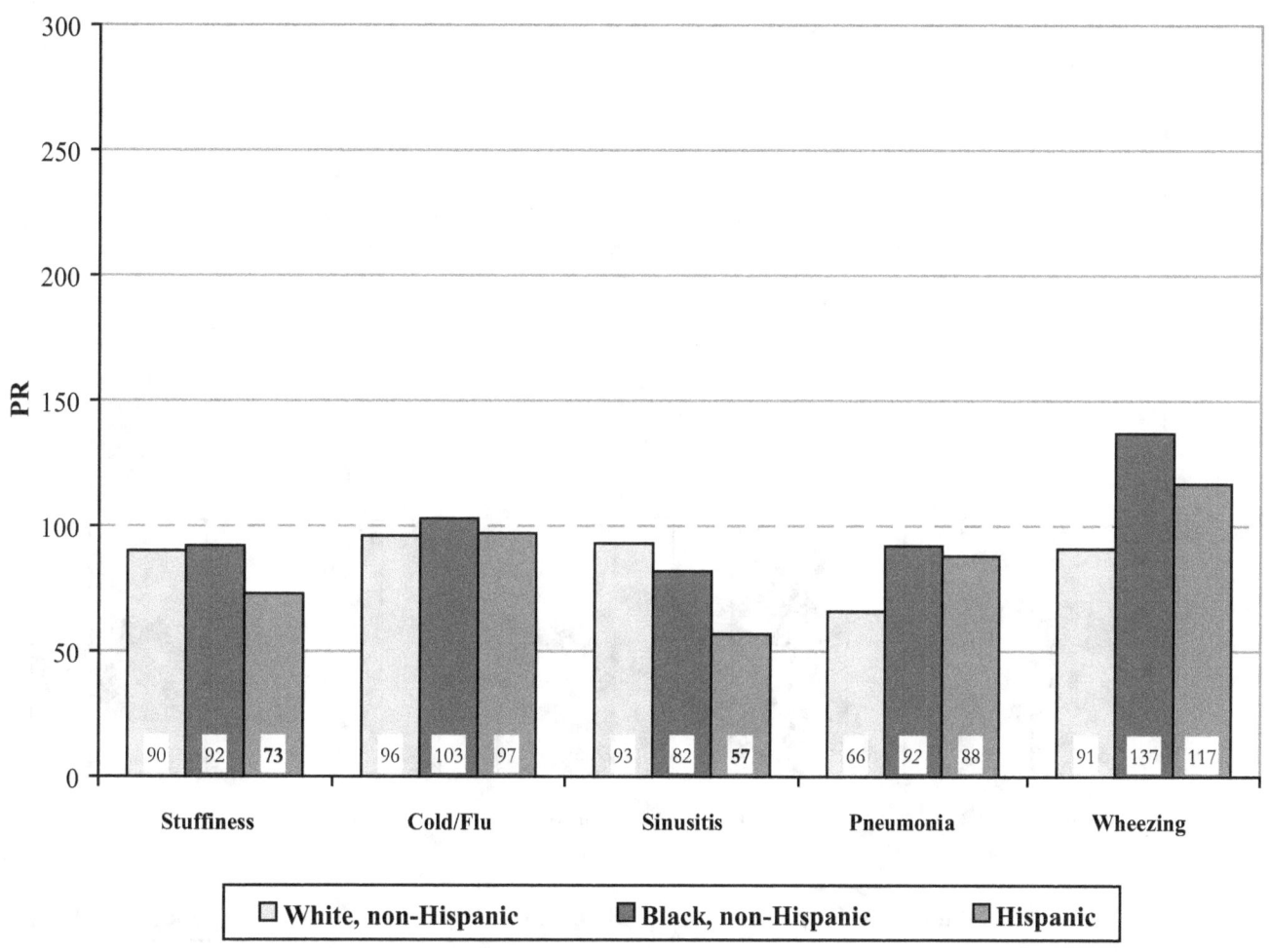

NOTE: Based on responses to the following questions:
"During the past 12 months, have you had any episodes of stuffy, itchy, or runny nose?"
"During the past 12 months, have you had a cold or the flu?"
"During the past 12 months, have you had sinusitis or sinus problems?"
"During the past 12 months, have you had pneumonia?"
"Have you had wheezing or whistling in your chest at any time in the past 12 months?"
PRs in **bold** are significantly different from 100 ($p<0.05$). PRs in *italics* are based on fewer than five observed cases. See appendices for source description and methods.
SOURCE: National Center for Health Statistics, Third National Health and Nutrition Examination Survey (NHANES III)

Morbidity by Race/Ethnicity within Respiratory Condition and Agricultural Group–NHANES III

Figure 3-27. Respiratory conditions (past year), farm managers: Prevalence ratio (PR) adjusted for age, sex, and smoking status by race/ethnicity, U.S. residents age 17 and over, 1988–1994

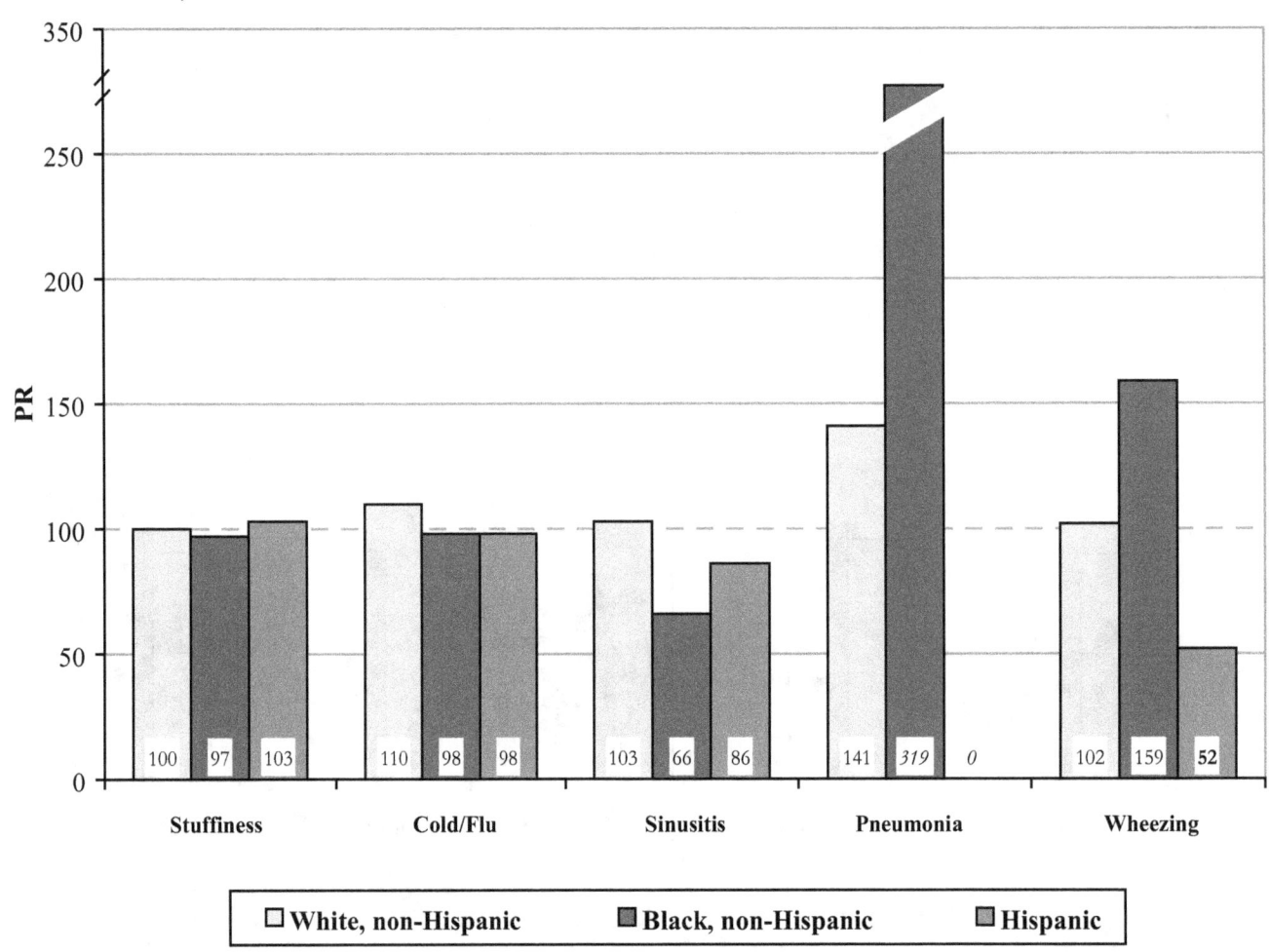

NOTE: Based on responses to the following questions:
 "During the past 12 months, have you had any episodes of stuffy, itchy, or runny nose?"
 "During the past 12 months, have you had a cold or the flu?"
 "During the past 12 months, have you had sinusitis or sinus problems?"
 "During the past 12 months, have you had pneumonia?"
 "Have you had wheezing or whistling in your chest at any time in the past 12 months?"
PRs in **bold** are significantly different from 100 (p<0.05). PRs in *italics* are based on fewer than five observed cases. See appendices for source description and methods.
SOURCE: National Center for Health Statistics, Third National Health and Nutrition Examination Survey (NHANES III)

Morbidity by Race/Ethnicity within Respiratory Condition and Agricultural Group–
NHANES III

Figure 3-28. Respiratory conditions (past year), other agricultural workers: Prevalence ratio (PR) adjusted for age, sex, and smoking status by race/ethnicity, U.S. residents age 17 and over, 1988–1994

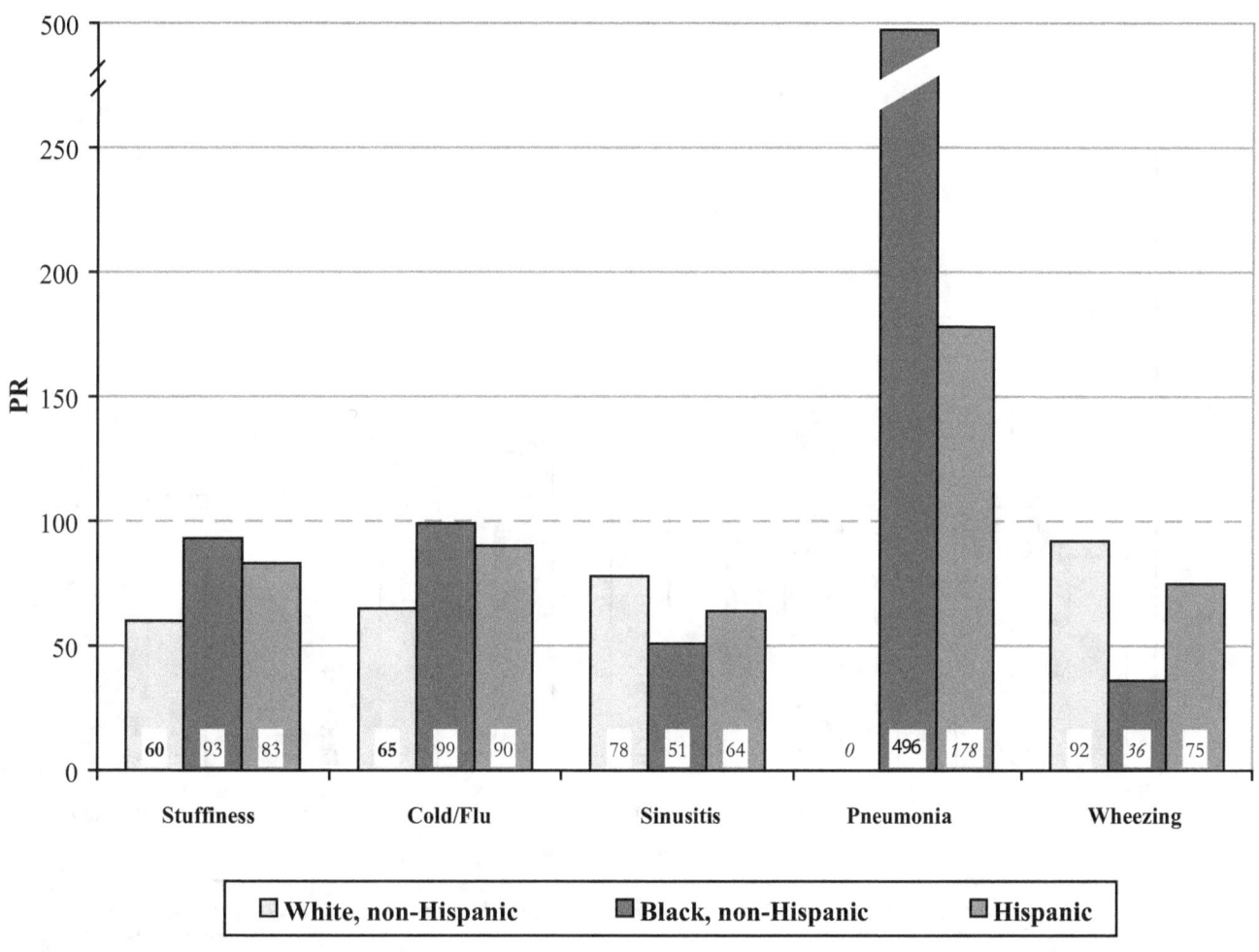

NOTE: Based on responses to the following questions:
"During the past 12 months, have you had any episodes of stuffy, itchy, or runny nose?"
"During the past 12 months, have you had a cold or the flu?"
"During the past 12 months, have you had sinusitis or sinus problems?"
"During the past 12 months, have you had pneumonia?"
"Have you had wheezing or whistling in your chest at any time in the past 12 months?"
PRs in **bold** are significantly different from 100 (p<0.05). PRs in *italics* are based on fewer than five observed cases. See appendices for source description and methods.
SOURCE: National Center for Health Statistics, Third National Health and Nutrition Examination Survey (NHANES III)

Morbidity by Race/Ethnicity within Respiratory Condition and Agricultural Group–NHANES III

Figure 3-29. Respiratory conditions (ever), farm workers: Prevalence ratio (PR) adjusted for age, sex, and smoking status by race/ethnicity, U.S. residents age 17 and over, 1988–1994

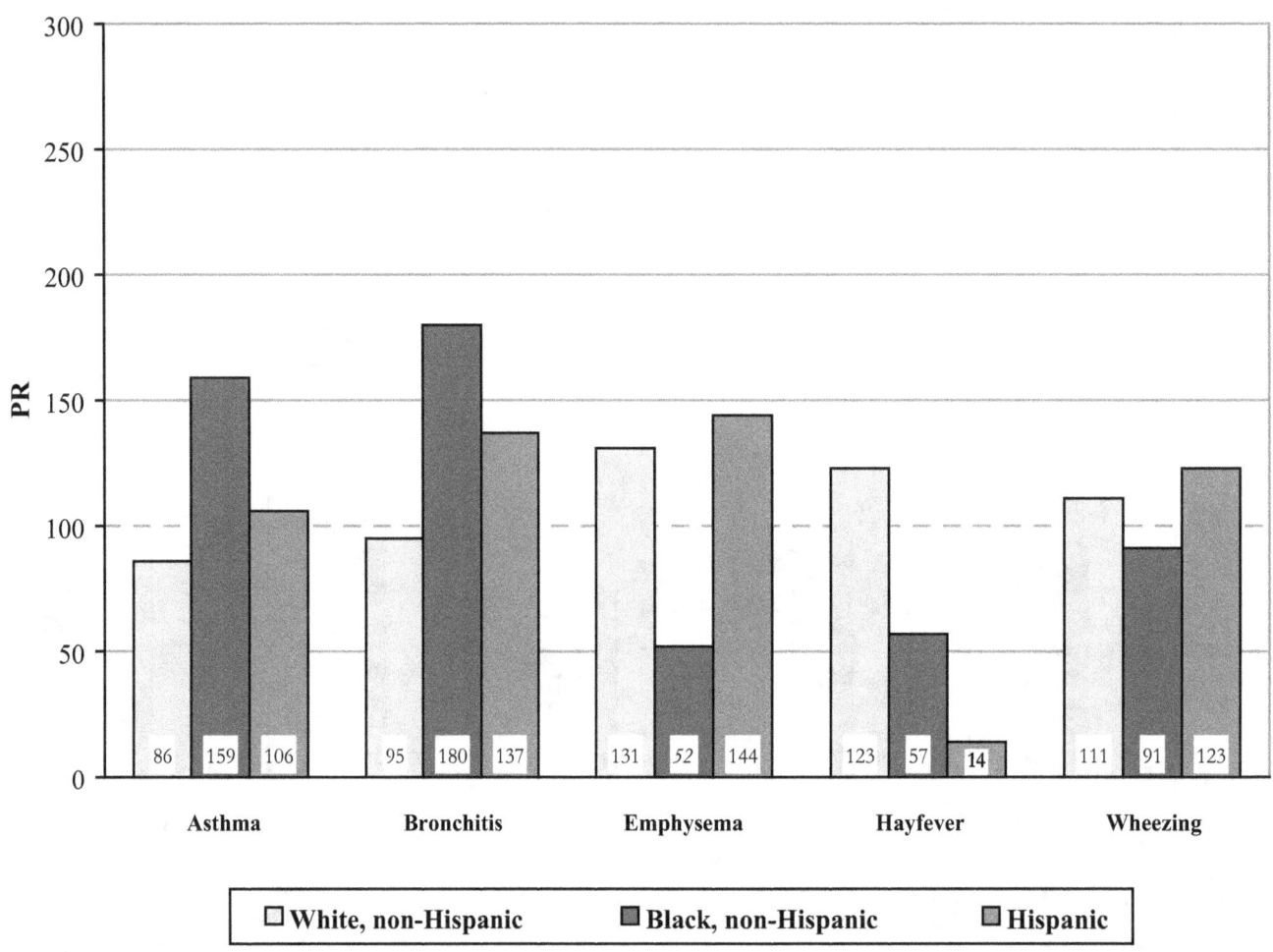

NOTE: Based on responses to the following questions:
"Has a doctor ever told you that you had asthma?"
"Has a doctor ever told you that you had chronic bronchitis?"
"Has a doctor ever told you that you had emphysema?"
"Has a doctor ever told you that you had hay fever?"
"Apart from when you have a cold, does your chest ever sound wheezy or whistling?"

PRs in **bold** are significantly different from 100 ($p<0.05$). PRs in *italics* are based on fewer than five observed cases. See appendices for source description and methods.
SOURCE: National Center for Health Statistics, Third National Health and Nutrition Examination Survey (NHANES III)

Morbidity by Race/Ethnicity within Respiratory Condition and Agricultural Group–NHANES III

Figure 3-30. Respiratory conditions (ever), farm managers: Prevalence ratio (PR) adjusted for age, sex, and smoking status by race/ethnicity, U.S. residents age 17 and over, 1988–1994

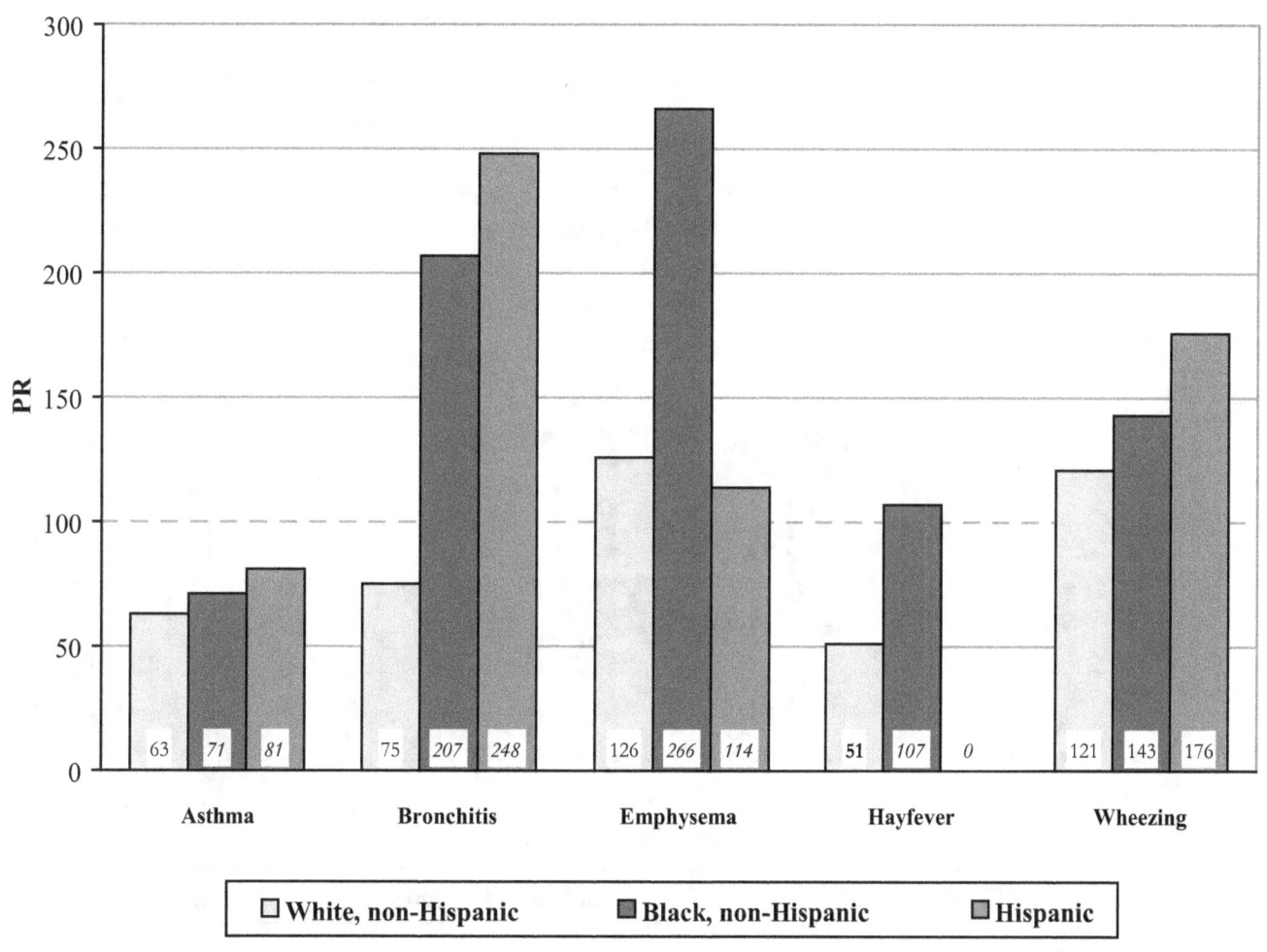

NOTE: Based on responses to the following questions:
 "Has a doctor ever told you that you had asthma?"
 "Has a doctor ever told you that you had chronic bronchitis?"
 "Has a doctor ever told you that you had emphysema?"
 "Has a doctor ever told you that you had hay fever?"
 "Apart from when you have a cold, does your chest ever sound wheezy or whistling?"
PRs in **bold** are significantly different from 100 (p<0.05). PRs in *italics* are based on fewer than five observed cases. See appendices for source description and methods.
SOURCE: National Center for Health Statistics, Third National Health and Nutrition Examination Survey (NHANES III)

227

Morbidity by Race/Ethnicity within Respiratory Condition and Agricultural Group–NHANES III

Figure 3-31. Respiratory conditions (ever), other agricultural workers: Prevalence ratio (PR) adjusted for age, sex, and smoking status by race/ethnicity, U.S. residents age 17 and over, 1988–1994

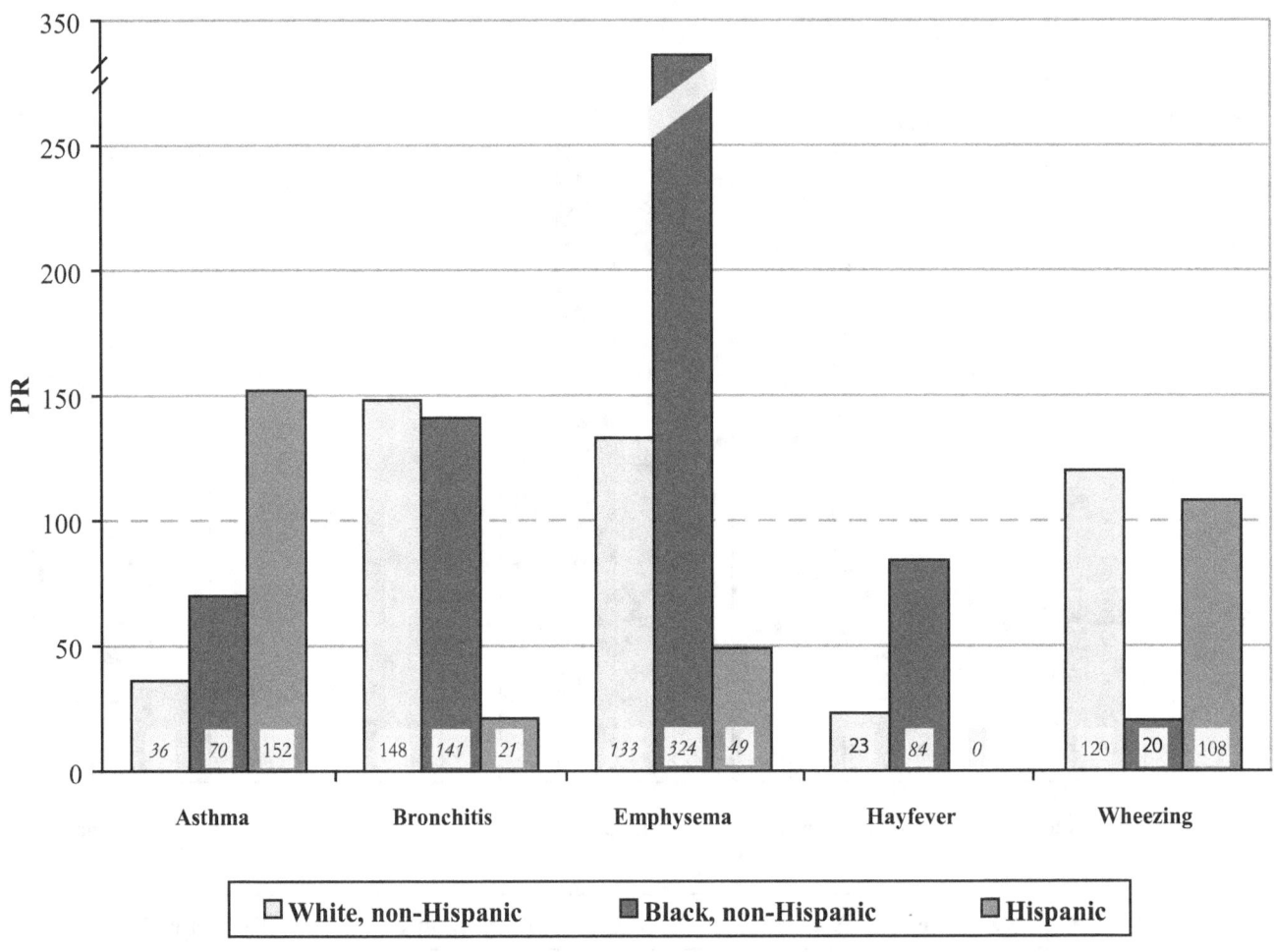

NOTE: Based on responses to the following questions:
 "Has a doctor ever told you that you had asthma?"
 "Has a doctor ever told you that you had chronic bronchitis?"
 "Has a doctor ever told you that you had emphysema?"
 "Has a doctor ever told you that you had hay fever?"
 "Apart from when you have a cold, does your chest ever sound wheezy or whistling?"

PRs in **bold** are significantly different from 100 (p<0.05). PRs in *italics* are based on fewer than five observed cases. See appendices for source description and methods.

SOURCE: National Center for Health Statistics, Third National Health and Nutrition Examination Survey (NHANES III)

Morbidity by Smoking Status within Respiratory Condition and Agricultural Group–NHANES III

Figure 3-32. Respiratory conditions (current), farm workers: Prevalence ratio (PR) adjusted for age, sex, and race/ethnicity by smoking status, U.S. residents age 17 and over, 1988–1994

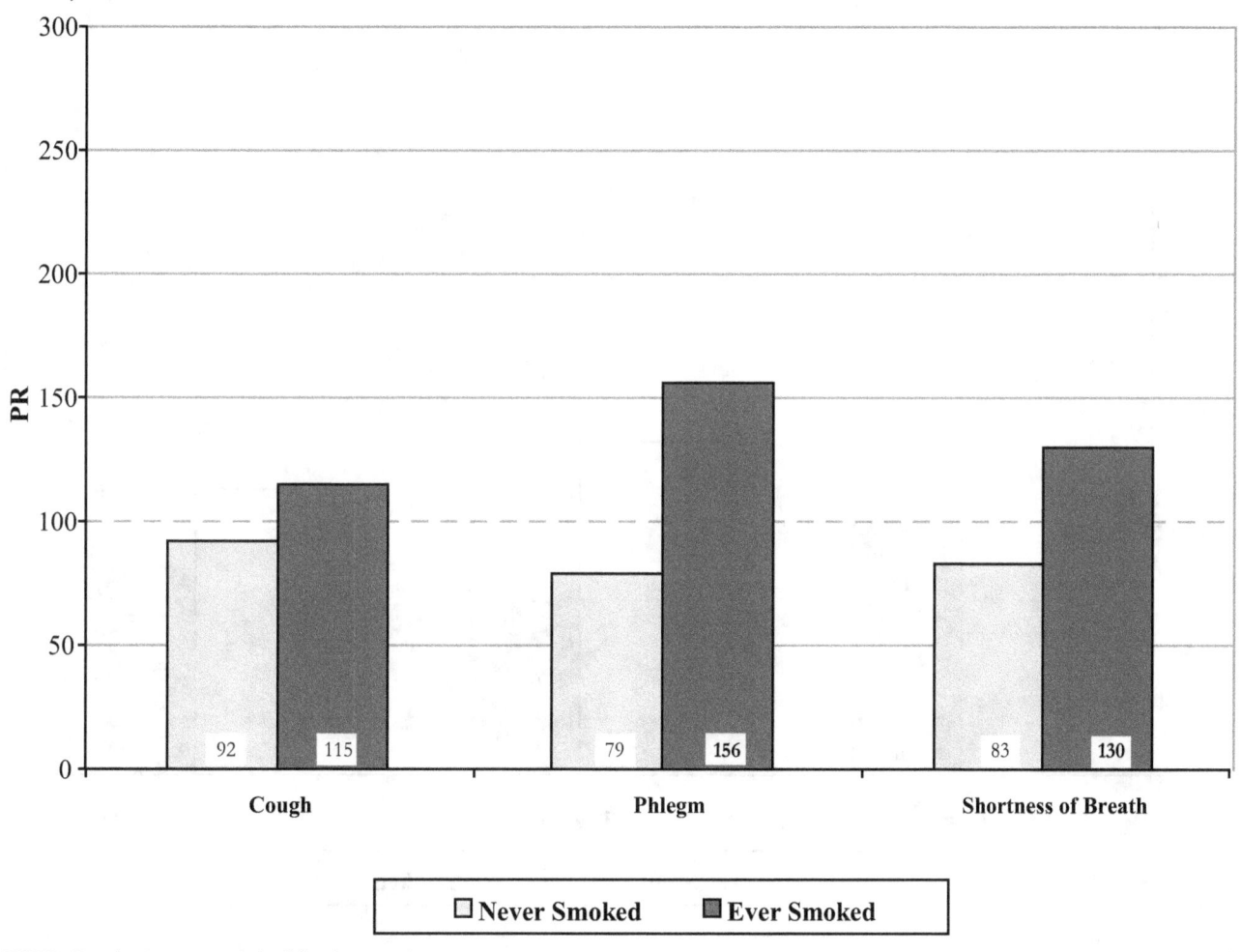

NOTE: Based on responses to the following questions:
"Do you usually cough on most days for 3 consecutive months or more during the year?"
"Do you bring up phlegm on most days for 3 consecutive months or more during the year?"
"Are you troubled by shortness of breath when hurrying on level ground or walking up a slight hill?"
PRs in **bold** are significantly different from 100 ($p<0.05$). PRs in *italics* are based on fewer than five observed cases. See appendices for source description and methods.
SOURCE: National Center for Health Statistics, Third National Health and Nutrition Examination Survey (NHANES III)

Morbidity by Smoking Status within Respiratory Condition and Agricultural Group–NHANES III

Figure 3-33. Respiratory conditions (current), farm managers: Prevalence ratio (PR) adjusted for age, sex, and race/ethnicity by smoking status, U.S. residents age 17 and over, 1988–1994

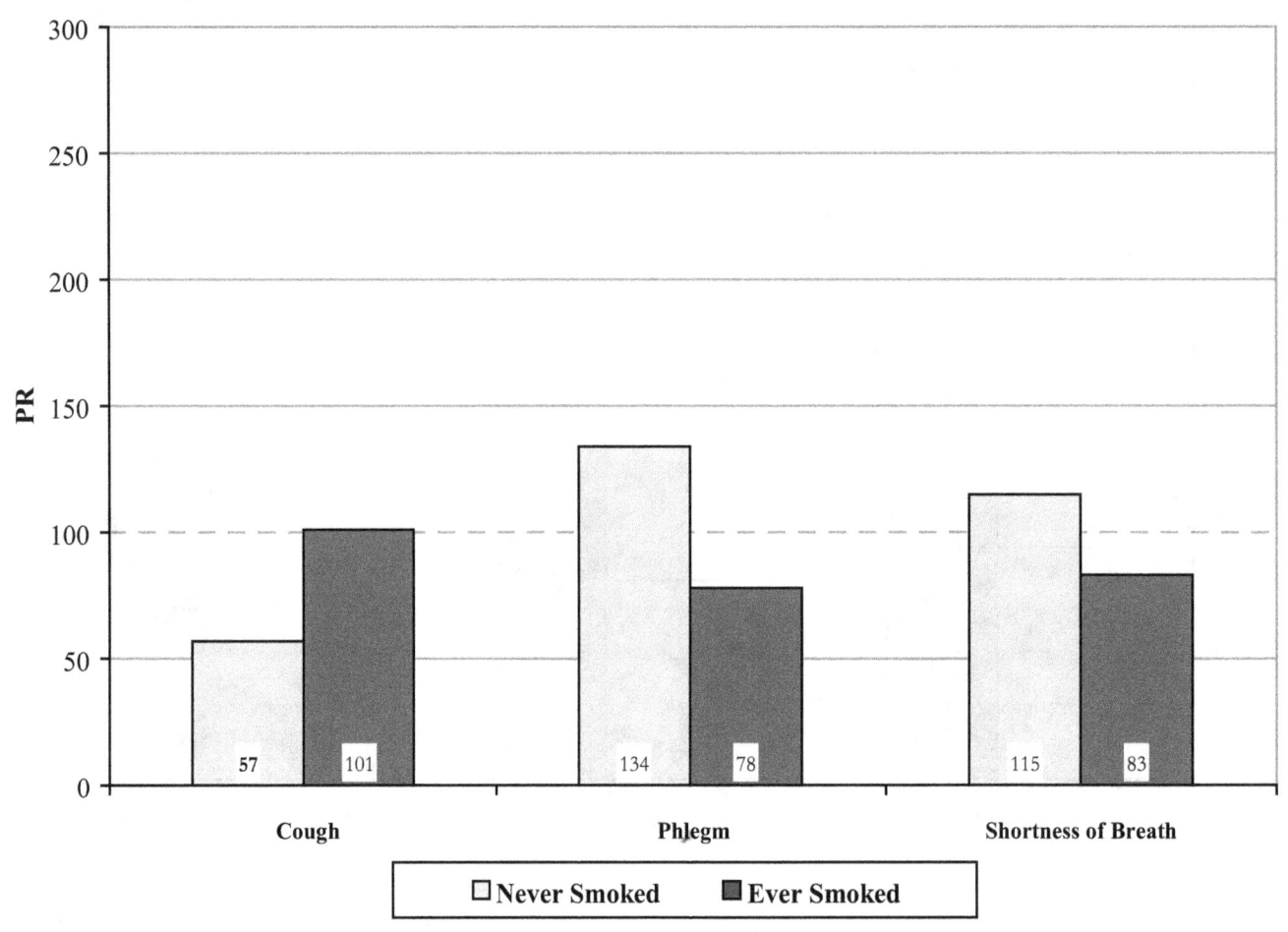

NOTE: Based on responses to the following questions:
 "Do you usually cough on most days for 3 consecutive months or more during the year?"
 "Do you bring up phlegm on most days for 3 consecutive months or more during the year?"
 "Are you troubled by shortness of breath when hurrying on level ground or walking up a slight hill?"
PRs in **bold** are significantly different from 100 (p<0.05). PRs in *italics* are based on fewer than five observed cases. See appendices for source description and methods.
SOURCE: National Center for Health Statistics, Third National Health and Nutrition Examination Survey (NHANES III)

Morbidity by Smoking Status within Respiratory Condition and Agricultural Group–
NHANES III

Figure 3-34. Respiratory conditions (current), other agricultural workers: Prevalence ratio (PR) adjusted for age, sex, and race/ethnicity by smoking status, U.S. residents age 17 and over, 1988–1994

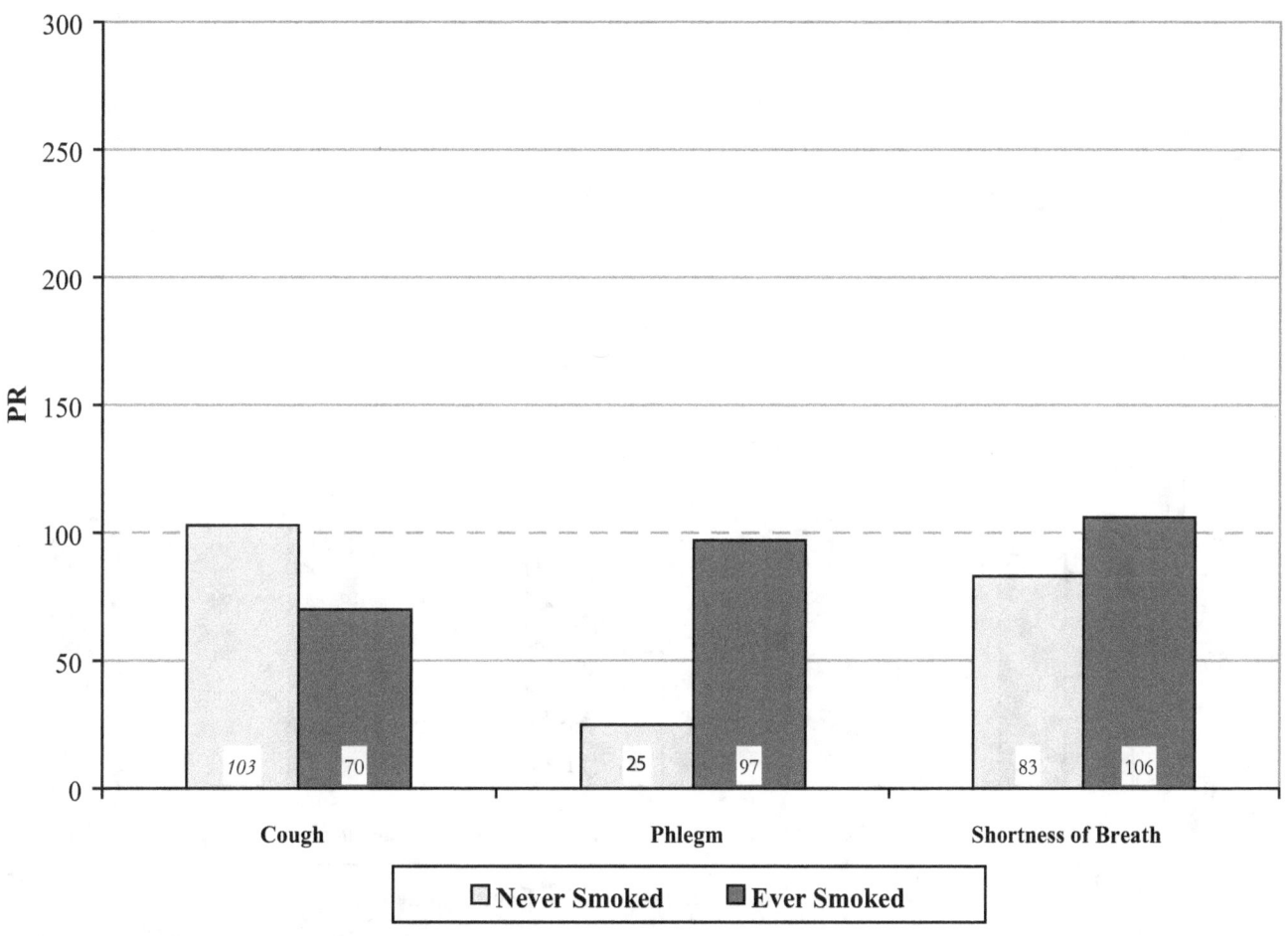

NOTE: Based on responses to the following questions:
 "Do you usually cough on most days for 3 consecutive months or more during the year?"
 "Do you bring up phlegm on most days for 3 consecutive months or more during the year?"
 "Are you troubled by shortness of breath when hurrying on level ground or walking up a slight hill?"
PRs in **bold** are significantly different from 100 (p<0.05). PRs in *italics* are based on fewer than five observed cases. See appendices for source description and methods.
SOURCE: National Center for Health Statistics, Third National Health and Nutrition Examination Survey (NHANES III)

Morbidity by Smoking Status within Respiratory Condition and Agricultural Group– NHANES III

Figure 3-35. Respiratory conditions (past year), farm workers: Prevalence ratio (PR) adjusted for age, sex, and race/ethnicity by smoking status, U.S. residents age 17 and over, 1988–1994

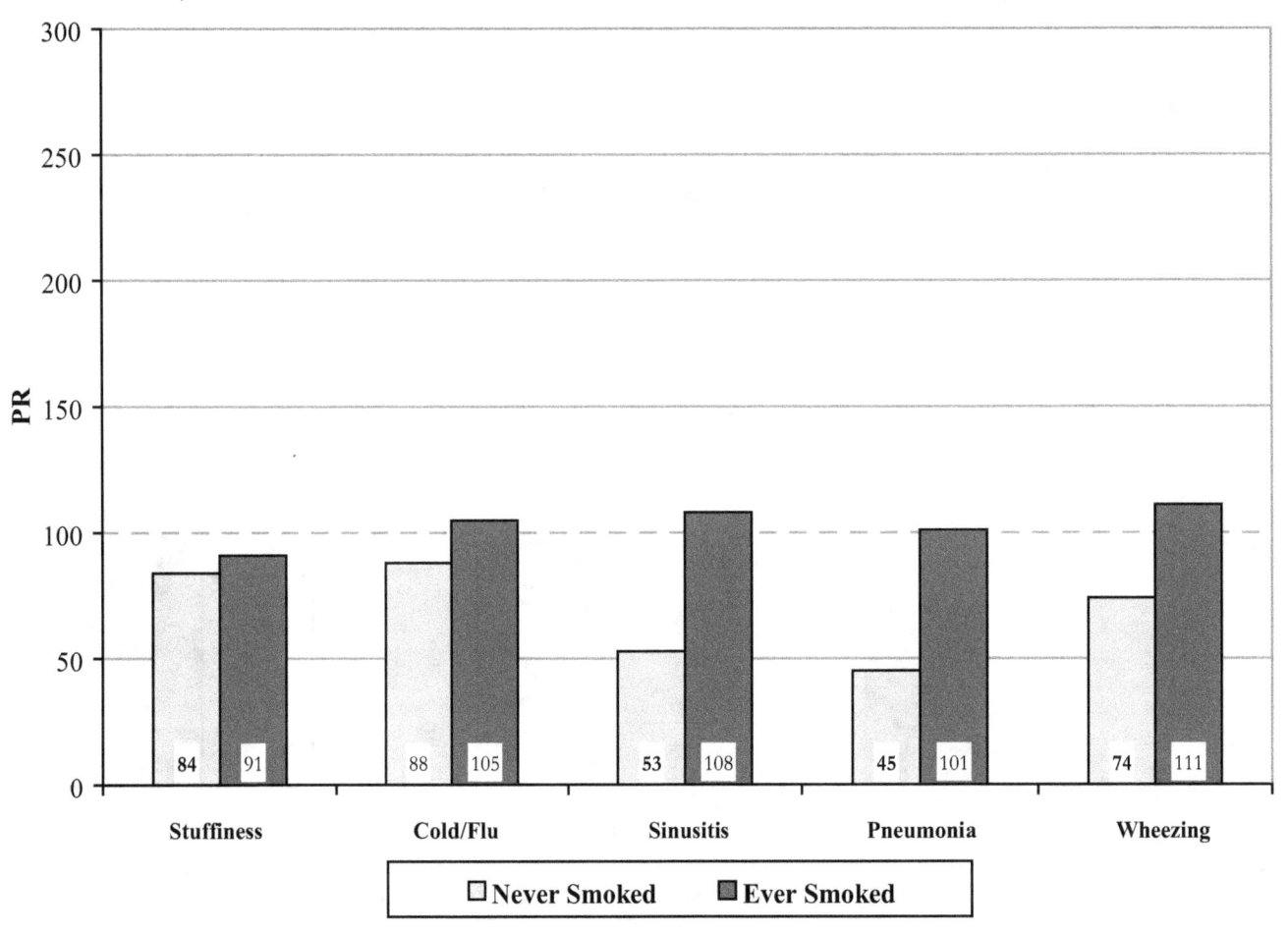

NOTE: Based on responses to the following questions:
"During the past 12 months, have you had any episodes of stuffy, itchy, or runny nose?"
"During the past 12 months, have you had a cold or the flu?"
"During the past 12 months, have you had sinusitis or sinus problems?"
"During the past 12 months, have you had pneumonia?"
"Have you had wheezing or whistling in your chest at any time in the past 12 months?"
PRs in **bold** are significantly different from 100 (p<0.05). PRs in *italics* are based on fewer than five observed cases. See appendices for source description and methods.
SOURCE: National Center for Health Statistics, Third National Health and Nutrition Examination Survey (NHANES III)

Morbidity by Smoking Status within Respiratory Condition and Agricultural Group–
NHANES III

Figure 3-36. Respiratory conditions (past year), farm managers: Prevalence ratio (PR) adjusted for age, sex, and race/ethnicity by smoking status, U.S. residents age 17 and over, 1988–1994

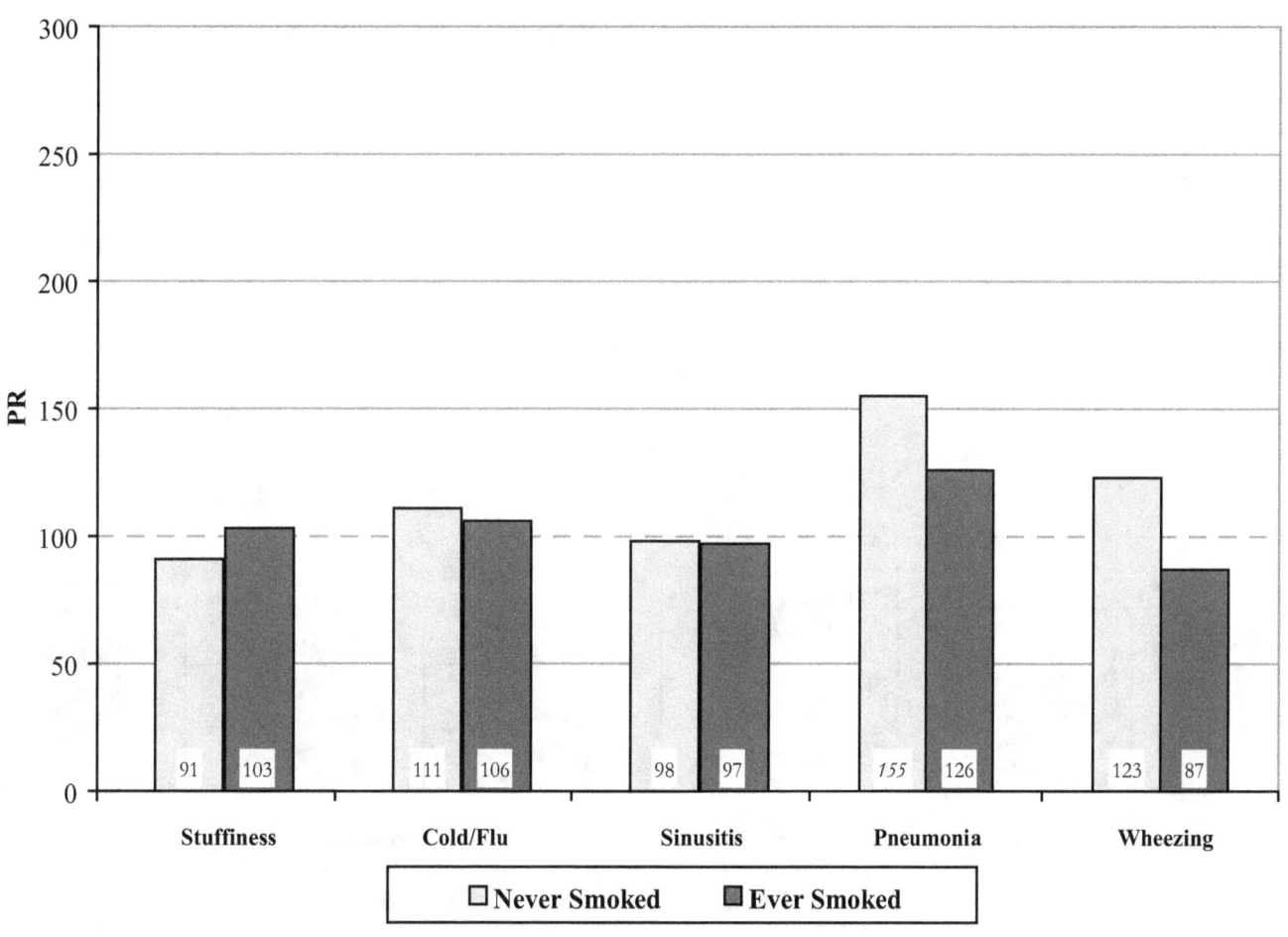

NOTE: Based on responses to the following questions:
 "During the past 12 months, have you had any episodes of stuffy, itchy, or runny nose?"
 "During the past 12 months, have you had a cold or the flu?"
 "During the past 12 months, have you had sinusitis or sinus problems?"
 "During the past 12 months, have you had pneumonia?"
 "Have you had wheezing or whistling in your chest at any time in the past 12 months?"
PRs in **bold** are significantly different from 100 (p<0.05). PRs in *italics* are based on fewer than five observed cases. See appendices for source description and methods.
SOURCE: National Center for Health Statistics, Third National Health and Nutrition Examination Survey (NHANES III)

Morbidity by Smoking Status within Respiratory Condition and Agricultural Group–NHANES III

Figure 3-37. Respiratory conditions (past year), other agricultural workers: Prevalence ratio (PR) adjusted for age, sex, and race/ethnicity by smoking status, U.S. residents age 17 and over, 1988–1994

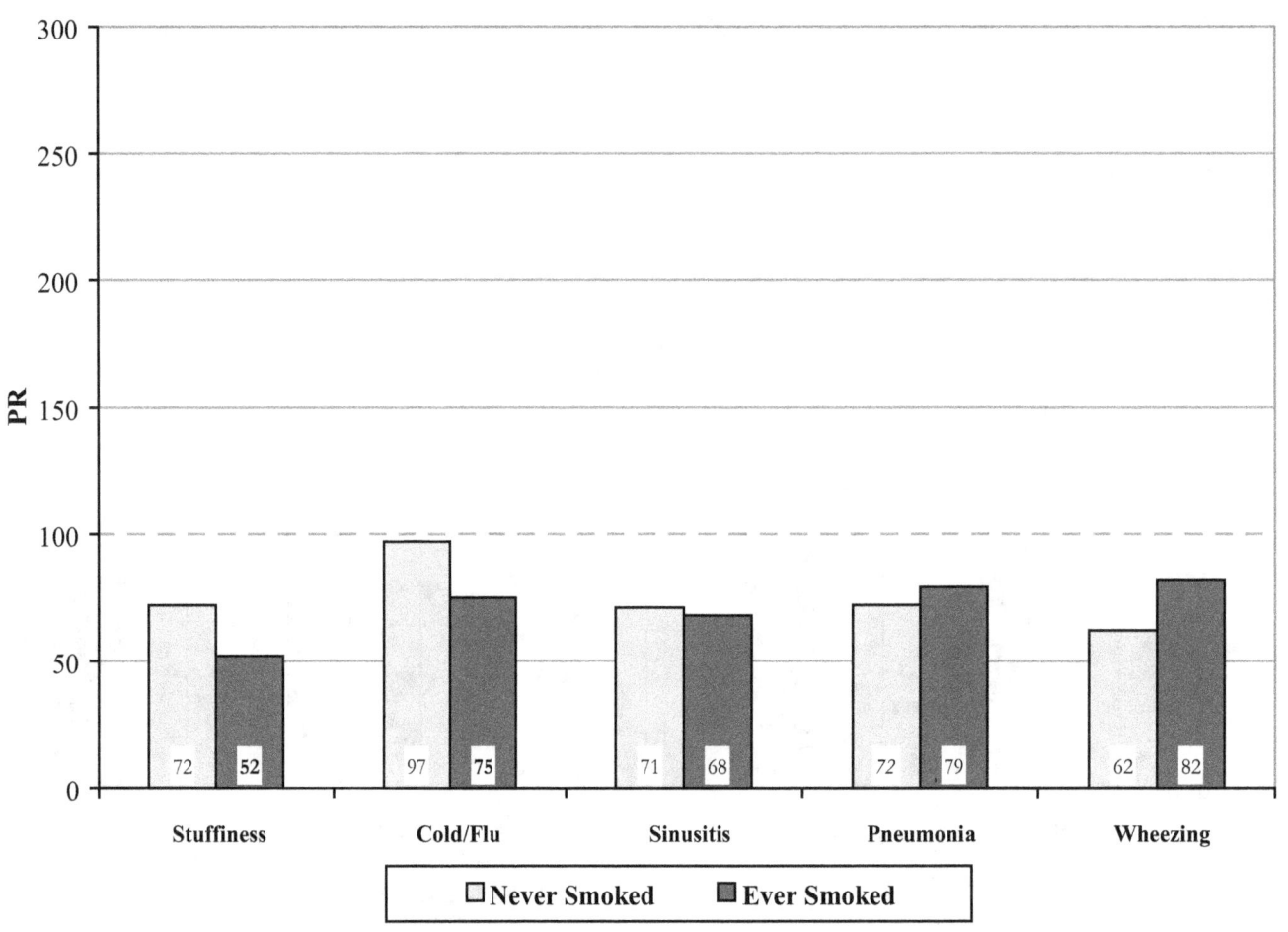

NOTE: Based on responses to the following questions:
 "During the past 12 months, have you had any episodes of stuffy, itchy, or runny nose?"
 "During the past 12 months, have you had a cold or the flu?"
 "During the past 12 months, have you had sinusitis or sinus problems?"
 "During the past 12 months, have you had pneumonia?"
 "Have you had wheezing or whistling in your chest at any time in the past 12 months?"
PRs in **bold** are significantly different from 100 ($p<0.05$). PRs in *italics* are based on fewer than five observed cases. See appendices for source description and methods.
SOURCE: National Center for Health Statistics, Third National Health and Nutrition Examination Survey (NHANES III)

Morbidity by Smoking Status within Respiratory Condition and Agricultural Group–
NHANES III

Figure 3-38. Respiratory conditions (ever), farm workers: Prevalence ratio (PR) adjusted for age, sex, and race/ethnicity by smoking status, U.S. residents age 17 and over, 1988–1994

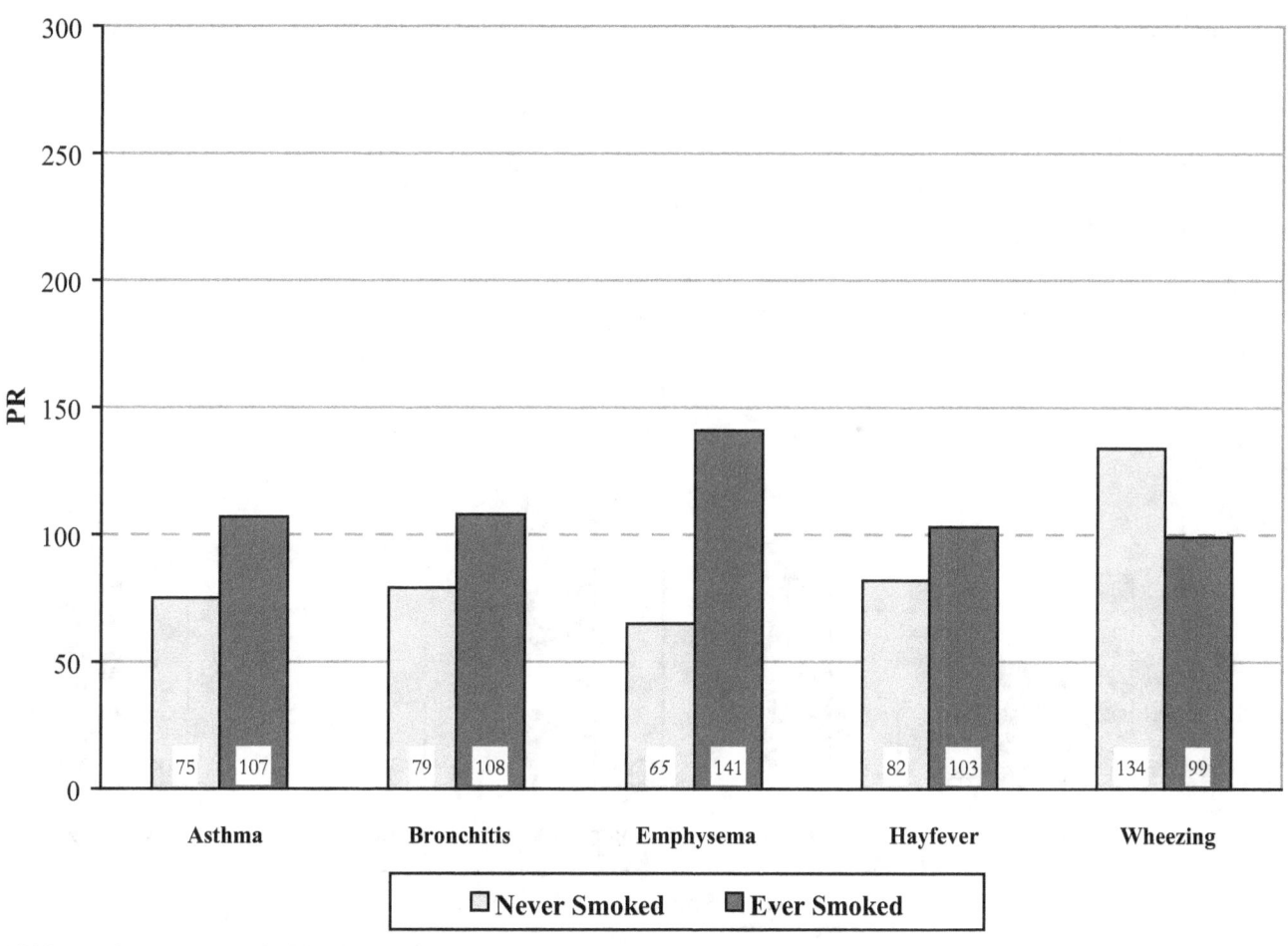

NOTE: Based on responses to the following questions:
 "Has a doctor ever told you that you had asthma?"
 "Has a doctor ever told you that you had chronic bronchitis?"
 "Has a doctor ever told you that you had emphysema?"
 "Has a doctor ever told you that you had hay fever?"
 "Apart from when you have a cold, does your chest ever sound wheezy or whistling?"
PRs in **bold** are significantly different from 100 (p<0.05). PRs in *italics* are based on fewer than five observed cases. See appendices for source description and methods.
SOURCE: National Center for Health Statistics, Third National Health and Nutrition Examination Survey (NHANES III)

Morbidity by Smoking Status within Respiratory Condition and Agricultural Group–NHANES III

Figure 3-39. Respiratory conditions (ever), farm managers: Prevalence ratio (PR) adjusted for age, sex, and race/ethnicity by smoking status, U.S. residents age 17 and over, 1988–1994

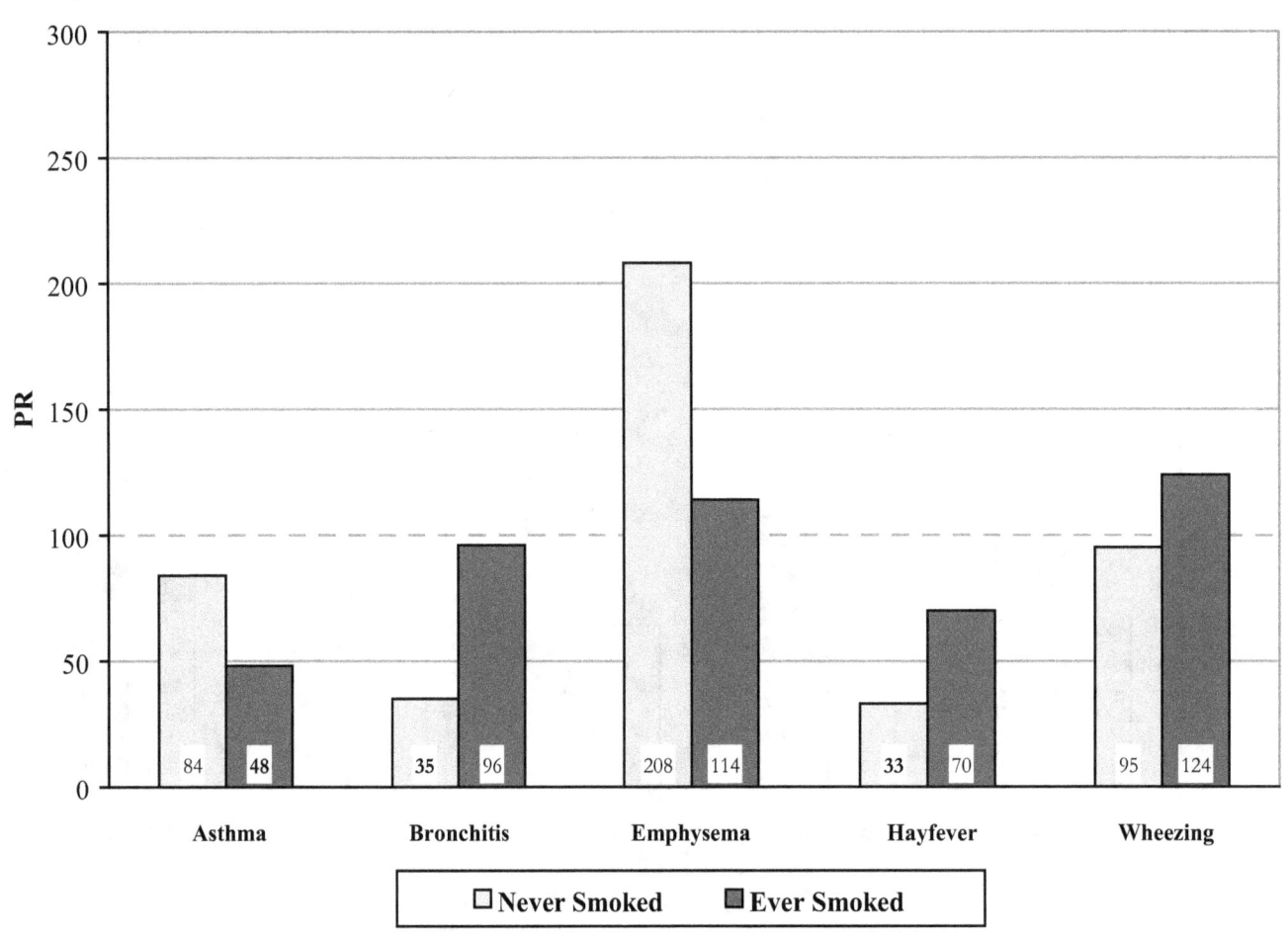

NOTE: Based on responses to the following questions:
 "Has a doctor ever told you that you had asthma?"
 "Has a doctor ever told you that you had chronic bronchitis?"
 "Has a doctor ever told you that you had emphysema?"
 "Has a doctor ever told you that you had hay fever?"
 "Apart from when you have a cold, does your chest ever sound wheezy or whistling?"
PRs in **bold** are significantly different from 100 (p<0.05). PRs in *italics* are based on fewer than five observed cases. See appendices for source description and methods.
SOURCE: National Center for Health Statistics, Third National Health and Nutrition Examination Survey (NHANES III)

Morbidity by Smoking Status within Respiratory Condition and Agricultural Group–NHANES III

Figure 3-40. Respiratory conditions (ever), other agricultural workers: Prevalence ratio (PR) adjusted for age, sex, and race/ethnicity by smoking status, U.S. residents age 17 and over, 1988–1994

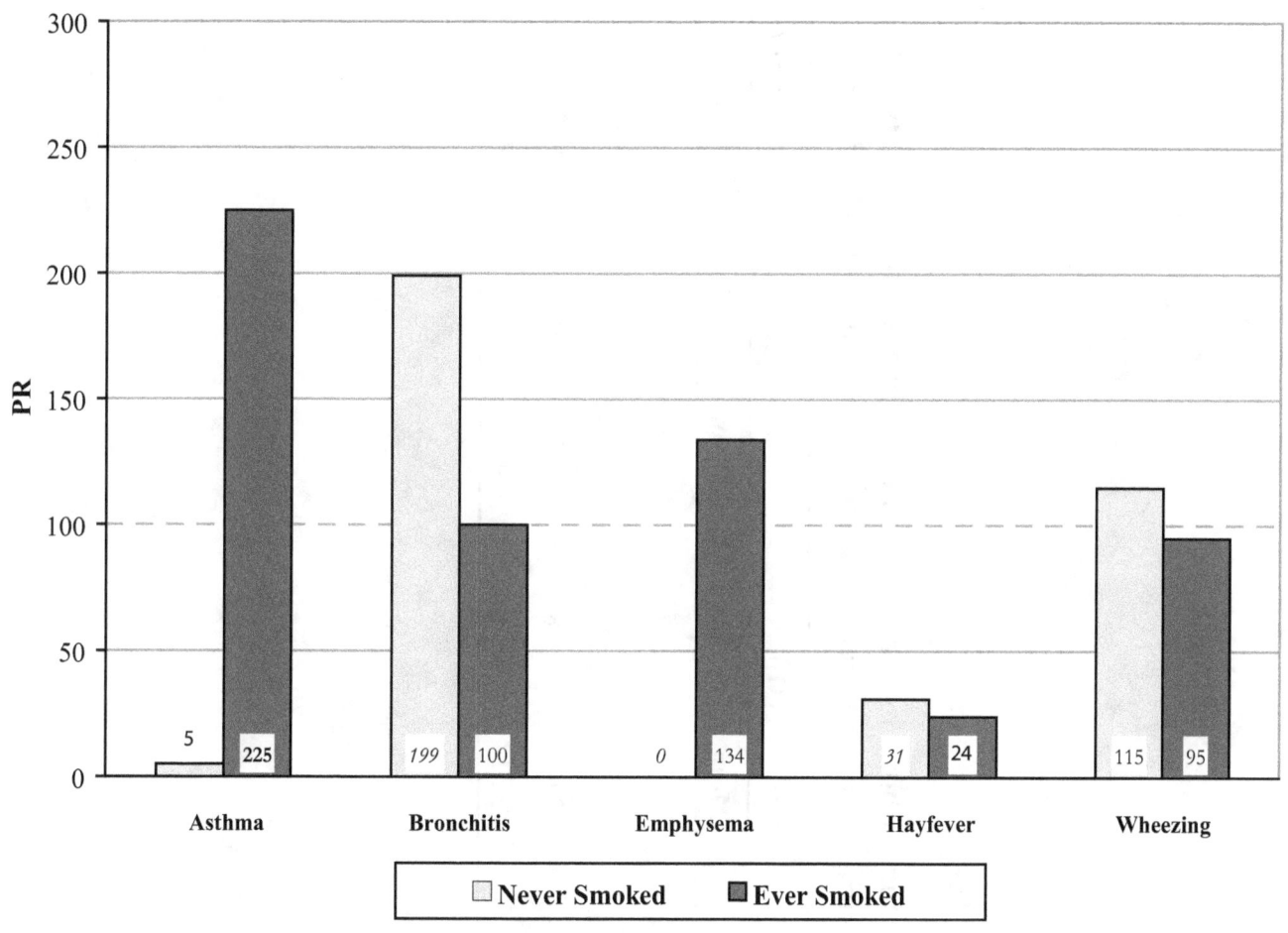

NOTE: Based on responses to the following questions:
"Has a doctor ever told you that you had asthma?"
"Has a doctor ever told you that you had chronic bronchitis?"
"Has a doctor ever told you that you had emphysema?"
"Has a doctor ever told you that you had hay fever?"
"Apart from when you have a cold, does your chest ever sound wheezy or whistling?"

PRs in **bold** are significantly different from 100 ($p<0.05$). PRs in *italics* are based on fewer than five observed cases. See appendices for source description and methods.
SOURCE: National Center for Health Statistics, Third National Health and Nutrition Examination Survey (NHANES III)

Table 3-20. Spirometry: Forced expiratory volume in one second (FEV$_1$), forced vital capacity (FVC), and peak expiratory flow (PEF) by agricultural group, U.S. residents age 17 and over, 1988–1994

Worker Group	Number Observed	FEV$_1$ (L)		FVC (L)		PEF (L/sec)	
		Mean	SD	Mean	SD	Mean	SD
Farm Workers	268	3.29	1.18	4.26	1.39	7.49	2.70
Farm Managers	610	3.28	1.00	4.30	1.11	7.77	2.34
Other Agricultural Workers	152	3.66	0.78	4.53	0.92	8.73	2.11
All Non-agricultural Workers	14,811	3.22	0.97	4.07	1.14	7.60	2.32

L - liters   L/sec - liters per second   SD - standard deviation
NOTE: See appendices for source description and methods.
SOURCE: National Center for Health Statistics, Third National Health and Nutrition Examination Survey (NHANES III)

Table 3-21. Spirometry: Percent predicted forced expiratory volume in one second (FEV$_1$), forced vital capacity (FVC), and peak expiratory flow (PEF) by agricultural group, U.S. residents age 17 and over, 1988–1994

Worker Group	Number Observed	FEV$_1$ (L) Mean	FEV$_1$ (L) SD	FVC (L) Mean	FVC (L) SD	PEF (L/sec) Mean	PEF (L/sec) SD
Farm Workers	268	95.2	18.6	99.0	16.3	92.1	23.0
Farm Managers	610	96.4	16.8	98.6	14.2	92.7	19.9
Other Agricultural Workers	152	99.1	12.8	101.3	12.2	98.7	20.5
All Non-agricultural Workers	14,811	95.6	15.9	98.1	13.9	95.6	19.6

L - liters   L/sec - liters per second   SD - standard deviation
NOTE: See appendices for source description and methods.
SOURCE: National Center for Health Statistics, Third National Health and Nutrition Examination Survey (NHANES III)

Morbidity by Agricultural Group within Spirometry Index: FEV$_1$, FVC, PEF–NHANES III

Table 3-22a. Obstructive abnormality: Estimated prevalence and prevalence ratio (PR) adjusted for age, sex, race/ethnicity, and smoking status by agricultural group, U.S. residents age 17 and over, 1988–1994

Worker Group	Number Observed	Estimated Prevalence of Condition in U.S. n	Estimated Prevalence of Condition in U.S. (%)	PR	95% Confidence Interval LCL	95% Confidence Interval UCL
Farm Workers	110	660,225	22.5	**173**	143	209
Farm Managers	52	240,430	10.8	83	63	109
Other Agricultural Workers	19	140,157	13.1	101	61	158
All Non-agricultural Workers	1,823	20,100,351	13.0	100		

n - estimated number   LCL - lower confidence limit   UCL - upper confidence limit
NOTE: Estimated number in U.S., estimated percent with condition, and PR are based on weighted sample results. PRs in **bold** are significantly different from 100 (p<0.05). PRs in *italics* are based on fewer than five observed cases. See appendices for source description and methods.
SOURCE: National Center for Health Statistics, Third National Health and Nutrition Examination Survey (NHANES III)

Table 3-22b. Restrictive abnormality: Estimated prevalence and prevalence ratio (PR) adjusted for age, sex, race/ethnicity, and smoking status by agricultural group, U.S. residents age 17 and over, 1988–1994

Worker Group	Number Observed	Estimated Prevalence of Condition in U.S. n	Estimated Prevalence of Condition in U.S. (%)	PR	95% Confidence Interval LCL	95% Confidence Interval UCL
Farm Workers	48	165,352	5.6	80	59	106
Farm Managers	20	117,745	5.3	75	46	116
Other Agricultural Workers	6	27,978	2.6	**37**	14	81
All Non-agricultural Workers	1,024	10,906,623	7.1	100		

n - estimated number   LCL - lower confidence limit   UCL - upper confidence limit
NOTE: Estimated number in U.S., estimated percent with condition, and PR are based on weighted sample results. PRs in **bold** are significantly different from 100 (p<0.05). PRs in *italics* are based on fewer than five observed cases. See appendices for source description and methods.
SOURCE: National Center for Health Statistics, Third National Health and Nutrition Examination Survey (NHANES III)

Morbidity by Agricultural Group and Sex within Spirometry Index:
FEV_1, FVC, PEF–NHANES III

Figure 3-41. Percent predicted forced expiratory volume in one second (FEV_1) by agricultural group and sex, U.S. residents age 17 and over, 1988–1994

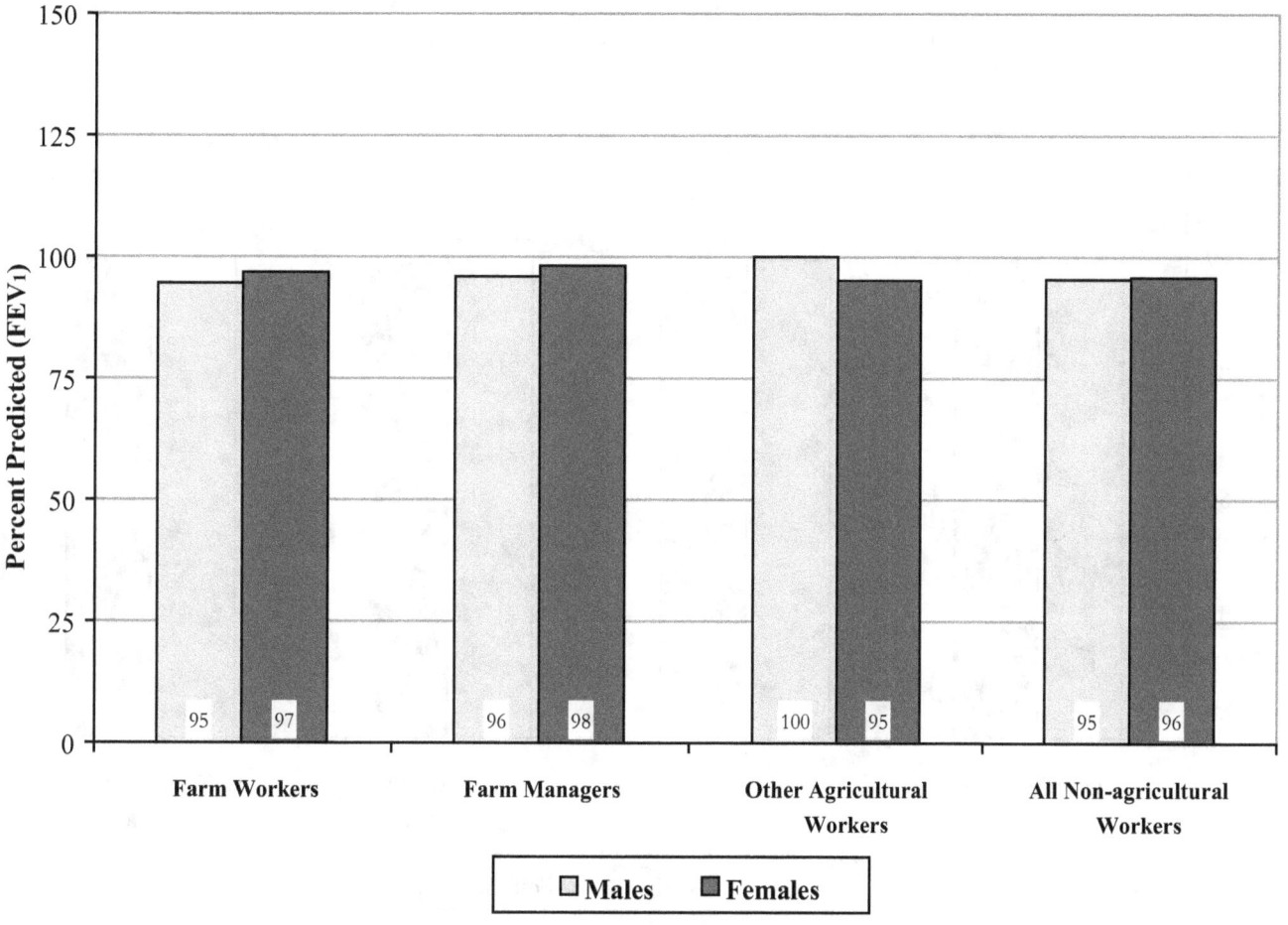

NOTE: See appendices for source description and methods.
SOURCE: National Center for Health Statistics, Third National Health and Nutrition Examination Survey (NHANES III)

Morbidity by Agricultural Group and Sex within Spirometry Index:
FEV$_1$, FVC, PEF–NHANES III

Figure 3-42. Percent predicted forced vital capacity (FVC) by agricultural group and sex, U.S. residents age 17 and over, 1988–1994

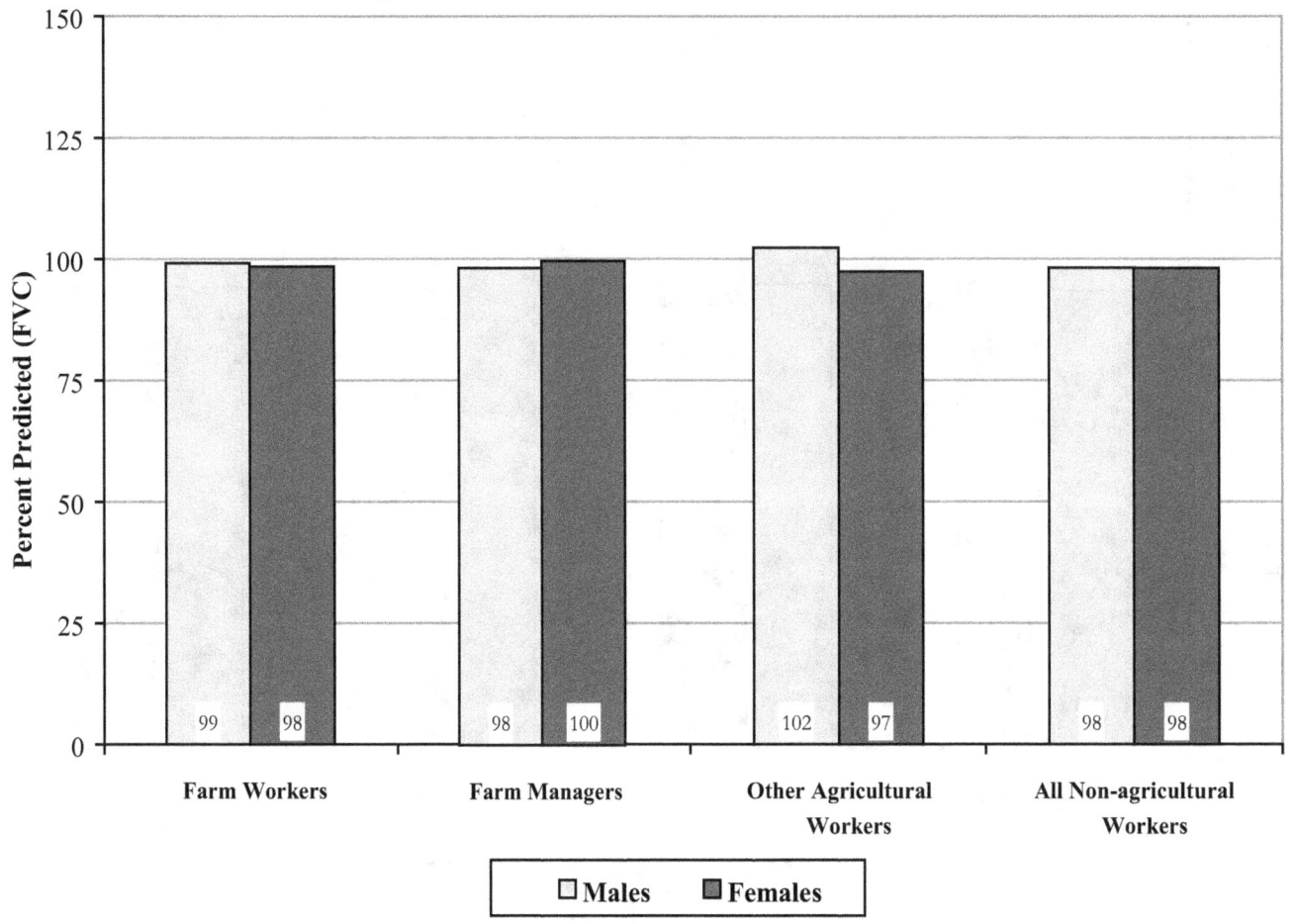

NOTE: See appendices for source description and methods.
SOURCE: National Center for Health Statistics, Third National Health and Nutrition Examination Survey (NHANES III)

Figure 3-43. Percent predicted peak expiratory flow (PEF) by agricultural group and sex, U.S. residents age 17 and over, 1988–1994

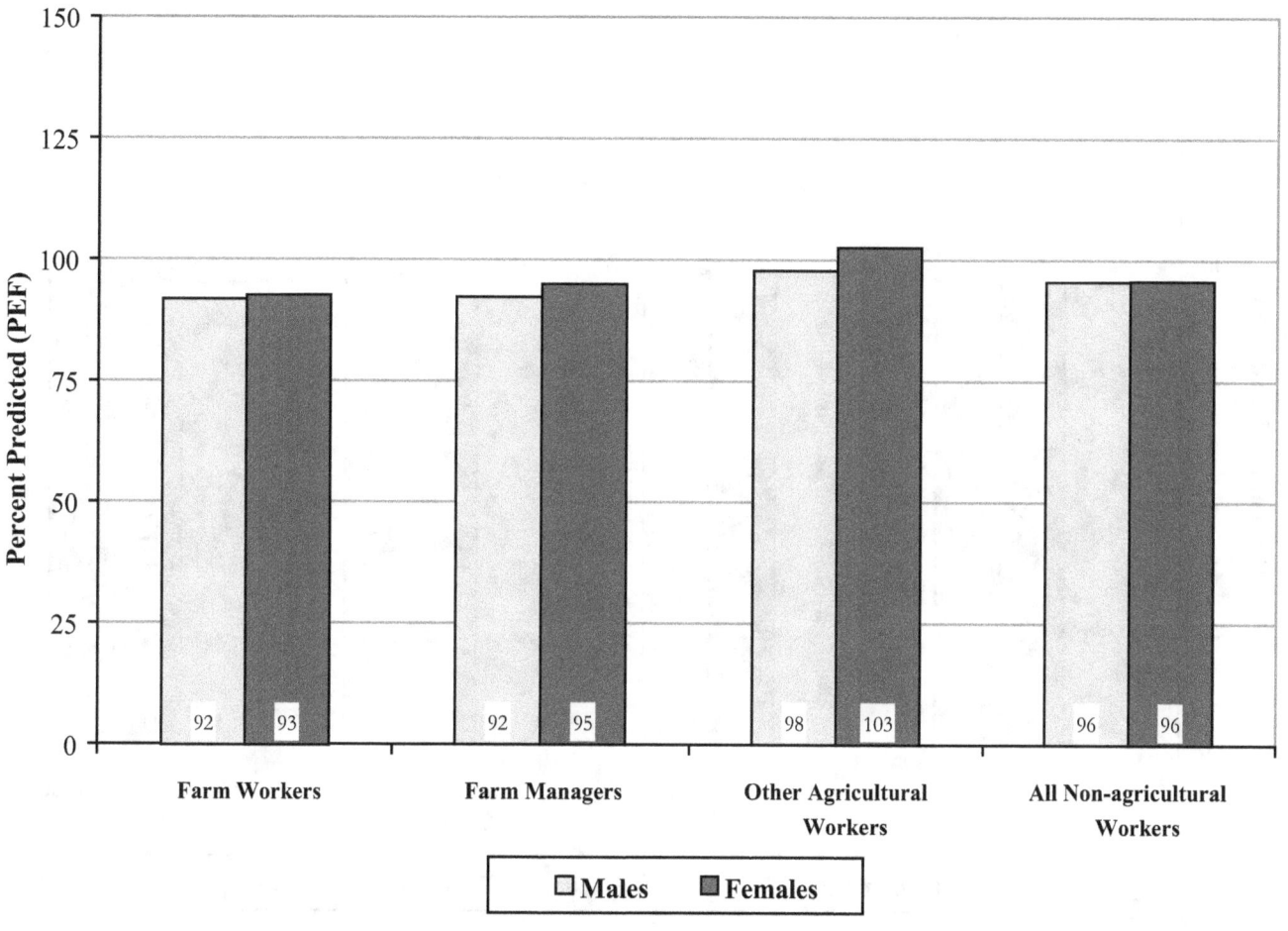

NOTE: See appendices for source description and methods.
SOURCE: National Center for Health Statistics, Third National Health and Nutrition Examination Survey (NHANES III)

Morbidity by Agricultural Group and Race/Ethnicity within Spirometry Index: FEV$_1$, FVC, PEF–NHANES III

Figure 3-44. Percent predicted forced expiratory volume in one second (FEV$_1$) by agricultural group and race/ethnicity, U.S. residents age 17 and over, 1988–1994

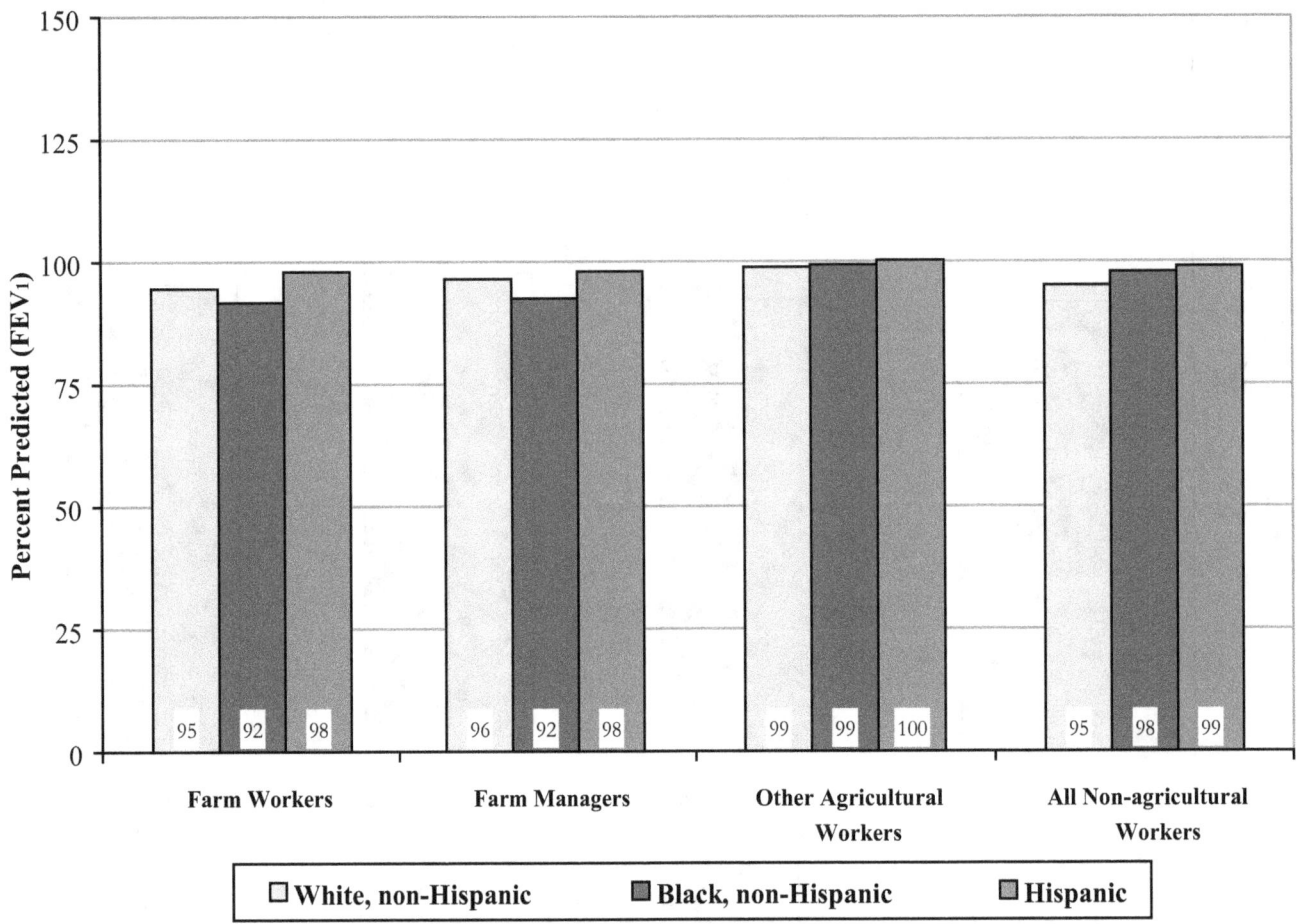

NOTE: See appendices for source description and methods.
SOURCE: National Center for Health Statistics, Third National Health and Nutrition Examination Survey (NHANES III)

Morbidity by Agricultural Group and Race/Ethnicity within Spirometry Index: FEV$_1$, FVC, PEF–NHANES III

Figure 3-45. Percent predicted forced vital capacity (FVC) by agricultural group and race/ethnicity, U.S. residents age 17 and over, 1988–1994

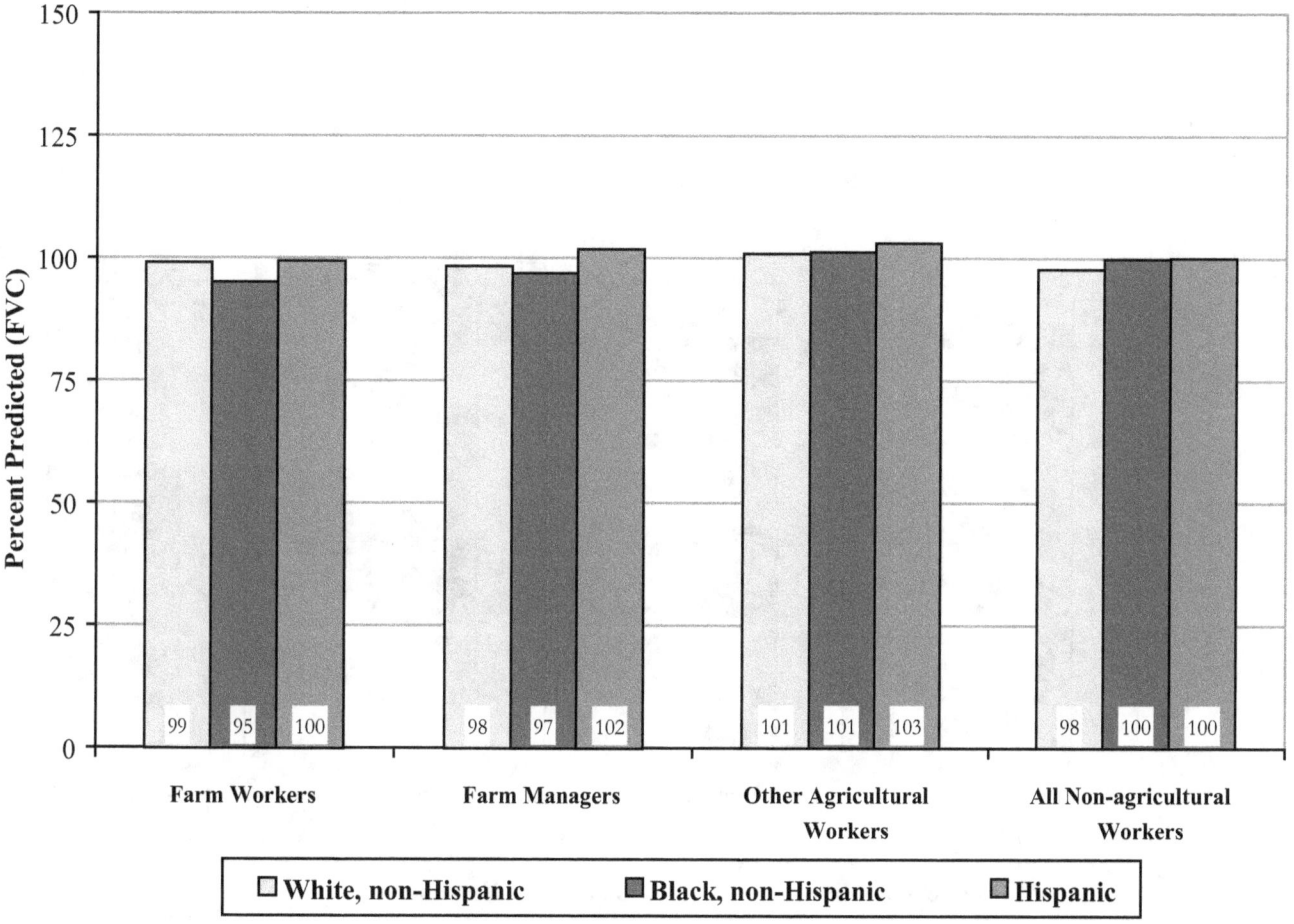

NOTE: See appendices for source description and methods.
SOURCE: National Center for Health Statistics, Third National Health and Nutrition Examination Survey (NHANES III)

Morbidity by Agricultural Group and Race/Ethnicity within Spirometry Index: FEV₁, FVC, PEF–NHANES III

Figure 3-46. Percent predicted peak expiratory flow (PEF) by agricultural group and race/ethnicity, U.S. residents age 17 and over, 1988–1994

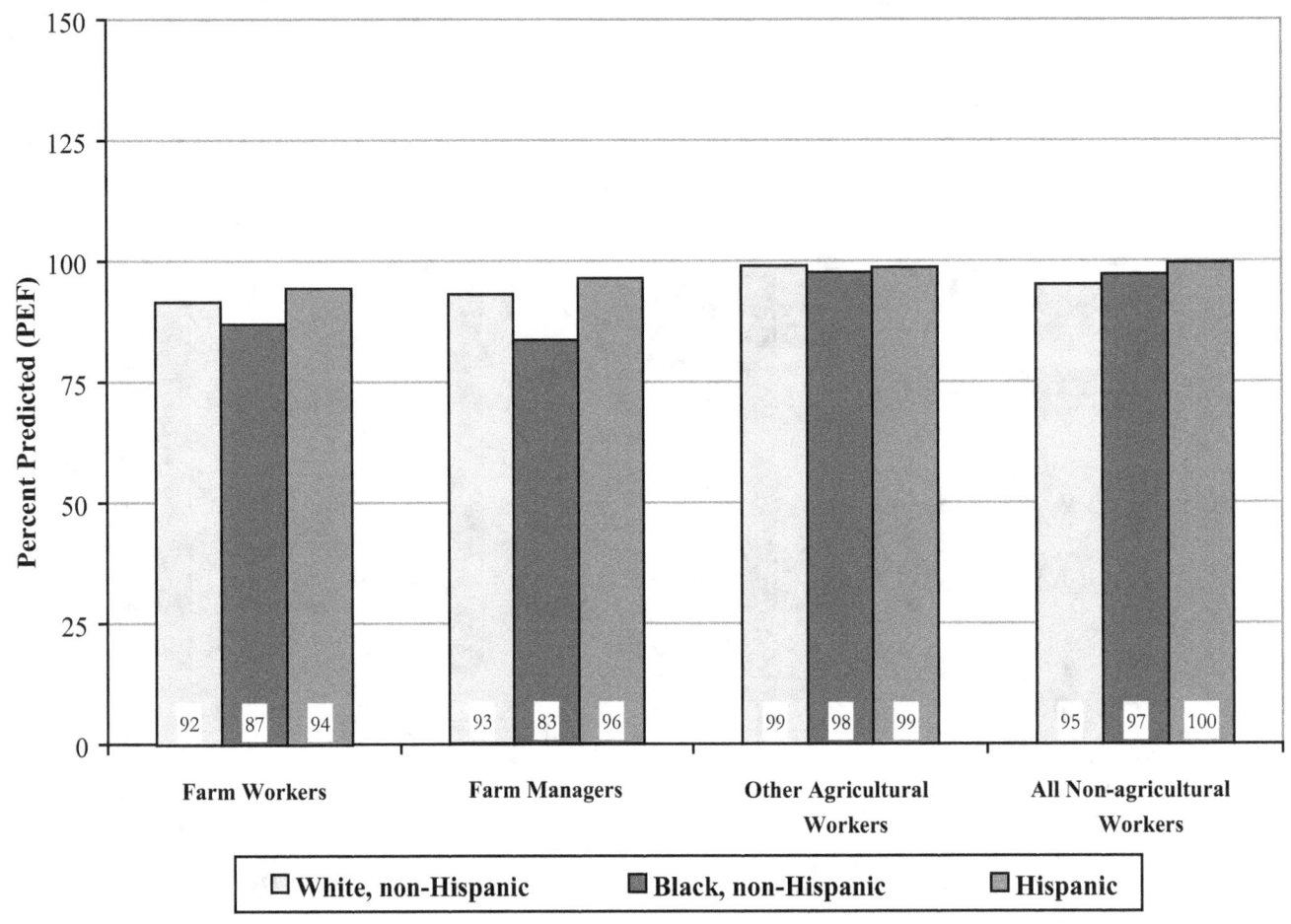

NOTE: See appendices for source description and methods.
SOURCE: National Center for Health Statistics, Third National Health and Nutrition Examination Survey (NHANES III)

Morbidity by Agricultural Group and Smoking Status within Spirometry Index: FEV$_1$, FVC, PEF–NHANES III

Figure 3-47. Percent predicted forced expiratory volume in one second (FEV$_1$) by agricultural group and smoking status, U.S. residents age 17 and over, 1988–1994

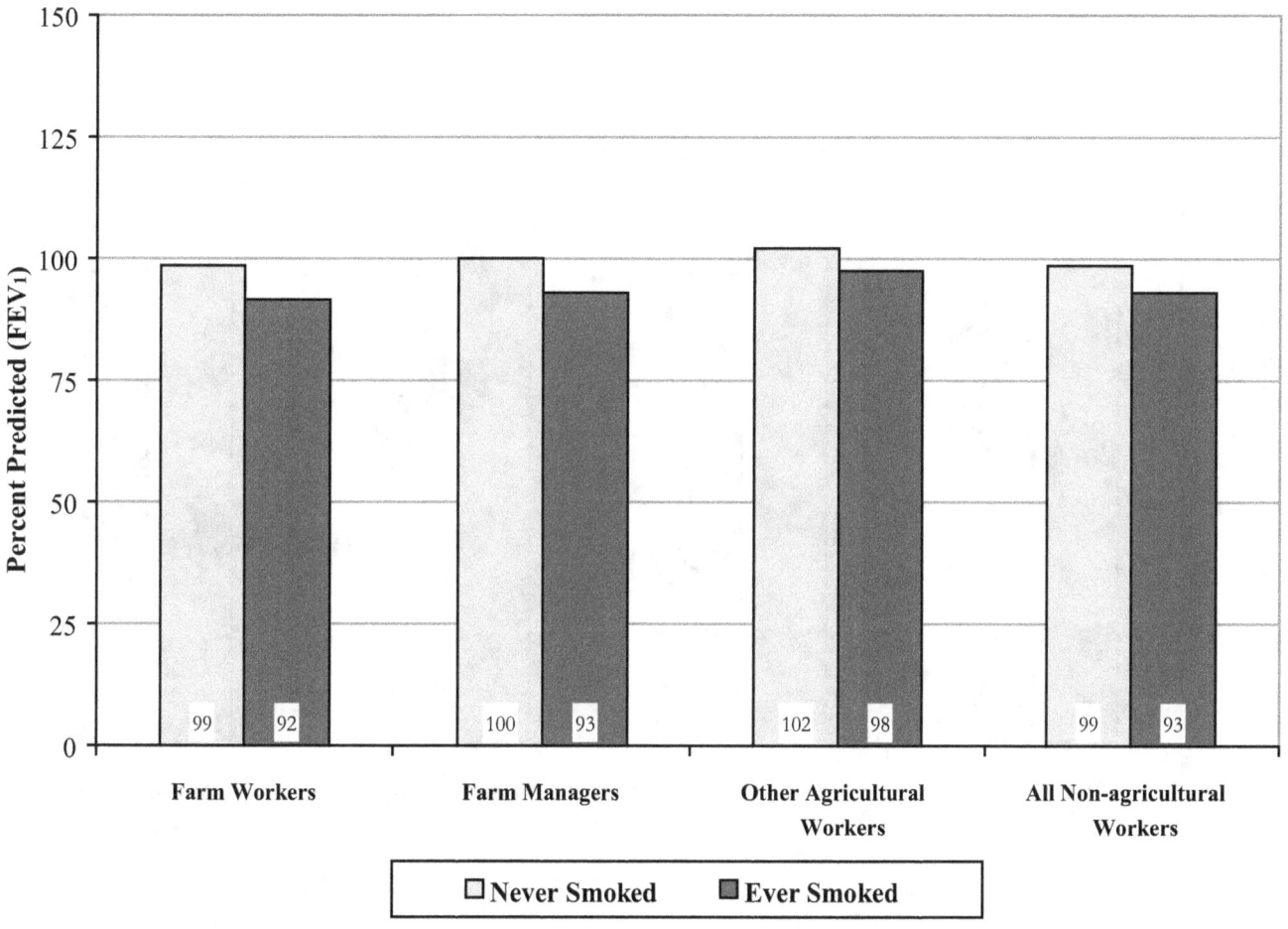

NOTE: See appendices for source description and methods.
SOURCE: National Center for Health Statistics, Third National Health and Nutrition Examination Survey (NHANES III)

Morbidity by Agricultural Group and Smoking Status within Spirometry Index: FEV₁, FVC, PEF–NHANES III

Figure 3-48. Percent predicted forced vital capacity (FVC) by agricultural group and smoking status, U.S. residents age 17 and over, 1988–1994

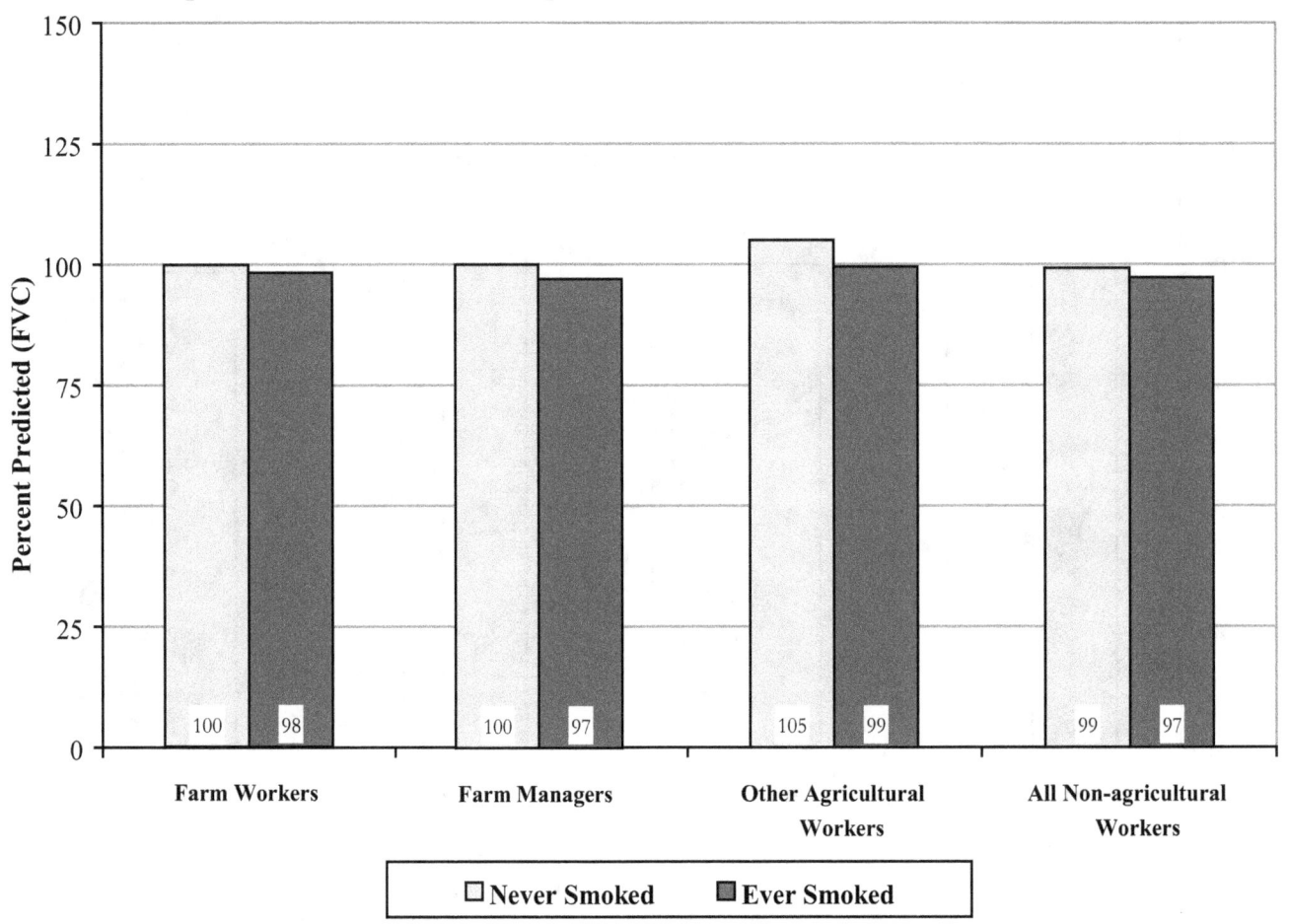

NOTE: See appendices for source description and methods.
SOURCE: National Center for Health Statistics, Third National Health and Nutrition Examination Survey (NHANES III)

Morbidity by Agricultural Group and Smoking Status within Spirometry Index:
FEV₁, FVC, PEF–NHANES III

Figure 3-49. Percent predicted peak expiratory flow (PEF) by agricultural group and smoking status, U.S. residents age 17 and over, 1988–1994

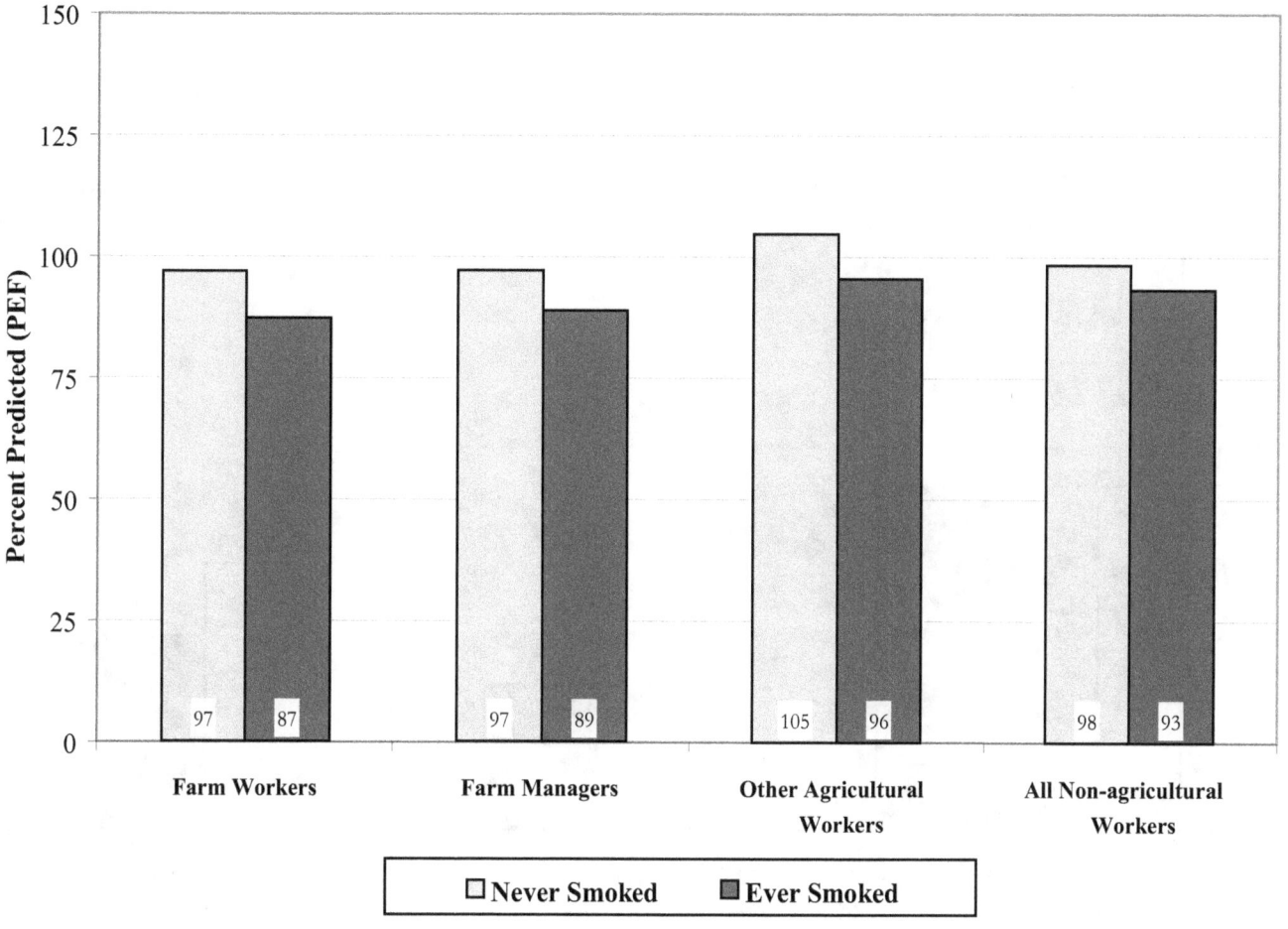

NOTE: See appendices for source description and methods.
SOURCE: National Center for Health Statistics, Third National Health and Nutrition Examination Survey (NHANES III)

Morbidity by Agricultural Group and Sex within Spirometry Index: Obstructive and Restrictive Abnormality–NHANES III

Figure 3-50. Spirometry, obstructive abnormality: Prevalence ratio (PR) adjusted for age and race/ethnicity by agricultural group and sex, U.S. residents age 17 and over, 1988–1994

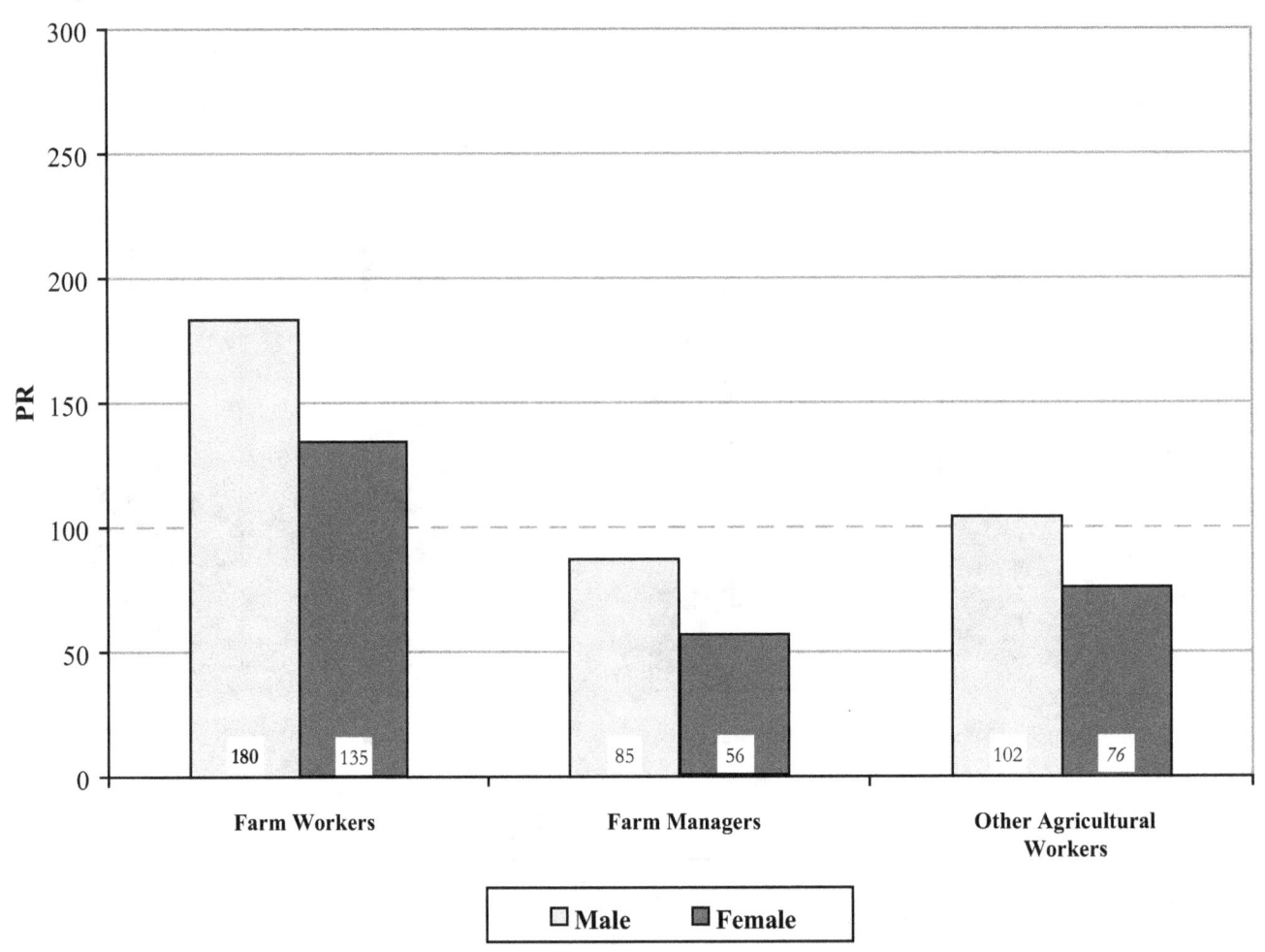

NOTE: PRs in **bold** are significantly different from 100 ($p<0.05$). PRs in *italics* are based on fewer than five observed cases. See appendices for source description and methods.
SOURCE: National Center for Health Statistics, Third National Health and Nutrition Examination Survey (NHANES III)

Morbidity by Agricultural Group and Sex within Spirometry Index:
Obstructive and Restrictive Abnormality–NHANES III

Figure 3-51. Spirometry, restrictive abnormality: Prevalence ratio (PR) adjusted for age and race/ethnicity by agricultural group and sex, U.S. residents age 17 and over, 1988–1994

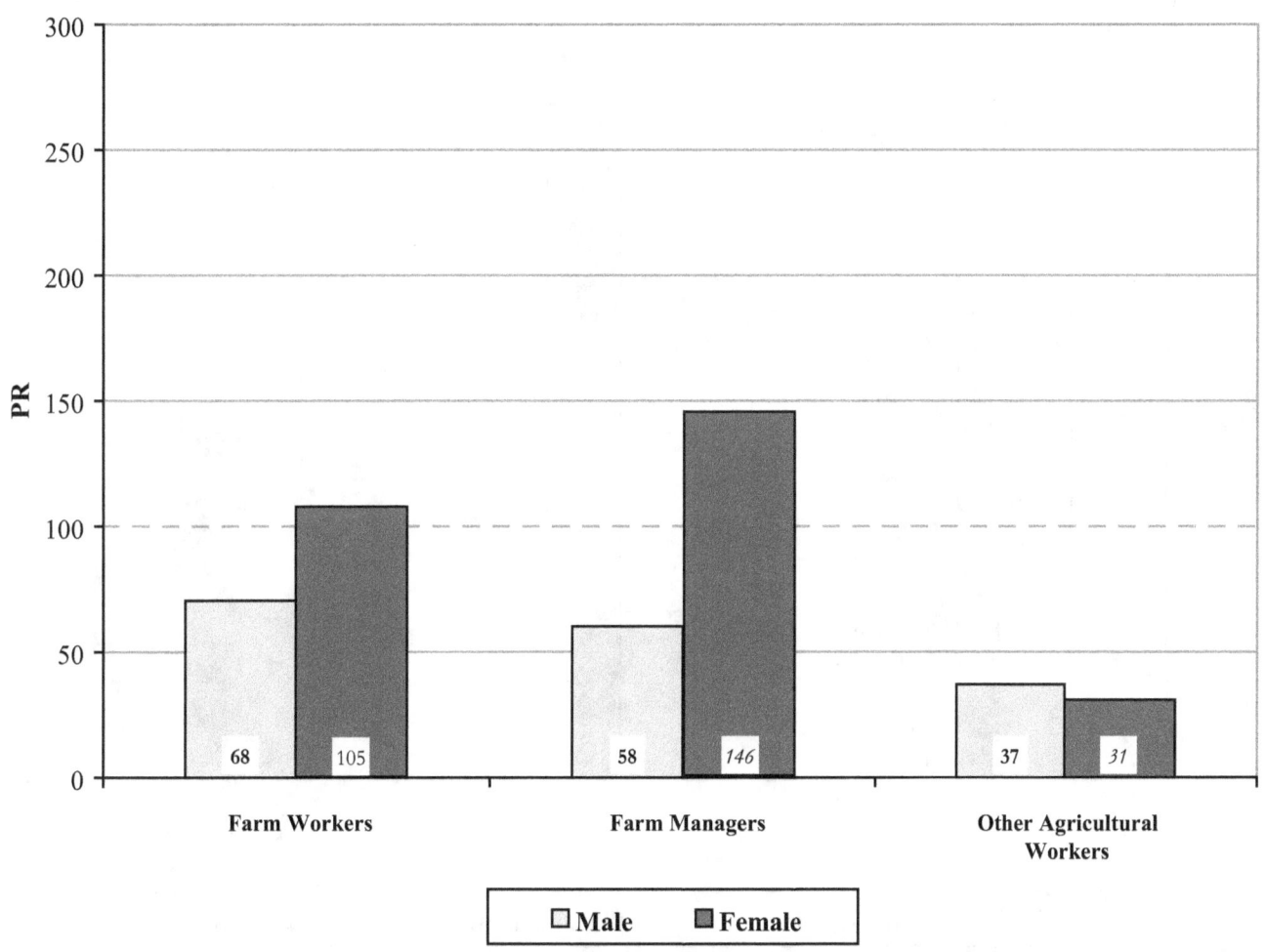

NOTE: PRs in **bold** are significantly different from 100 (p<0.05). PRs in *italics* are based on fewer than five observed cases. See appendices for source description and methods.
SOURCE: National Center for Health Statistics, Third National Health and Nutrition Examination Survey (NHANES III)

Morbidity by Agricultural Group and Race/Ethnicity within Spirometry Index: Obstructive and Restrictive Abnormality–NHANES III

Figure 3-52. Spirometry, obstructive abnormality: Prevalence ratio (PR) adjusted for age and sex by agricultural group and race/ethnicity, U.S. residents age 17 and over, 1988–1994

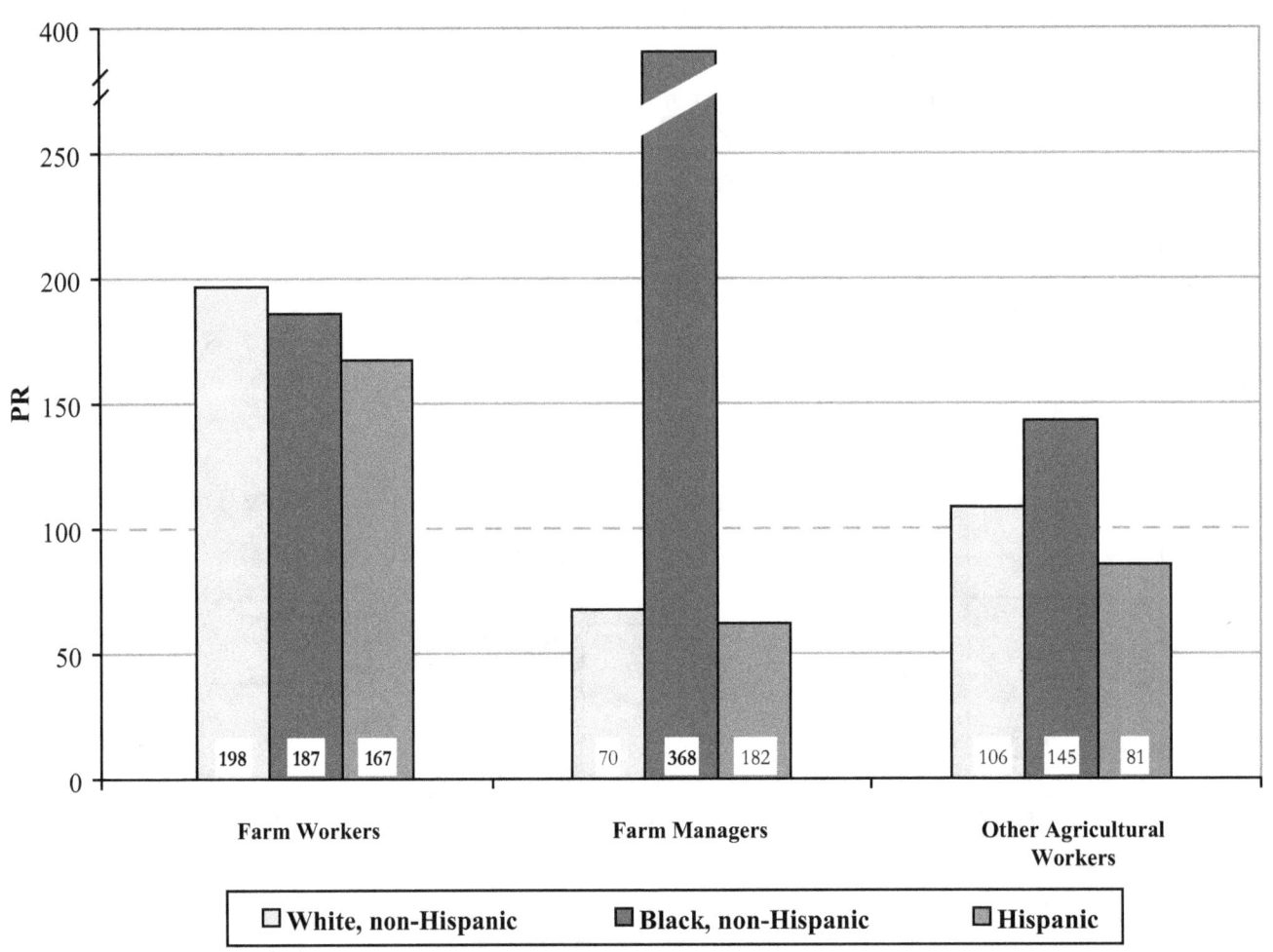

NOTE: PRs in **bold** are significantly different from 100 (p<0.05). PRs in *italics* are based on fewer than five observed cases. See appendices for source description and methods.
SOURCE: National Center for Health Statistics, Third National Health and Nutrition Examination Survey (NHANES III)

Morbidity by Agricultural Group and Race/Ethnicity within Spirometry Index: Obstructive and Restrictive Abnormality–NHANES III

Figure 3-53. Spirometry, restrictive abnormality: Prevalence ratio (PR) adjusted for age and sex by agricultural group and race/ethnicity, U.S. residents age 17 and over, 1988–1994

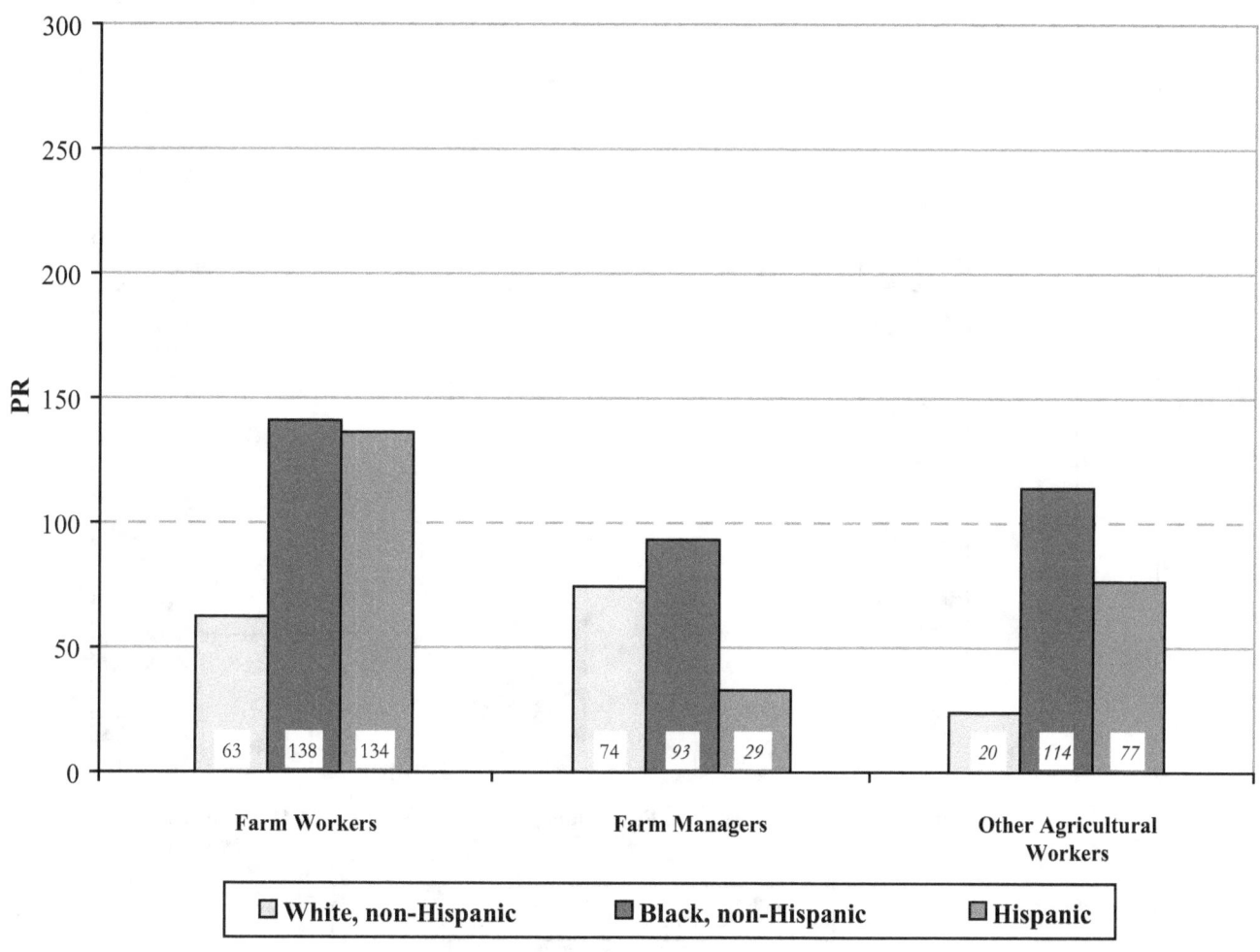

NOTE: PRs in **bold** are significantly different from 100 (p<0.05). PRs in *italics* are based on fewer than five observed cases. See appendices for source description and methods.
SOURCE: National Center for Health Statistics, Third National Health and Nutrition Examination Survey (NHANES III)

Morbidity by Agricultural Group and Smoking Status within Spirometry Index: Obstructive and Restrictive Abnormality–NHANES III

Figure 3-54. Spirometry, obstructive abnormality: Prevalence ratio (PR) adjusted for age, sex, and race/ethnicity by agricultural group and smoking status, U.S. residents age 17 and over, 1988–1994

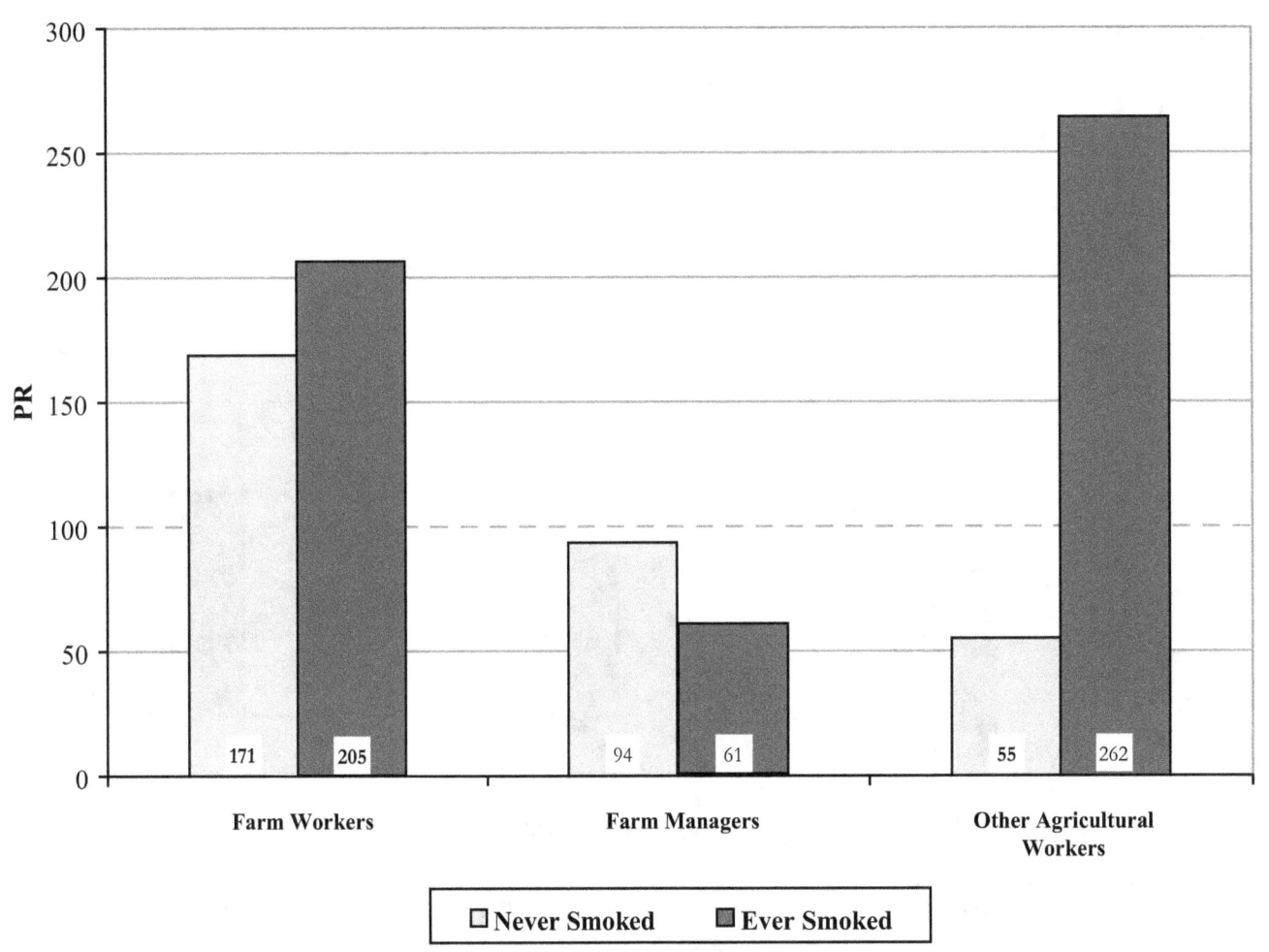

NOTE: PRs in **bold** are significantly different from 100 (p<0.05). PRs in *italics* are based on fewer than five observed cases. See appendices for source description and methods.
SOURCE: National Center for Health Statistics, Third National Health and Nutrition Examination Survey (NHANES III)

Morbidity by Agricultural Group and Smoking Status within Spirometry Index: Obstructive and Restrictive Abnormality–NHANES III

Figure 3-55. Spirometry, restrictive abnormality: Prevalence ratio (PR) adjusted for age, sex, and race/ethnicity by agricultural group and smoking status, U.S. residents age 17 and over, 1988–1994

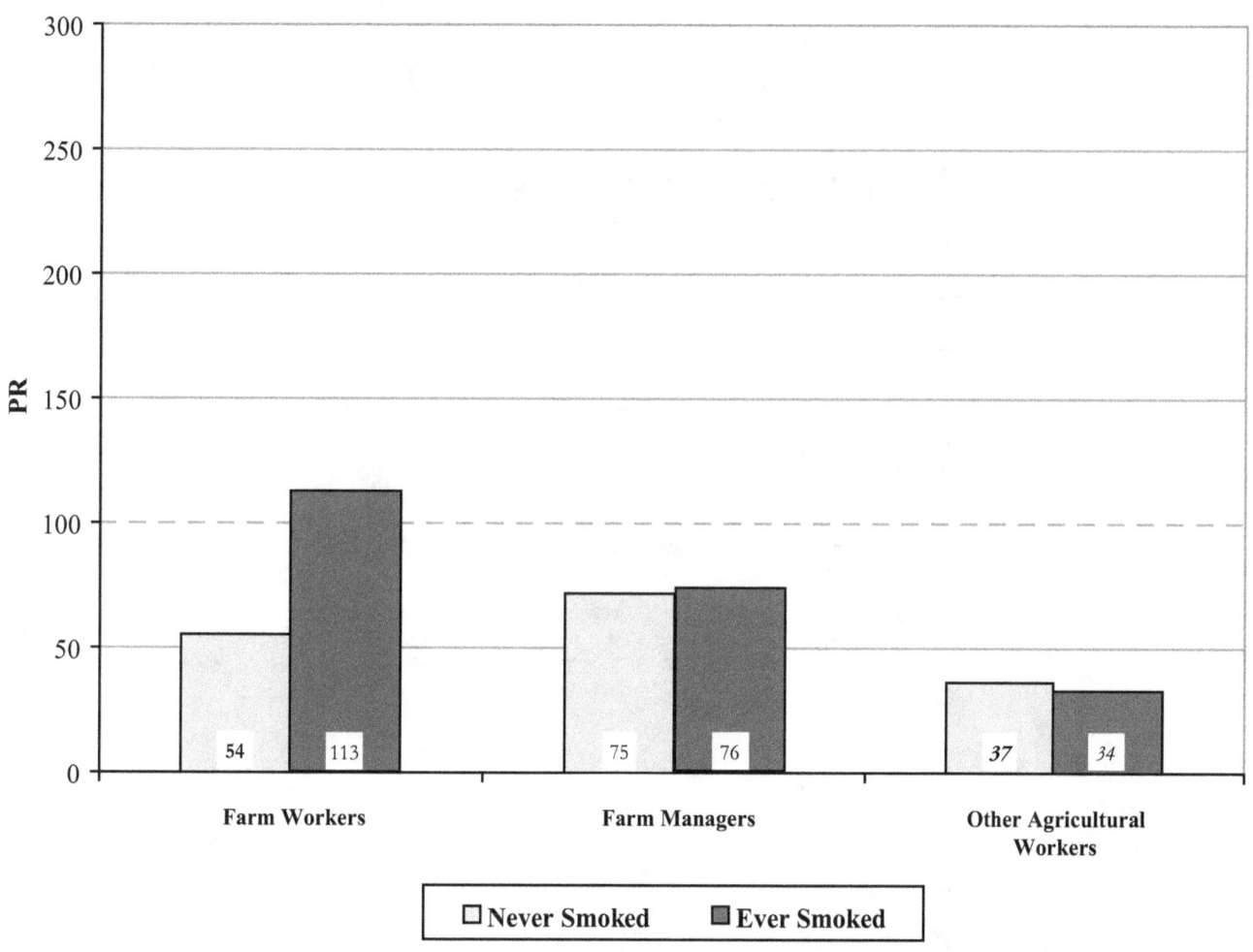

NOTE: PRs in **bold** are significantly different from 100 (p<0.05). PRs in *italics* are based on fewer than five observed cases. See appendices for source description and methods.
SOURCE: National Center for Health Statistics, Third National Health and Nutrition Examination Survey (NHANES III)

Morbidity by Agricultural Group within Dust Diseases of the Lung and Respiratory Conditions Due to Toxic Agents–SOII

Table 3-23. Dust diseases of the lung: Estimated incidence per 10,000 workers by agricultural group, 1995–2001

Agricultural Group	1995	1996	1997	1998	1999	2000	2001
Agriculture/Forestry/Fishing*	0.1	0.1	0.3	0.2	0.2	0.2	0.1
Agricultural Production*	0.2	0.1	0.4	0.4	0.3	<0.05	0.2
Agricultural Production, Crops*	0.2	0.1	0.3	0.3	0.5	<0.05	0.2
Agricultural Production, Livestock*	0.2	0.1	0.7	0.9	<0.05	<0.05	0.3
Landscape/Horticultural Services	<0.05	<0.05	0.3	<0.05	0.1	<0.05	<0.05
Forestry	<0.05	<0.05	<0.05	<0.05	<0.05	10.2	<0.05
Fishing/Hunting/Trapping	<0.05	<0.05	7.8	<0.05	<0.05	<0.05	<0.05
All Private Industry*	0.3	0.4	0.3	0.2	0.2	0.2	0.1

*Excludes farms with less than 11 employees.
SOURCE: Bureau of Labor Statistics: Survey of Occupational Injuries and Illnesses

Morbidity by Agricultural Group within Dust Diseases of the Lung and Respiratory Conditions Due to Toxic Agents–SOII

Table 3-24. Respiratory conditions due to toxic agents: Estimated incidence per 10,000 workers by agricultural group, 1995–2001

Agricultural Group	1995	1996	1997	1998	1999	2000	2001
Agriculture/Forestry/Fishing*	1.4	1.7	2.7	3.7	1.8	0.8	0.4
Agricultural Production*	1.8	1.9	3.0	4.4	1.6	1.2	0.9
Agricultural Production, Crops*	1.8	1.5	3.2	2.3	1.4	0.6	0.6
Agricultural Production, Livestock*	2.1	3.1	2.1	11.2	1.9	3.0	1.7
Landscape/Horticultural Services	0.5	<0.05	0.7	2.4	1.3	0.2	<0.05
Forestry	<0.05	<0.05	<0.05	<0.05	1.8	1.6	0.8
Fishing/Hunting/Trapping	3.3	7.4	15.7	5.4	<0.05	2.4	<0.05
All Private Industry*	3.0	2.6	2.4	2.0	1.8	1.6	1.6

*Excludes farms with less than 11 employees.
SOURCE: Bureau of Labor Statistics: Survey of Occupational Injuries and Illnesses

Section 4

Recommendations for Future Studies

Recommendations for Future Studies

As noted in the *Limitations* section of this report, the results in this report are subject to various constraints on their interpretation. The following are recommendations for future study that would help fill gaps and improve data quality.

- **For the mortality analysis, increase the number of states having reliable industry and occupation data.** The PMR analysis in this report relies on information from only 24 states. These were the states that coded both industry and occupation on death certificates and supplied sufficiently reliable data for analysis. Collectively, these states account for 32% of the U.S. agricultural worker population. That fraction could be doubled, to about 60% of the national agricultural worker population, if three additional states–California, Texas, and Florida– were to supply reliable industry and occupation information. Inclusion of further states would progressively increase the representativeness of the findings. Furthermore, the inclusion of additional states would enable more reliable estimation of the PMRs for diseases that are rare.

- **Expand temporal coverage of mortality analyses.** Another approach to increasing the reliability of the findings for rare mortality outcomes would be to expand the temporal coverage beyond the range 1988–1998. However, including data for the years 1998 to the present would require reconciling respiratory disease codes across the 9th and 10th ICD revisions. Because of the transition from the 9th to the 10th revision of the ICD in 1999, a comparability study on the respiratory diseases would be necessary to evaluate any apparent changes in disease frequency causes as a result of the ICD revision.

- **For the morbidity analysis, add further years or cycles of the NHIS or NHANES survey data.** Addition of further years from the NHIS and NHANES would enable more reliable estimation of results, particularly when the data are disaggregated by worker group, sex, and race.

- **Undertake comprehensive industrial hygiene surveys of worker exposures.** Although no exposure databases were identified for application to this report, good exposure data are needed for assessment of work-related respiratory disease for agricultural workers. The best means of filling the gap would be through special-purpose surveys targeting agricultural workers. For the results of such surveys to be meaningful, in terms of the ability to generalize results, they would need to be reasonably broad in coverage–at least statewide, preferably for states with a significant agricultural worker population such as California, Texas or Florida. Regional studies would also be useful.

Appendices

Appendix A
Sources of Data

Two main types of data sources were sought for this surveillance report: those that were medical-outcome related (i.e., mortality, morbidity) and those that were exposure related. However, no major databases of exposure data pertinent to respiratory disease in agricultural workers were identified. As a consequence, this report is restricted to health outcomes only.

Multiple-Cause-of-Death Data

The National Center for Health Statistics (NCHS) has made available annual multiple-cause-of-death data files for public use since 1968. These files contain records of all deaths in the United States (approximately two million annually) that are reported to state vital statistics offices. Each death record includes codes for up to 20 conditions listed on the death certificate, including both underlying and contributing causes of death in two fields: the entity axis, which preserves diagnostic detail for all listed conditions and their placement on the death certificate; and the record axis, which reorders the codes alphanumerically, removes redundancies, and occasionally combines some associated conditions. Other data include age, race/ethnicity, sex, and state and county of residence at time of death. In addition, usual industry and occupation codes are available for decedents from some states since 1985. NCHS annually determines that certain quality criteria have been met by usual industry and occupation data from individual states (see Appendix D). Multiple-cause-of-death data for 1988–1998 were used in this report.

For more information: http://www.cdc.gov/nchs/about/major/dvs/mortdata.htm

National Health Interview Survey Data

NCHS makes available public-use data from the National Health Interview Survey (NHIS), an annual health survey conducted since 1960. The NHIS is a cross-sectional household interview survey on the health of the U.S. civilian, non-institutionalized population. The main objective of the NHIS is to monitor the health of the population of the United States through the collection and analysis of survey information on a broad range of health topics. NHIS data are collected annually by personal interview from approximately 40,000 households and include about 100,000 persons, with over-sampling of blacks and Hispanics. Through weighting procedures, estimates can be derived that are representative of the target population. The annual response rate of the NHIS is near 90% of eligible households in the sample. Data from the 1997, 1998 and 1999 surveys were used for this report.

For more information: http://www.cdc.gov/nchs/nhis.htm

The Third National Health and Nutrition Examination Survey Data

NCHS makes available public-use data from the third National Health and Nutrition Examination Survey (NHANES III) conducted from 1988 through 1994. NHANES III was designed to provide national estimates of the health and nutritional status of the U.S. civilian, non-institutionalized population. The NHANES III was a complex, multi-stage, stratified, clustered interview and medical survey of about 5,000 individuals per year, with over-sampling of blacks and Hispanics (and certain other groups). Through weighting procedures, estimates can be derived that are representative of the target population. The NHANES III elicited information on demography, chest symptoms, smoking history, industry and occupation, as well as deriving information on many other medical and health-related variables. Of the 39,695 individuals selected in NHANES III, 33,994 (86%) were interviewed and 20,492 undertook spirometry.

For more information: http://www.cdc.gov/nchs/nhanes.htm

Appendix A: Sources of Data

Survey of Occupational Injuries and Illnesses Data

The Bureau of Labor Statistics (BLS) Survey of Occupational Injuries and Illnesses (SOII), done in cooperation with participating state agencies, involves data collection by mail from a sample of approximately 250,000 establishments each calendar year. Nearly all industries in the private sector (employers covered by the Occupational Safety and Health Act of 1970) are included. Annual BLS reports of these data incorporate corresponding data from mine operators, provided to BLS by the Mine Safety and Health Administration (MSHA), and from railroad transportation employers, provided to BLS by the Federal Railroad Administration. National estimates of injury and illness incidence rates by industry are developed from the survey data.

Beginning in 1992, the survey was expanded to provide more information on illnesses resulting in days away from work, allowing for more detailed classification of respiratory system diseases. For this report, annual summary data on respiratory illnesses were extracted from BLS annual reports on occupational injuries and illnesses. Data from 1995–2002 SOII surveys were used for this report.

For more information: http://www.bls.gov/iif/oshsum.htm

Demographic Data

Statistics on the distribution of agricultural workers by occupation for 1997 and 2002 were obtained from the Bureau of Labor Statistics *Current Population Survey*.

For more information: ftp://ftp.bls.gov/pub/special.requests/lf/aa97/aat11.txt and ftp://ftp.bls.gov/pub/special.requests/lf/aa2002/aat11.txt

Appendix B
Methods

Mortality Analyses of NCHS Multiple-Cause-of-Death Data

For this report, the number of deaths for each respiratory condition was defined as either (1) the number of decedents for which the condition was coded as the underlying cause of death, or (2) the number of decedents for which the condition was coded as one of the multiple causes of death (i.e., either the underlying or contributing cause of death). For the years 1988–1998, these numbers were tabulated from the record axis of the NCHS multiple-cause-of-death data files. See Appendix C for a listing of the ICD-9 codes that were used in this analysis. The tables in Section 2 of this report are based solely on multiple causes of death whereas the figures in Section 2 are based on both underlying cause and multiple causes of death.

Appendix D shows the states and years with industry and occupation data on death certificates that were used for the mortality analysis.

Deaths for the analysis also were restricted to persons 15 years of age or older, appropriate when examining worker populations. Five age categories were used for the analysis: 15-54, 55-64, 65-74, 75-84, and 85 years or older. Because the age-at-death distribution is slanted toward older ages, there was a fairly even distribution across the five age categories that were used. Race and ethnicity were combined into a single variable for the analysis, categorized as follows: (1) white, non-Hispanic; (2) black, non-Hispanic; (3) other, non-Hispanic; (4) Hispanic; or (5) unknown race/ethnicity.

Deaths were tabulated by agricultural groups and by sex, age, and race/ethnicity. The agricultural groups were defined based on industry and occupation codes shown in Appendix E. Six agricultural groups were defined: (1) crop farm workers, (2) livestock farm workers, (3) farm managers, (4) landscape and horticultural workers, (5) forestry workers, and (6) fishery workers. The remaining non-agricultural workers were used as a comparison group for the analysis.

Combinations of occupation and industry codes that were used to define agricultural groups are listed in Appendix E.

Although most ICD-9 codes used in the analysis clearly are respiratory diseases, a few might be considered only marginally related. The rationale for including the marginally related diseases was as follows:

- Tuberculosis (010-018): Tuberculosis is an infectious disease caused by *Mycobacterium tuberculosis*. It mainly involves the respiratory tract. Some of the ICD-9 codes explicitly specify other organ systems – for example, code 013 (tuberculosis of the meninges and central nervous system; code 014 (tuberculosis of intestines, peritoneum, and mesenteric glands); code 015 (tuberculosis of bones and joints); code 016 (tuberculosis of genitourinary system); and code 017 (tuberculosis of other organs). However, pulmonary tuberculosis (and other respiratory tuberculosis) predominates in terms of tuberculosis deaths in the United States.

- Mycoses (110-118): Mycoses are fungal infections that can affect various organs, including the lungs and other respiratory organs. Many, but not all of the serious and sometimes fatal mycotic infections do involve the lungs.

- Sarcoidosis (135): Sarcoidosis is a systemic granulomatous inflammatory disease of unknown etiology that typically involves the lungs.

The analysis was accomplished primarily by calculating a proportionate mortality ratio (PMR) for each worker group, for selected respiratory conditions. The PMR is defined as the observed number of deaths with the condition of interest

Appendix B: Methods

(mentioned as either the underlying cause of death or a contributing cause) among all deaths in a specified worker group, divided by the expected number of deaths among those decedents for that condition. For this analysis, PMRs were calculated based on both the underlying cause of death and multiple causes of death (i.e., either underlying or contributing).

For calculating the PMRs, first, deaths from each condition of interest were tabulated by worker group, and for 50 demographic groups (i.e., all combinations of two sex categories, five race/ethnicity categories, and five age categories) within each worker group. This tabulation was performed separately for each of the years 1988–1998. These results were then summed across years to get totals for all demographic groups within each worker group.

The tabulation of observed deaths was performed separately for the underlying cause of death (at the 3-digit level of detail for ICD-9 codes) and for multiple causes of death. Further tabulations were performed for groupings of the 3-digit ICD codes (see Appendix C for a listing of all groupings).

The expected number of deaths for any worker group, for a specific condition, is the number that would have occurred if that worker group had the same proportion of deaths for that condition as did the comparison group. The expected numbers of deaths were calculated by disease and by demographic group for the six worker groups of interest, by multiplying the total number of observed deaths for each worker group by the fraction of deaths for that disease that occurred in the comparison group. The expected deaths then were summed for each worker group across the 50 demographic groups. The number of observed deaths was divided by the sum of expected deaths and then multiplied by 100 to obtain the PMR. A PMR greater than 100 indicates that there were more deaths associated with the condition in a specified agricultural group than expected.

Lower and upper confidence limits (LCLs and UCLs) for the PMR, at a 95% level of statistical confidence, were calculated in accordance with a method described by Bailar and Ederer.[1] The method applies to the ratio of a Poisson variable to its expectation, and is appropriate for this analysis involving diseases for which the fraction of deaths attributable is relatively small. A PMR was considered to be different from 100 at the 95% level of statistical significance (i.e., $p<0.05$) if the 95% confidence interval did not overlap 100.

Morbidity Analyses of National Health Interview Survey Data

Because the data from the National Health Interview Survey (NHIS) are based on a sample of the U.S. population, the number of data points can be relatively small when the analysis is restricted to a subgroup such as agricultural workers. Consequently, the three most recent years (1997–1999) for which NHIS results were available in the form of public-use files were combined in the analysis, to obtain a relatively greater statistical stability.

Weights that are inverse to the probability of selection for each respondent are provided with each yearly NHIS data set to enable development of national estimates from the sample data. These weights were applied separately to each year of data.

The estimates derived from the NHIS data sets concerned the number (and percent) of respondents with specific conditions. More specifically, responses were analyzed for the following questions:

---

[1] Bailar JC, Ederer F [1964]. Significance factors for the ratio of a Poisson variable to its expectation. Biometrics *20*: 639-643.

- Have you EVER been told by a doctor or other health professional that you had emphysema?
- Have you EVER been told by a doctor or other health professional that you had asthma?
- Have you EVER been told by a doctor or other health professional that you had cancer or a malignancy of any kind? What kind of cancer was it? ... lung?
- During the past 12 months, have you been told by a doctor or other health professional that you had hayfever?
- During the past 12 months, have you been told by a doctor or other health professional that you had sinusitis?
- During the past 12 months, have you been told by a doctor or other health professional that you had chronic bronchitis?

The industry and occupation codes used by NCHS for the NHIS data sets are shown in Appendix F. Three agricultural groups were defined for the NHIS data sets, based on a combination of occupation/industry codes for a respondent's current job as shown in Appendix F. The remaining respondents were classified as non-agricultural workers. The occupation code of 6 (natural mathematical and computer scientists) was included for the forestry and fishery agricultural group because of relatively high proportion of the respondents were identified in the industry code of 2 (forestry and fisheries).

The number (and percent) of individuals in each worker group of interest with each of the above respiratory conditions was calculated. As with the mortality data, a comparison group (non-agricultural workers) was used as a basis for calculating the expected number of workers with each condition. The expected numbers were calculated separately within each of 80 categories representing combinations of sex (male, female), race/ethnicity (white, non-Hispanic; black, non-Hispanic; other, non-Hispanic, Hispanic), age (18-25, 25-34, 35-44, 45-64, 65+), and smoking status (never smoked or ever smoked, based on the question "Have you smoked at least 100 cigarettes in your entire life?"). The observed and expected numbers then were summed across the 80 demographic categories. Prevalence ratios (PRs), or ratios of summed observed to expected numbers, were calculated and then multiplied by 100 to obtain a convenient reference point, and 95% LCLs and UCLs were calculated according to the method described by Bailar and Ederer[1] for mortality data. (Strictly speaking, the method may not apply directly for some conditions that are not considered rare, but it should provide an adequate approximation for purposes of screening the results to discount those based on very small numbers of observations). A PR was considered to be different from 100 at the 95% level of statistical significance (i.e., $p<0.05$) if the 95% confidence interval did not overlap 100.

Morbidity Analyses of the third National Health and Nutritional Examination Survey

Results from the third National Health and Nutritional Examination Survey (NHANES III) also are based on a statistical sample of the U.S. population, and weights are provided for each respondent in the public-use data files to enable development of national estimates from the sample data. As with the NHIS data, much of the analysis with the NHANES III data set concerned the number (and percent) of respondents with specific conditions that could be considered respiratory in nature. Although the NHANES data set had fewer respondents overall than NHIS (with one round of survey results rather than three available for analysis), there were more questions for NHANES III that concerned respiratory conditions. Responses were analyzed for the following questions:

- Has a doctor ever told you that you had asthma?
- Has a doctor ever told you that you had chronic bronchitis?
- Has a doctor ever told you that you had emphysema?
- Has a doctor ever told you that you had hay fever?

Appendix B: Methods

- Apart from when you have a cold, does your chest ever sound wheezy or whistling?
- Do you usually cough on most days for 3 consecutive months or more during the year?
- Do you bring up phlegm on most days for 3 consecutive months or more during the year?
- Are you troubled by shortness of breath when hurrying on level ground or walking up a slight hill?
- During the past 12 months, have you had any episodes of stuffy, itchy, or runny nose?
- During the past 12 months, have you had a cold or the flu?
- During the past 12 months, have you had sinusitis or sinus problems?
- During the past 12 months, have you had pneumonia?
- Have you had wheezing or whistling in your chest at any time in the past 12 months?

Industry and occupation codes from NHANES III for the worker groups of interest are similar to one another. The occupation code (for longest job held) was used to define three worker groups for the NHANES III data set as shown in Appendix G.

As with the NHIS data, the number (and percent) of individuals with each of the above respiratory conditions was calculated for each worker group, and expected numbers were calculated separately within each of 80 categories for combinations of sex, race/ethnicity, age, and smoking status. The categories used for sex, race/ethnicity, age, and smoking status were the same as those used for the NHIS data sets. Similarly, the question used for determining smoking status for NHANES III participants was the same as that used for the NHIS —"Have you smoked at least 100 cigarettes in your entire life?"

Prevalence ratios (PRs), or ratios of summed observed to expected numbers, were calculated in the same manner as described for the NHIS data and were then multiplied by 100 to obtain a convenient reference point. Similarly, 95% LCLs and UCLs were calculated according to the method described for mortality data. (Strictly speaking, the method may not apply directly for some conditions that are not considered rare, but it should provide an adequate approximation for purposes of screening the results to discount those based on very small numbers of observations. A PR was considered to be different from 100 at the 95% level of statistical significance (i.e., $p<0.05$) if the 95% confidence interval did not overlap 100.

A unique feature of the NHANES data set is the inclusion of spirometry data. The following spirometric parameters were used in the analysis: forced expiratory volume in one second (FEV_1); forced vital capacity (FVC); and peak expiratory flow (PEF). Expected values for each of these measures were obtained on an individual-respondent basis, using prediction equations developed by Hankinson et al.[2] These equations provide expected values for each of the three spirometric parameters based on the subject's sex, race/ethnicity, age, and height. Percent predicted ratios were calculated for each subject for each parameter, and resulting distributions were summarized for each worker group (and a comparison group) in terms of the mean and standard deviation of the distribution.

In addition to the summary statistics described above, prevalence ratios based on the fraction of individuals with obstructive or restrictive abnormalities (using the American Thoracic Society criteria[3]) were calculated. Individuals with

---

[2] Hankinson JL, Odencrantz JR, Fedan KB [1999]. Spriometric reference values from a sample of the general U.S. population. Am J Respir Crit Care Med *159*: 179-187.

[3] American Thoracic Society Statement [1991]. Lung function testing: Selection of reference values and interpretative strategies. Am Rev Respir Dis *144*:1202-1218.

Appendix B: Methods

obstructive abnormalities were defined as those for whom the FEV_1/FVC ratio was below the lower limit of normal (LLN), again using prediction equations provided by Hankinson et al.[2] Subjects with restrictive abnormalities were defined as those with an FEV_1/FVC ratio above the LLN but with an FVC value that was below the LLN.

BLS Data

Unlike the NHIS and NHANES data, public-use data files are not available for the injury and illness data reported by BLS. Consequently, incidence rates summarized by industry for selected types of illness (dust diseases of the lung and respiratory conditions due to toxic agents) were extracted from BLS reports for the most recent years available: 1995-2001.

Demographic Data

Estimates extracted from selected BLS web sites or publications were used to develop the demographic statistics for agricultural workers shown in Section 1 of this report.

Statistics on the distribution of agricultural workers by occupation, for the year 2002 (most recent available), were taken from the *Current Population Survey*. Statistics on the distribution of agricultural workers by the state in which they worked, for the year 2002, also were taken from the *Occupational Employment Survey* and were retrieved (state by state) from the same web site (Table 1-2).

Statistics on the distribution of agricultural groups by sex and race/ethnicity, for the years 1997 and 2002, were taken from the *Current Population Survey* (Figures 1-1 through 1-3).

Appendix C
ICD-9 Codes and Descriptions for Respiratory Diseases Included in the Mortality Analysis

ICD Code	Description
Tuberculosis (010-018)	
010*	Primary tuberculous infection
011	Pulmonary tuberculosis
012	Other respiratory tuberculosis
013	Tuberculosis of meninges and central nervous system
014	Tuberculosis of intestines, peritoneum, and mesenteric glands
015	Tuberculosis of bones and joints
016	Tuberculosis of genitourinary system
017	Tuberculosis of other organs
018	Miliary tuberculosis
Mycoses (110-118)	
110	Dermatophytosis
111	Dermatomycosis, other and unspecified
112	Candidiasis
114	Coccidioidomycosis
115	Histoplasmosis
116	Blastomycotic infection
117	Other mycoses
118*	Opportunistic mycoses
Sarcoidosis (135)	
135	Sarcoidosis
Malignant Neoplasms of Trachea/Bronchus/Lung/Pleura (162-163)	
162	Malignant neoplasm of trachea, bronchus, and Lung
163	Malignant neoplasm of pleura
Acute Respiratory Infections (460-466)	
460	Acute nasopharyngitis [common cold]
461*	Acute sinusitis
462	Acute pharyngitis
463	Acute tonsillitis
464	Acute laryngitis and tracheitis
465	Acute upper respiratory infections of multiple or unspecified sites
466	Acute bronchitis and bronchiolitis
Other Diseases of Upper Respiratory Tract (470-478)	
470*	Deflected nasal septum
471	Nasal polyps
472	Chronic pharyngitis and nasopharyngitis
473	Chronic sinusitis
474	Chronic disease of tonsils and adenoids
475	Peritonsillar abscess
476*	Chronic laryngitis and laryngotracheitis
Other Diseases of Upper Respiratory Tract (cont'd))	
477	Allergic rhinitis
478	Other diseases of upper respiratory tract
Pneumonia and Influenza (480-487)	
480	Viral pneumonia
481	Pneumococcal pneumonia
482	Other bacterial pneumonia
483	Pneumonia due to other specified organism
485	Bronchopneumonia, organism unspecified
486	Pneumonia, organism unspecified
487	Influenza
Chronic Obstructive Pulmonary Disease and Allied Conditions (490-496)	
490	Bronchitis, not specified as acute or chronic
491	Chronic bronchitis
492	Emphysema
493	Asthma
494	Bronchiectasis
495	Extrinsic allergic alveolitis (hypersensitivity pneumonitis)
496	Chronic airway obstruction, not elsewhere classified
Pneumoconiosis and Other Lung Diseases - External Agents (500-508)	
500	Coal workers' pneumoconiosis
501	Asbestosis
502	Pneumoconiosis due to other silica or silicates
503	Pneumoconiosis due to other inorganic dust
504	Pneumonopathy due to inhalation of other dust
505	Pneumoconiosis, unspecified
506	Respiratory conditions due to chemical fumes and vapors
507	Pneumonitis due to solids and liquids
508	Respiratory conditions due to other and unspecified external agents
Other Diseases of Respiratory System (510-519)	
510	Empyema
511	Pleurisy
512	Pneumothorax
513	Abscess of lung and mediastinum
514	Pulmonary congestion and hypostasis
515	Postinflammatory pulmonary fibrosis
516	Other alveolar and parietoalveolar pneumonopathy
518	Other diseases of lung
519	Other diseases of respiratory system

*ICD code had no observed deaths for each of the agricultural groups defined in Appendix D.

Appendix D
States and Years with Industry and Occupation Codes from Death Certificates Used in the Mortality Analysis, 1988–1998

State	1988	1989	1990	1991	1992	1993	1994	1995	1996	1997	1998
Alaska	X										
Colorado	X	X	X	X	X	X	X	X	X	X	X
Georgia	X	X	X	X	X	X	X	X	X	X	X
Hawaii						X	X	x	X		X
Idaho	X	X	X	X	X	X	X	X	X	X	X
Indiana	X	X	X	X	X	X		X	x		X
Kansas	X	X	X	X	X	X	X	X	X	X	X
Kentucky	X	X	X	X	X	X	X	X	X	X	X
Maine	X	X	X	X	X	X	X	X	X		X
Nevada	X	X	X	X	X	X	X	X	X	X	X
New Hampshire	X	X	X	X	X	X	X	X	X		X
New Jersey	X	X	X	X	X	X	X	X	X	X	X
New Mexico	X	X	X	X	X	X	X	X	X	X	X
North Carolina	X	X	X	X	X	X	X	X	X	X	X
Ohio	X	X	X	X	X	X		X	X	X	X
Oklahoma	X	X	X	X	X	X		x	x		
Rhode Island	X	X	X	X	X	X	X	X	X	X	X
South Carolina	X	X	X	X	X	X	X	X	X	X	X
Tennessee	X										
Utah	X	X	X	X	X	X	X	X	X	X	X
Vermont	X	X	X	X	X	X	X	X	X	X	X
Washington		X	X	X	X	x	x				
West Virginia	X	X	X	X	X	X	X	X	X	X	X
Wisconsin	X	X	X	X	X	X	X	X	X	X	X

NOTE: Upper case 'X' means the occupation/industry data coded from state death certificates met NCHS quality criteria; lower case 'x' means the data did not meet NCHS quality criteria. Data that did not meet NCHS quality criteria comprised 1.8% of the total deaths represented in the mortality analysis.

SOURCE: National Center for Health Statiscs multiple cause-of-death data

Appendix E
Agricultural Groups Used in the Mortality Analysis and Their Derivation from the U.S. Bureau of Census Industry and Occupation Codes

Table E-1. Derivation of agricultural groups from U.S. Bureau of Census industry and occupation codes

Census Occupation Code (See Table E-2.)	Census Industry Code (See Table E-2.)				
	010	011	020	031	032
473	crop farm workers	livestock farm workers			
474	landscape and horticultural workers	landscape and horticultural workers	landscape and horticultural workers		
475	farm managers	farm managers	landscape and horticultural workers		
476	landscape and horticultural workers	landscape and horticultural workers	landscape and horticultural workers		
477	farm managers	farm managers		forestry workers	fishery workers
479	crop farm workers	livestock farm workers	landscape and horticultural workers	forestry workers	
483		livestock farm workers			fishery workers
484	crop farm workers	crop farm workers	landscape and horticultural workers		
485	landscape and horticultural workers	landscape and horticultural workers	landscape and horticultural workers	forestry workers	fishery workers
486	landscape and horticultural workers	landscape and horticultural workers	landscape and horticultural workers	forestry workers	
494				forestry workers	
495				forestry workers	
496	forestry workers	forestry workers	forestry workers	forestry workers	
497					fishery workers
498					fishery workers
499					fishery workers

SOURCE: U.S. Bureau of the Census: Classified Index of Industries and Occupations. 1990 Census of Population and Housing, first edition

Appendix E: Agricultural Groups Used in the Mortality Analysis

Table E-2. U.S. Bureau of Census industry and occupation codes used in the mortality data analyses

Industry Codes

Agriculture, Forestry, and Fisheries

- 010 Agricultural production, crops
- 011 Agricultural production, livestock
- 020 Landscape and horticultural services
- 031 Forestry
- 032 Fishing, hunting, and trapping

Occupation Codes

Farming, Forestry, and Fishing Occupations

- 473 Farmers, except horticultural
- 474 Horticultural specialty farmers
- 475 Managers, farms, except horticultural
- 476 Managers, horticultural specialty farms
- 477 Supervisors, farm workers
- 479 Farm workers
- 483 Marine life cultivation workers
- 484 Nursery workers
- 485 Supervisors, related agricultural occupations
- 486 Groundskeepers and gardeners, except farm
- 494 Supervisors, forestry, and logging workers
- 495 Forestry workers, except logging
- 496 Timber cutting and logging occupations
- 497 Captains and other officers, fishing vessels
- 498 Fishers
- 499 Hunters and trappers

SOURCE: U.S. Bureau of the Census: Classified Index of Industries and Occupations. 1990 Census of Population and Housing, first edition

Appendix F
Agricultural Groups Used in the Morbidity Analysis and Their Derivation from the National Health and Interview Survey (NHIS) Industry and Occupation Codes

		NHIS Industry Code	
		1 Agriculture	2 Forestry and Fisheries
NHIS Occupation Code			
6	Natural mathematical and computer scientists		forestry and fishery workers
29	Farm operators and managers	farm managers	
30	Farm workers and other agricultural workers	farm workers	forestry and fishery workers
31	Forestry and fishing occupations		forestry and fishery workers

SOURCE: 1997/1998/1999 National Health Interview Surveys, Sample Adult Person Section – Public Use (pdf files, available from www.cdc.gov/nchs/about/major/nhis/quest_data_related_1997_forward.htm).

Appendix B

Appendix G
Agricultural Groups Used in the Morbidity Analysis and Their Derivation from the Third National Health and Nutrition Examination Survey (NHANES III) Industry and Occupation Codes

		NHANES III Industry Code	
NHANES III Occupation Code		1 Agricultural Production	2 Agricultural Services, Forestry, Fishing
25	Farm operators, managers, and supervisors	farm managers	
26	Farm and nursery workers	farm workers	
27	Related agricultural, forestry, fishing	other agricultural workers	other agricultural workers

SOURCE: Third National Health and Nutrition Examination Survey, Household Adult and Examination Data File Documentation (http://www.cdc.gov/nchs/about/major/nhanes/datalink.htm).